C0-AVJ-391

Semiconductor Device Physics and Simulation

MICRODEVICES
Physics and Fabrication Technologies

Series Editors: Ivor Brodie and Arden Sher
SRI International
Menlo Park, California

Recent volumes in the series:

COMPOUND AND JOSEPHSON HIGH-SPEED DEVICES
Edited by Takahiko Misugi and Akihiro Shibatomi

ELECTRON AND ION OPTICS
Miklos Szilagyi

ELECTRON BEAM TESTING TECHNOLOGY
Edited by John T. L. Thong

ORIENTED CRYSTALLIZATION ON AMORPHOUS SUBSTRATES
E. I. Givargizov

PHYSICS OF HIGH-SPEED TRANSISTORS
Juras Požela

THE PHYSICS OF MICRO/NANO-FABRICATION
Ivor Brodie and Julius J. Muray

PHYSICS OF SUBMICRON DEVICES
David K. Ferry and Robert O. Grondin

THE PHYSICS OF SUBMICRON LITHOGRAPHY
Kamil A. Valiev

RAPID THERMAL PROCESSING OF SEMICONDUCTORS
Victor E. Borisenko and Peter J. Hesketh

SEMICONDUCTOR ALLOYS
Physics and Materials Engineering
An-Ban Chen and Arden Sher

SEMICONDUCTOR DEVICE PHYSICS AND SIMULATION
J. S. Yuan and J. J. Liou

SEMICONDUCTOR LITHOGRAPHY
Principles, Practices, and Materials
Wayne M. Moreau

SEMICONDUCTOR PHYSICAL ELECTRONICS
Sheng S. Li

A Continuation Order Plan is available for this series. A continuation order will bring delivery of each new volume immediately upon publication. Volumes are billed only upon actual shipment. For further information please contact the publisher.

Semiconductor Device Physics and Simulation

J. S. Yuan and
J. J. Liou
University of Central Florida
Orlando, Florida

Plenum Press • New York and London

Library of Congress Cataloging-in-Publication Data

Yuan, J. S.
Semiconductor device physics and simulation / J.S. Yuan and J.J. Liou.
p. cm. -- (Microdevices)
Includes bibliographical references and index.
ISBN 0-306-45724-5
1. Semiconductors--Computer simulation. 2. Junction transistors--Computer simulation. I. Liou, Juin J. II. Title. III. Series.
TK7871.85.Y83 1998
621.3815'2'0113--dc21 98-18553
CIP

ISBN 0-306-45724-5

A Division of Plenum Publishing Corporation
233 Spring Street, New York, N.Y. 10013

http://www.plenum.com

10 9 8 7 6 5 4 3 2 1

Printed in the United States of America

To my late father and my mother

—J. S. Yuan

Preface

The advent of the microelectronics technology has made ever-increasing numbers of small devices on a same chip. The rapid emergence of ultra-large-scaled-integrated (ULSI) technology has moved device dimension into the sub-quarter-micron regime and put more than 10 million transistors on a single chip. While traditional closed-form analytical models furnish useful intuition into how semiconductor devices behave, they no longer provide consistently accurate results for all modes of operation of these very small devices. The reason is that, in such devices, various physical mechanisms affect the device performance in a complex manner, and the conventional assumptions (i.e., one-dimensional treatment, low-level injection, quasi-static approximation, etc.) employed in developing analytical models become questionable. Thus, the use of numerical device simulation becomes important in device modeling. Researchers and engineers will rely even more on device simulation for device design and analysis in the future.

This book provides comprehensive coverage of device simulation and analysis for various modern semiconductor devices. It will serve as a reference for researchers, engineers, and students who require in-depth, up-to-date information and understanding of semiconductor device physics and characteristics. The materials of the book are limited to conventional and mainstream semiconductor devices; photonic devices such as light-emitting and laser diodes are not included, nor does the book cover device modeling, device fabrication, and circuit applications. It is assumed that the reader has already acquired a basic understanding of device structures and operations, such as those given in *Solid-State Electronic Devices*, 4th ed. (Prentice-Hall, 1995) or *Device Electronics for Integrated Circuits*, 2nd ed. (Wiley, 1986).

A two-dimensional device simulator called MEDICI, originally developed by Stanford University (PISCES II) and supported by Technology Modeling Associates, Inc., CA, is used to perform the analysis and to generate the simulation results. Basically, MEDICI solves numerically the five classical semiconductor device equations: Poisson's equation, electron and hole continuity equations, and electron and hole drift–diffusion current equations. Relevant device physics, such as heavy doping effects, concentration, and field-dependent free-carrier mobilities, and concentration-dependent free-carrier lifetimes, are incorporated in the program. The capability of MEDICI goes beyond the traditional device simulator, however, in that it also contains two optional modules which allow one to simulate the device performance, including the effects of lattice heating and heterostructure. A detailed description of MEDICI is given in Chapter 1.

The book is organized into nine chapters. Except for the first chapter, which discusses semiconductor fundamentals and the capabilities and features of device simulators, each chapter covers a semiconductor device, providing dc, ac, and transient simulation results, discussions of relevant device physics, and their implications to device design and analysis. The devices covered, in the same order as the chapters, are *p–n* junction, bipolar junction transistors, junction field-effect transistors, metal–oxide–semiconductor field-effect transistors, bipolar/CMOS devices, metal–semiconductor field-effect transistors, heterojunction bipolar transistors, and photoconductive diodes.

All chapters contain useful figures to illustrate the physical mechanisms and characteristics of the semiconductor devices simulated by MEDICI and, in some cases, those observed in measurements. Extensive references have also been given as an aid to the reader who wishes to carry out an in-depth study on a particular topic.

J. S. Yuan and J. J. Liou
Department of Electrical
and Computer Engineering
University of Central Florida

Contents

1

Introduction

Since the discovery of the bipolar junction transistor (BJT) (Bardeen and Brattain, 1948) considerable effort has been devoted to understanding semiconductor devices and electron device modeling. Due to the complexity of the semiconductor equations, however, closed-form solutions are seldom possible. Two typical approaches to the problem were developed: (1) The regional approach involves very simplified assumptions on device geometry and mathematical models. Devices are usually treated as one-dimensional structures having individual segments with constant impurity concentration. They are further split into quasi-neutral and space-charge regions where different assumptions are made. Local solutions are then linked at the boundaries between two adjacent regions, often leading to closed-form expressions for the device terminal characteristics. (2) The second approach attempts a fully numerical solution of the basic semiconductor equations by using a realistic device description and allowing for all relevant physical effects. Two-dimensional models comprising Poisson and continuity equations first appeared in 1969 for the junction field-effect transistor (JFET) (Kennedy and O'Brien, 1969) and the bipolar junction transistor (Slotboom, 1969). Two-dimensional metal–oxide field-effect transistor (MOSFET) models soon followed (Vandorpe and Xuong, 1971; Kennedy and Murley, 1973; Mock, 1973; Bube, 1974). More recently, two-dimensional programs have appeared with varying degrees of generality and sophistication. With the increasing complexity of integrated circuit (IC) fabrication processes and diminishing feature sizes of IC device structures, computer programs have proven valuable in the development and characterization of new IC technologies.

1.1. SEMICONDUCTOR DEVICE FUNDAMENTALS

Electrons and holes in semiconductors are known to have both wave and particle properties (wave–particle duality). The analysis treating free carriers as waves is referred to as the quantum approach, whereas the analysis treating free carriers as particles is normally called the classical approach. According to quantum mechanics the behavior of a particle is described by the wave function $\Theta(r,t)$, which is a solution of the Schrödinger wave equation

$$\frac{-\hbar^2}{2m}\nabla^2\Theta + V(r,t)\Theta = i\hbar\frac{\partial\Theta}{\partial t} \tag{1.1.1}$$

where $V(r,t)$ is the potential energy function of the particle and m is the effective mass.

A brief comparison of particle-like and wavelike properties is given as follows. A particle-like electron has a position in space; its location can be specified by giving its spatial coordinates. It has momentum mv and energy $E = mv^2/2 + V$. A wavelike electron is extended in space; spatial characteristics are specified by a wavelength λ. Momentum is described in terms of a wave number $\kappa = 2\pi/\lambda$. It can be present at all frequencies ω if the "wave" is effectively infinite (unconfined) or only at a set of discrete frequencies if the "wave" is finite (confined to a specific region of space) (Bube, 1974). The correlation between the particle-like and wavelike properties is as follows: if particle-like momentum is mv, then wavelike wavelength is $\lambda = h/mv$. If particle-like energy is E, then wavelike frequency is $\omega = E/\hbar\omega$. Particle-like properties can exhibit all energies $E = \hbar\omega \geq 0$ if the "wave" is effectively infinite but only a set of discrete energy levels $E_j = \hbar\omega_j$ if the "wave" is confined.

For typical silicon semiconductor devices today, the particle-like property is valid. In classical particle theory, free-carrier transport in a semiconductor can normally be described with the concept of drift and diffusion. The drift mechanism describes the unilateral flow of electrons and holes driven by the electric field induced in the semiconductor, whereas the diffusion mechanism describes the unilateral flow of electrons and holes as a result of the spatial gradient of electron and hole concentration in the semiconductor.

1.1.1. Mobility and Carrier Scattering

Consider a small electric field applied to the lattice. The electrons are accelerated in a direction opposite to the field during the time between collisions. Electrons tend to move in the direction opposite to the electric field because of drift. The electrons exchange energy during collisions with the lattice and move toward their thermal equilibrium positions. If the E field is small, the energy exchanged is also small, and the lattice is not appreciably heated by the current.

A perfectly periodic lattice would not scatter free carriers. However, at any temperature above absolute zero the lattice atoms vibrate. These vibrations disturb periodicity and allow energy to be transferred between the carriers and lattice. The interactions with lattice vibrations can be viewed as collisions with energetic particles called phonons. Theoretical analysis indicates that the carrier mobility should decrease with temperature in proportion to T^{η}, with η between 1.5 and 2.5, when lattice scattering is dominant. Experimental values of η range from 1.66 to 3.

In addition to lattice vibration, dopant impurities can also cause local distortions in the lattice and scattering of free carriers. However, unlike scattering caused by lattice vibration, scattering from ionized impurities becomes less significant at high temperatures, depending on doping concentration.

For direct-band-gap semiconductors such as GaAs, optical phonon scattering is significant. Optical phonon scattering becomes the predominant scattering source at high temperatures or at high electric fields. Both polar and nonpolar optical phonons are responsible for this type of scattering. Polar optical phonon scattering is the predominant scattering mechanism for ionic or polar crystals such as II–VI and III–V compound semiconductors. Polar optical phonon scattering is associated with atomic polarization arising from displacement caused by optical phonons. Nonpolar optical phonon scattering becomes important for silicon and germanium crystals above room temperatures in which intervalley scattering becomes the dominant process.

In general, electron mobility in a semiconductor due to multiple scattering processes can be calculated from the expression

$$\mu_n = \frac{q\langle\tau\rangle}{m_n^*} \tag{1.1.2}$$

where m_n^* is the electron effective mass and τ is the relaxation time:

$$\frac{1}{\tau} = \sum_i \frac{1}{\tau_i} \tag{1.1.3}$$

τ_i denotes the relaxation time due to a particular scattering process. The electron mobility for multiple scattering processes can be calculated by finding the total relaxation time due to different scattering mechanisms and calculating the average relaxation time and the total electron mobility using (1.1.2).

If the relaxation time due to different scattering mechanisms is independent of energy, then one can use the simplified reciprocal sum formula to obtain the total electron mobility as follows:

$$\frac{1}{\mu} = \sum_{i=1}^{n} \frac{1}{\mu_i} \tag{1.1.4}$$

In classical semiconductor theory, electrons and holes are treated as particles. Current in a semiconductor consists of drift and diffusion currents. Drift current is due to an applied electric field E and is determined by the mobility of the carriers:

$$J_{n\text{drift}} = q\mu_n nE \tag{1.1.5}$$

$$J_{p\text{drift}} = q\mu_p pE \tag{1.1.6}$$

where q is the electron charge, μ_n is the electron mobility, μ_p is the hole mobility, n is the electron concentration, and p is the hole concentration. It is clear from (1.1.5) and (1.1.6) that the mobility significantly affects the device current characteristics. It is desirable to have a formula for mobility as a function of doping level, temperature, and materials that can be used to predict device characteristics in semiconductor device modeling and simulation. The mobility in phosphorous-doped silicon as a function of doping derived from a combination of experimental and theoretical data is (Arora *et al.*, 1982).

$$\mu_n = 88.0\, T_n^{-0.57} + \frac{7.4 \times 10^8\, T^{-2.33}}{1 + \left(\dfrac{N}{1.26 \times 10^{17}\, T_n^{2.4}}\right) 0.88\, T_n^{-0.146}} \tag{1.1.7}$$

$$\mu_p = 54.3\, T_n^{-0.57} + \frac{1.4 \times 10^8\, T^{-2.33}}{1 + \left(\dfrac{N}{2.35 \times 10^{17}\, T_n^{2.4}}\right) 0.88\, T_n^{-0.146}} \tag{1.1.8}$$

where N is the total ionized impurity concentration, T is the absolute temperature, and $T_n = T/300$.

1.1.2. Carrier Transport by Diffusion and the Einstein Relation

Diffusion is the natural result of the random motion of carriers. Each particle moves in an arbitrary direction until it collides with another particle or gets scattered from the lattice or impurity, after which it moves in a new direction. If a gradient of carrier concentrations exists, carriers diffuse from the higher concentration to the lower concentration in the semiconductor. This process continues until the carriers are uniformly distributed. Diffusion current due to the gradient of the carrier concentrations is

$$J_{n\text{diff}} = qD_n \frac{dn}{dx} \tag{1.1.9}$$

$$J_{p\text{diff}} = -qD_p \frac{dp}{dx} \tag{1.1.10}$$

where D_n is the electron diffusion coefficient and D_p is the hole diffusion coefficient. For nondegenerate semiconductors the mobility and diffusion coefficients are related by the Einstein relation:

$$D_n = \frac{kT}{q} \mu_n \tag{1.1.11}$$

$$D_p = \frac{kT}{q} \mu_p \tag{1.1.12}$$

where k is Boltzmann's constant.

1.1.3. Recombination and Generation

When nonequilibrium carrier densities are induced in a semiconductor through the action of an external stimulus, the density will return to equilibrium after removal of the external influence. Carriers present above equilibrium values are termed excess carriers, and the opposite condition can be considered to involve a negative excess carrier density. In the former case, carriers are annihilated by recombination; in the latter case, they are created by generation.

Recombination can occur through a band-to-band transition in which a conduction electron recombines with a hole in the valence band. This process is important for direct-band-gap semiconductors such as GaAs. A second recombination process, which is important for indirect semiconductors such as Si, occurs through the medium of a recombination center. The single-level recombination can be described by four processes: electron capture, electron emission, hole capture, and hole emission. The recombination rate approaches a maximum as the energy level of the recombination center approaches midgap. Thus, the most effective recombination centers are those located near the middle of the band gap. For multiple-level traps the recombination process has gross qualitative features that are similar to those of the single-level case. The behavioral details are different, however, particularly at high levels of injection, where the asymptotic lifetime is an average of the lifetimes associated with all the positively charged, negatively charged, and neutral trapping levels.

Electron and hole lifetimes are concentration dependent and can be modeled as (Roulston *et al.*, 1982)

$$\tau_n = \frac{\tau_{n0}}{1 + N/N_{\mathrm{SRH}n}} \tag{1.1.13}$$

$$\tau_p = \frac{\tau_{p0}}{1 + N/N_{\mathrm{SRH}p}} \tag{1.1.14}$$

where τ_{n0} and τ_{p0} are the electron and hole lifetimes at low doping concentration, and $N_{\mathrm{SRH}n}$ and $N_{\mathrm{SRH}p}$ are the Shockley–Read–Hall concentration coefficients for electrons and holes, respectively. The default value for $N_{\mathrm{SRH}n}$ and $N_{\mathrm{SRH}p}$ used in many device simulators is 5×10^{16} cm^{-3}.

1.1.4. Heavy Doping Effects and Band-Gap Narrowing

When the doping level of semiconductors exceeds about 10^{18} cm^{-3}, heavy doping effects such as band-gap narrowing, Auger recombination, and minority carrier mobility come into play. For instance, as the doping level increases, the impurity band becomes skewed. A downshift of conduction-band edge with nearly parabolic density of states occurs due to electron–donor interaction. The valence band also moves up by approximately the same amount. Furthermore, a nearly equal shift of the two band edges occurs due to many-body effects, i.e., carrier–carrier interactions. In addition, local fluctuation of electrostatic potential due to nonuniform spatial distribution of impurity atoms introduces band tails. The effective energy band gap due to heavy doping effects is thus reduced.

The band-gap narrowing is important for modeling the current gain of the bipolar transistor. Until recently there was no universal agreement about the experimental values of band-gap narrowing on the method that gives the most reliable values. It has been shown that the earlier values derived from electrical measurements were too high at doping levels above 3×10^{19} cm^{-3} (del Alamo and Swanson, 1984; del Alamo *et al.*, 1985). The theoretical results of Bergren and Sernelius (1985) agree with the values for band-gap narrowing obtained from the luminescence data (Dumke, 1983) as well as the electrical data (del Alamo and Swanson, 1984; del Alamo *et al.*, 1985). The theoretical data do not differ much from those given by Slotboom and de Graaff (1976) as follows:

$$\Delta E_G = 0.009\left[\ln\left(\frac{N}{10^{17}}\right) + \sqrt{\ln^2\left(\frac{N}{10^{17}}\right) + 0.5}\right] \tag{1.1.15}$$

A second very important effect that occurs at high doping levels is due to Auger recombination. This is due to the direct band-to-band recombination between an electron and a hole across the forbidden gap, accompanied by the transfer of energy to another free electron or hole. The recombination lifetime for the Auger process is given by

$$\tau_{An} = \frac{1}{c_n N^2} \tag{1.1.16}$$

$$\tau_{Ap} = \frac{1}{c_p N^2} \tag{1.1.17}$$

where the Auger recombination coefficient c_n has been measured for electrons in silicon to be in the range 0.4×10^{-31} to 6×10^{-31} cm^6 s^{-1} (Roulston *et al.*, 1982; Dziewior and Schmid, 1977), and the Auger recombination coefficient c_p has been measured for holes to be 1×10^{-31} cm^6 s^{-1}.

Normally it is assumed that the minority carrier mobility (for example, hole mobility in n-type material) is the same as that given for the same carrier type when it is a majority carrier (for example, hole mobility in p-type material of the same doping level). Recently, minority carrier transport parameters have been assessed experimentally by using chopped optical excitation (Dziewior and Silber, 1979), photoluminescence decay (Swirhun *et al.*, 1986), and direct transient measurements (Misiakos *et al.*, 1989). Significant discrepancies between the minority hole mobility (μ_p in n-type Si) and the majority hole mobility (μ_p in p-type Si) in the degenerate material have been found (Dziewior and Silber, 1979; Swirhun *et al.*, 1986; Misiakos *et al.*, 1989; Green, 1990). For example, the degenerate Si doping of 7.2×10^{19} cm^{-3}, the minority hole mobility is 2.8 times larger than the majority hole mobility at room temperature. Thus, using μ_n (in p-type Si) for a heavily doped n^+ emitter in device simulation can result in significant errors in predicting base current and current gain for the bipolar transistor. An empirical equation from curve fitting of experimental data for the minority hole mobility is (Swirhun *et al.*, 1986)

$$\mu_p = 130 + \frac{370}{1 + \left(\frac{N_D}{8 \times 10^{17}}\right)^{1.25}} \tag{1.1.18}$$

1.1.5. Carrier Concentration in Semiconductors and Fermi–Dirac Statistics

When the semiconductor is lightly doped, the use of Boltzmann statistics is adequate in determining the Fermi level and carrier concentration:

$$n = N_C e^{(E_F - E_C)/kT} \tag{1.1.19}$$

$$p = N_V e^{(E_V - E_F)/kT} \tag{1.1.20}$$

where N_C and N_V are the density of states for the conduction and valence bands, E_C and E_V are the energy levels at the bottom of the conduction band and at the top of the valence band, respectively, and E_F is the Fermi level.

It is, however, necessary to use Fermi–Dirac statistics to calculate the Fermi level when the semiconductor is heavily doped ($>10^{18}$ cm^{-3}):

$$n = N_c F_{1/2}\left(\frac{E_F - E_C}{kT}\right) \tag{1.1.21}$$

$$p = N_V F_{1/2}\left(\frac{E_V - E_F}{kT}\right) \tag{1.1.22}$$

where $F_{1/2}$ is the Fermi–Dirac integral of order one-half:

$$F_{1/2}(\eta_{n,p}) = \frac{2}{\sqrt{\pi}} \int_0^\infty \frac{x^{1/2}\, dx}{1 + e^{x - \eta_{n,p}}}$$

and $\eta_n = (E_{Fn} - E_C)/kT$ and $\eta_p = (E_V - E_{Fp})/kT$. The effective intrinsic carrier concentration taking into account the heavy doping effects is

$$n_{ie}^2 = n_i^2 e^{\Delta E_G/kT} \tag{1.1.23}$$

where n_i is the intrinsic carrier concentration.

For degenerate semiconductors the mobility and diffusion coefficients are related by (Sze, 1981)

$$D_n = \frac{kT}{q}\mu_n \frac{F_{1/2}(\eta_n)}{F_{-1/2}(\eta_n)} \tag{1.1.24}$$

$$D_p = \frac{kT}{q}\mu_p \frac{F_{1/2}(\eta_p)}{F_{-1/2}(\eta_p)} \tag{1.1.25}$$

1.2. BASIC SEMICONDUCTOR EQUATIONS USED IN DEVICE SIMULATORS

The device simulators developed by universities and industry include SEDAN (1984), BIPOLE (1993), BAMBI (Franz *et al.*, 1989), MINIMOS (Fischer *et al.*, 1994), PISCES (1989), MEDICI (1993), DAVINCI (1993), ATLAS (1996), and FIELDY (Buturla *et al.*, 1988). SEDAN (SEmiconductor Device ANalysis), developed at Stanford University, is a one-dimensional device simulator. Devices consisting of up to five metallurgical junctions can be analyzed. SEDAN performs device analysis by solving Poisson's equation or simultaneously solving Poisson's equation and transient current continuity equations to obtain the potential and carrier concentration distributions.

BIPOLE is a quasi-two-dimensional device simulator. It was developed at the University of Waterloo, Canada, to provide a convenient and rapid means of predicting terminal electrical characteristics of bipolar transistors from input consisting of fabrication data such as mask dimensions, impurity profiles, and recombination data. The calculation is based on the variable boundary regional approach using one-dimensional transport equations. Two-dimensional and quasi-cylindrical edge effects are handled by combining the vertical one-dimensional analysis with a coupled one-dimensional horizontal analysis of the transport equations in the base region.

BAMBI (Basic Analyzer of MOS and BIpolar devices) is a two-dimensional device simulator developed at Technishe Universitat Wien, Austria. It solves Poisson's and continuity equations simultaneously. Arbitrary shapes can be simulated. The process of grid generation and management in BAMBI is automated. The program also includes an automatic time-step control algorithm. The program does not include models for the physical parameters such as μ_n and μ_p. Instead, they are defined by external functions. This allows general use of the package.

MINIMOS, developed at the Technical University Vienna, Austria, is a two- and three-dimensional numerical simulator of semiconductor field-effect transistors. The fundamental semiconductor equations solved in MINIMOS numerically comprise Poisson's equation and two carrier continuity equations for electrons and holes. MINIMOS builds the consistent solution of the semiconductor equations in a hierarchical manner, starting with a relatively simple model that is subsequently refined by taking into account more complicated physical mechanisms. The numerical schemes of MINIMOS are designed to handle planar and nonplanar device structures in two and three dimensions.

PISCES, a powerful two-dimensional device simulation program to develop MOS and bipolar integrated circuits, models the two-dimensional distributions of potential and carrier concentrations in a device to predict its electrical characteristics for any bias condition. PISCES was developed at Stanford University and improved by Technology Modeling Associates (TMA).

MEDICI is an enhanced version of PISCES. The MEDICI program, developed by TMA, allows all model and material parameters that are accessible to the user to be modified on a region-to-region basis. With its nonuniform triangular simulation grid, MEDICI can model arbitrary device geometries with planar and nonplanar surface topologies. MEDICI solves the semiconductor equations at each node within the device. From the grid structure of the device and impurity doping profile, MEDICI can provide two-dimensional contours and vectors.

DAVINCI is a three-dimensional device simulator that solves three-dimensional distributions of potential and carrier concentrations in a device to predict its electrical characteristics for any bias condition. Poisson's equation and current continuity equations for electrons and holes in three dimensions are solved to analyze minority and majority carrier devices.

ATLAS provides general capabilities for numerical, physics-based, two- and three-dimensional simulation of semiconductor devices. It predicts the electrical behavior of specified semiconductor structures and provides insight into the internal physical mechanisms associated with device operation. This device simulation software was developed by Silvaco International.

Overall, these device simulators solve the following five basic semiconductor equations:

Poisson's equation:

$$\varepsilon \nabla^2 \Phi = -q(p - n + N_D^+ - N_A^-) - \rho_F \tag{1.2.1}$$

the continuity equations for electrons and holes:

$$\frac{\partial n}{\partial t} = \frac{1}{q} \nabla J_n - U_n \tag{1.2.2}$$

$$\frac{\partial p}{\partial t} = -\frac{1}{q} \nabla J_p - U_p \tag{1.2.3}$$

the electron and hole current density equations:

$$J_n = q\mu_n nE + qD_n \nabla n \tag{1.2.4}$$

$$J_p = q\mu_p nE - qD_p \nabla p \tag{1.2.5}$$

In (1.2.1)–(1.2.5) ε is the permittivity, Φ is the electrostatic potential, N_D^+ is the ionized donor concentration, N_A^- is the ionized acceptor concentration, ρ_F is the fixed charge density, which may be present due to fixed charge in insulating materials or charged interface states, t is the time, J_n is the electron current density, J_p is the hole current density, U_n is the electron recombination rate, and U_p is the hole recombination rate.

Although for most practical cases, full impurity ionization is assumed (i.e., $N_D^+ = N_D$ and $N_A^- = N_A$), the device simulators are able to model the incomplete ionization due to the carrier freeze-out effect:

$$N_D^+ = \frac{N_D}{1 + g_c e^{(E_{Fn} - E_D)/kT}} \tag{1.2.6}$$

$$N_A^- = \frac{N_A}{1 + g_d e^{(E_A - E_{Fp})/kT}} \tag{1.2.7}$$

where g_c is the degeneracy factor for electrons, g_d is the degeneracy factor for holes, E_{Fn} is the electron quasi-Fermi level, E_{Fp} is the hole quasi-Fermi level, E_D is the donor energy level, and E_A is the acceptor energy level.

In the continuity equations for electrons and holes, Shockley–Read–Hall and Auger recombination models are used

$$U_{\text{SRH}} = \frac{pn - n_{ie}^2}{\tau_n[n + n_{ie}e^{-(E_T - E_i)/kT}] + \tau_p[p + n_{ie}e^{(E_T - E_i)/kT}]} \tag{1.2.8}$$

$$U_{\text{Auger}} = c_n(pn^2 - nn_{ie}^2) + c_p(np^2 - pn_{ie}^2) \tag{1.2.9}$$

where E_i is the intrinsic energy level, E_T is the trap density energy level, τ_n is the electron lifetime, and τ_p is the hole lifetime.

The intrinsic concentration, energy band gap, effective density of states for the conduction band, and density of states for the valence band are temperature dependent:

$$n_i^2 = N_C(T)N_V(T)e^{-E_G(T)/kT} \tag{1.2.10}$$

$$N_C(T) = N_C(300)\left(\frac{T}{300}\right)^{3/2} \tag{1.2.11}$$

$$N_V(T) = N_V(300)\left(\frac{T}{300}\right)^{3/2} \tag{1.2.12}$$

$$E_G(T) = E_G(300) + \alpha\left(\frac{300^2}{300 + \beta} - \frac{T^2}{T + \beta}\right) \tag{1.2.13}$$

where $\alpha = 4.73 \times 10^{-4}$ for Si and 5.405×10^{-4} for GaAs and $\beta = 636$ for Si and 204 for GaAs (Sze, 1981).

1.3. NUMERICAL TECHNIQUES USED IN DEVICE SIMULATORS

Most device simulation programs use finite difference or finite element methods for solving the five basic semiconductor equations. In the finite difference method, the derivatives of the semiconductor equations are replaced by finite difference formulas at each node in a finite difference mesh. The boundary conditions are normally "Dirichlet" (values fixed) at the contacts and "Neumann" (gradient fixed) at other surfaces. The system of equations may be solved by direct methods or iterative schemes. A commonly used method is successive overrelaxation (SOR). The finite difference method is relatively easy to set up. It is important to employ small mesh spacing in the vicinity of junctions where the potential is changing rapidly.

The finite element method lends itself to structures that are far from rectangular. Since they employ elements that can be quite different in size for different regions, the method is also ideal for having small elements in the vicinity of junctions where the potential is rapidly changing. The boundary conditions are incorporated as integrals in a function that is minimized, and the scheme is independent of the specific boundary conditions; this provides a good degree of flexibility, particularly since elements of different sizes may be added without increasing the complexity.

BIPOLE uses the variable boundary regional approach. Boundaries are established between the space-charge layers and the neutral regions. Application of the Boltzmann relations provides a relationship between the injected carrier concentrations for holes and electrons and the applied bias. It is then used to solve the transport equations for diffusion and drift in the presence of arbitrary recombination in both quasi-neutral regions. This scheme is extremely fast in terms of computing time.

BAMBI is capable of simulating the behavior of arbitrarily shaped device domains including dielectrics. The exact numerical model accomplishes locally refined grid structures with automatic setup and adaptive to the different stationary and transient operating conditions. BAMBI has been designed with a three-pass concept: an input processor (PASS 1), a numeric processor (PASS 2), and an output processor (PASS 3). Each pass can be executed separately. PASS 1 writes the main program. PASS 2 initializes constants and provides the input data such as geometry, boundary conditions, etc. The main program and subroutines are then compiled and linked with the BAMBI library and the user-defined external functions for the physical parameters. The numeric processor outputs the results in packed form and creates the main program (PASS 3). The output processor converts the results into easily readable tables, calculates all physical quantities specified in the input deck, and furnishes plot files. BAMBI can be handled easily without further knowledge of the internal structure and its algorithms.

MINIMOS at first minimizes the error norms iteratively by accounting for the simplest physical model. Thus, after making a guess for the initial solution, MINIMOS starts with the internal model option. In this model MINIMOS does not solve the majority continuity equations, and generation–recombination effects are neglected. Poisson's equation is solved in the full simulation domain, whereas the minority continuity equation is solved only in the partial domain from the gate insulator to about half of the p–n junction depth. Within the deep bulk region the minority carriers are computed by extrapolation, assuming a constant quasi-Fermi level. This model offers fast execution since the computational demand for the costly solution of the carrier continuity equations is minimized. Within model 1-D MINIMOS builds its grid through an adaptive grid refinement loop. Each refinement step consists of a grid update followed by a Gummel iteration loop, which terminates when all error norms lie within the threshold set. After completing the grip loop (when the mesh weighting function is smaller than unity for all mesh intervals), MINIMOS switches its internal mode to 2-D. Now the minority continuity equation is solved in the full simulation domain; for the majority carriers there is still a constant quasi-Fermi level assumed. No grid updates are carried out. Since the deviation from thermal equilibrium in the deep bulk is usually small, only a few Gummel iteration steps are necessary.

MINIMOS 3D relies on a similar model structure as MINIMOS 2D. The key MODEL on the OPTION directive controls the computation of the three-dimensional solution in an equivalent way. At first it is assumed that M3MODE = 1 was set on the OPTION directive. M3MODE = 1 leads MINIMOS 3D to build the mesh in the z direction together with a coarse three-dimensional solution by solely solving the Poisson equation in three dimensions. The carrier concentrations are calculated from their quasi-Fermi level, assuming vanishing current densities in the z direction. The solution of the previously solved two-dimensional problem, corresponding to the setting of the MODEL key, is taken as the Dirichlet boundary condition in the first x–y plane. This first x–y plane is assumed to be the center of symmetry with respect to the z direction. This model offers comparatively fast execution since no solution of the carrier continuity equations is carried out.

With its nonuniform triangular tetrahedra simulation grid, DAVINCI can model arbitrary device geometries. It can also refine the simulation grid automatically during the solution process. Additional modes and elements can be added where a user-specified quantity, such as potential or impurity concentration, varies by more than a specified tolerance over existing mesh elements.

MEDICI uses the box method to discretize the differential operators on a general triangular grid. Each equation is integrated over a small polygon enclosing each node. The integration equates the flux into the polygon with the sources and sinks inside it so that conservation current and electric flux are built into the solution. The integral involved is performed on a triangle-by-triangle basis, which leads to a simple and elegant way of handling general surface and boundary conditions.

The discretization of the semiconductor device equations gives rise to a set of coupled nonlinear algebraic equations that must be solved by a nonlinear iteration method starting from an initial guess. Two methods are widely used: Gummel's and Newton's. In Gummel's method, the equations are solved sequentially. Poisson's equation is solved assuming fixed quasi-Fermi potentials; since the Poisson equation is nonlinear it is solved by an inner Newton loop. Then the new potential is substituted into the continuity equations, which are linear and can be solved directly. The new carrier concentrations are substituted back into the charge term of Poisson's equation and another cycle begins. At each state only one equation is being solved, so the matrix has *N* rows and *N* columns regardless of the number of carriers being solved. This is a decoupled method; one set of variables is held fixed while another set is solved. The success of the method depends, therefore, on the degree of coupling between the equations. In Newton's method, all of the variables in the problem are allowed to change during each iteration, and all of the coupling between variables is taken into account. Hence, the Newton method is very stable and the solution time is nearly independent of bias condition even at high levels of injection. Each approach involves solving several large linear systems of equations. The number of equations in each system is about one to three times the number of grid points, depending on the number of carriers being solved for. The nonlinear iteration usually converges at a linear or quadratic rate. In the former, the error decreases by about the same factor at each iteration. In a quadratic method, the error should go down at each iteration, giving rise to rapid convergence. For accurate solutions, it is advantageous to use a quadratic method. Newton's method is quadratic, while Gummel's method is linear in most cases.

A solution is considered converged and iterations will terminate when either the *X* norm falls below a certain tolerance or the right-hand-side (RHS) norm falls below a certain tolerance. The *X* norm measures the error as the size of the updates to the device variables at each iteration. For the *X* norm, the default tolerance is 10^{-5} kT/q for the potentials and 10^{-5} relative change in the concentrations. The RHS norm measures the error as a function of the difference between the left- and right-hand sides of the equations. For the RHS norm, the tolerance is 10^{-26} C/μm for the continuity equations. By default, MEDICI uses a combination of the *X* and RHS norms to determine convergence and, therefore, alleviates the difficulty of choosing one method or the other. Basically, the program assumes that a solution is converged when the *X* or RHS norm tolerances are satisfied at every node in the device. This often greatly reduces the number of iterations required to obtain a solution, compared to when the *X* or RHS norm is used alone, without sacrificing the accuracy of the solution.

The correct allocation of grid is a crucial issue in device simulation. The number of nodes in the grid has a direct influence on the simulation time. Because the different parts of a device have very different electrical behavior, it is usually necessary to allocate fine grid in some regions and coarse grid in others to maintain reasonable simulation time. Another aspect of grid allocation is the accurate representation of small device geometries. In order to model the carrier flow correctly, the grid must be a reasonable fit to the device shape. This consideration becomes more and more important as smaller, more nonplanar devices are simulated. MEDICI supports a general irregular grid structure. This permits the analysis of arbitrarily shaped devices and allows the refinement of particular regions with minimum impact to others.

Because of the extremely rigid mathematical nature of Poisson's equation and the electron and hole current continuity equations, strong stability requirements are placed on any proposed transient integration scheme. Additionally, it is most convenient to use one-step integration methods so that only the solution at the most recent time step is required. MEDICI has used the first-order (implicit) backward difference in time formula; i.e., electron and hole concentrations are discretized as

$$\frac{n_j - n_{j-1}}{\Delta t_j} = F_n(\psi_n, n_k, p_k) = F_n(j) \tag{1.3.1}$$

$$\frac{p_j - p_{j-1}}{\Delta t_j} = F_p(\psi_p, n_k, p_k) = F_p(j) \tag{1.3.2}$$

where $\Delta t_j = t_j - t_{j-1}$ and ψ_j, n_j, and p_j denotes the potential, electron concentration, and hole concentration at time t_j, respectively. The backward Euler method is known to be both *A*- and *L*-stable, but suffers from a large local truncation error (LTE) which is proportional to the size of the time steps taken.

The backward Euler method is a one-step method. An alternative is the second-order backward difference formula

$$\frac{1}{t_j - t_{j-2}}\left[\frac{(2-\gamma)n_j}{1-\gamma} - \frac{n_{j-1}}{1-\gamma} + \frac{(1-\gamma)n_{j-2}}{\gamma}\right] = F_n(j) \tag{1.3.3}$$

$$\frac{1}{t_j - t_{j-2}}\left[\frac{(2-\gamma)p_j}{1-\gamma} - \frac{p_{j-1}}{1-\gamma} + \frac{(1-\gamma)p_{j-2}}{\gamma}\right] = F_p(j) \tag{1.3.4}$$

where $\gamma = (t_{j-1} - t_{j-2})/(t_j - t_{j-2})$.

At a given point of time, MEDICI checks the local truncation error for the first- and second-order backward difference methods and selects the method that yields the largest time step. By default MEDICI will select time steps so that the local truncation error matches the user-specified criteria. Specifying a larger tolerance will result in a quicker, but less accurate, simulation. MEDICI also places additional restrictions on the time steps:

1. The time step size is allowed to increase at most by a factor of 2.
2. If the new time step is less than half of the previous step, the previous time step is recalculated.
3. If a time point fails to converge, the time step is reduced by a factor of 2 and the point is recalculated.

In addition to steady-state and transient simulation, MEDICI also allows small-signal analysis to be performed. An input of given amplitude and frequency can be applied to a device structure from which sinusoidal terminal currents and voltages are calculated. In MEDICI, the use of small-signal device analysis published by Laux (1985) is adopted. An ac sinusoidal voltage bias is applied to an electrode i such that

$$V_i = V_{i0} + \tilde{v}_i e^{j\omega t} \tag{1.3.5}$$

where V_{i0} is the dc bias, $\tilde{v}_i$ is the small-signal amplitude, ω is the angular frequency, and V_i is the actual bias (sum) to be simulated. Rearranging the basic partial differential equations in (1.2.1)–(1.2.5), we obtain

$$F_\psi(\psi,n,p) = \varepsilon\nabla^2\psi + q(p - n + N_D^+ - N_A^-) + \rho_F = 0 \tag{1.3.6}$$

$$F_n(\psi,n,p) = \frac{\vec{\nabla}\vec{J}_n}{q} - U_n = \frac{\partial n}{\partial t} \tag{1.3.7}$$

$$F_p(\psi,n,p) = -\frac{\vec{\nabla}\vec{J}_p}{q} - U_p = \frac{\partial p}{\partial t} \tag{1.3.8}$$

The ac solution to these equations can be written as

$$\psi_i = \psi_{i0} + \tilde{\psi}_i e^{j\omega t} \tag{1.3.9}$$

$$n_i = n_{i0} + \tilde{n}_i e^{j\omega t} \tag{1.3.10}$$

$$p_i = p_{i0} + \tilde{p}_i e^{j\omega t} \tag{1.3.11}$$

where ψ_{i0}, n_{i0}, and p_{i0} are the dc potential and carrier concentrations at node i while ψ_i, $\tilde{n}_i$, and $\tilde{p}_i$ are the respective ac values. By substituting (1.3.9)–(1.3.11) into (1.3.6)–(1.3.8) and expanding as a Taylor series of first order only, one obtains nonlinear equations of the following form for each of the three partial differential equations:

$$F(\psi,n,p) = F(\psi_0,n_0,p_0) + \tilde{\psi} e^{j\omega t}\frac{\partial F}{\partial \omega} + \tilde{n} e^{j\omega t}\frac{\partial F}{\partial n} + \tilde{p} e^{j\omega t}\frac{\partial F}{\partial p} \tag{1.3.12}$$

1.4. CAPABILITY AND LIMITATIONS OF DEVICE SIMULATORS

Device simulators are power tools to model the two-dimensional or three-dimensional distributions of potential and carrier concentrations as well as current vectors in a device in order to predict its electrical characteristics for any bias condition. Simulators are useful not only for majority carrier devices involving a single carrier type, such as MOSFETs, JFETs, and MESFET, but also for minority carrier devices involving both carriers (electrons and holes) such as *p–n* junction diodes, *n–p–n* bipolar transistors, and *p–n–p–n* thyristors. Device simulation can also be used to study devices under steady-state, small-signal, and transient operating conditions.

The capabilities of each device simulator are described as follows: In SEDAN, once the potential and carrier concentration distributions are evaluated, a variety of electrical parameters could be derived; they are

1. Layer conductances and resistances (all devices)
2. Low-frequency capacitance, flat-band voltage, and threshold voltage (MOS capacitor)
3. Junction capacitances (diode and transistor)
4. Large and small-signal current gains, base and emitter Gummel numbers, Early voltage, and cutoff frequency (transistor)

The SEDAN III version simulates Si, SiGe, GaAs, and AlGaAs devices.

BIPOLE can be used to account for tunneling and thermionic emission across the thin interfacial oxide layer in a polysilicon emitter (it is capable of simulating nonequilibrium transport due to hydrodynamic effects). This can be used for accurate avalanche multiplication and breakdown studies in high-speed transistors with shallow collector regions and heterojunction structures, including graded base Ge fraction in SiGe devices. It can also be used for BiCMOS devices with graded N or P well collectors. BIPOLE also offers terminal characteristics such as f_T and $f_{\max}$ versus I_C and has automatic parameter extraction for SPICE models.

MINIMOS can perform Monte Carlo simulation (Littlejohn *et al.*, 1983) of carrier transport. A Monte Carlo module couples self-consistently to the Poisson equation and replaces the drift–diffusion approximation in critical device areas. It is capable of simulating nonplanar structures in two and three dimensions. The model cards account for avalanche multiplication and hot carrier temperature effects. MINIMOS is also capable of simulating MESFETs on a silicon or GaAs substrate.

MEDICI is able to simulate CMOS latchup, avalanche breakdown, and single-event upset. The simulator accurately describes the behavior of deep submicron devices by providing the ability to solve the electron and hole energy balance equations consistently with the other device equations. Effects such as carrier heating and velocity overshoot are accounted for in MEDICI, and their influence on device behavior can be analyzed. Many physical models are incorporated into the program. Among these are models for recombination, mobility, and lifetime. In the low-field-mobility models, for example, the user has several options, including constant mobility, concentration-dependent mobility, analytical mobility, Arora mobility (Dziewior and Schmid, 1977), carrier–carrier scattering mobility (Dorkel and Letureq, 1981), and Phillips unified mobility (Klaasen, 1992) model. MEDICI also incorporates Boltzmann and Fermi–Dirac statistics, including the incomplete ionization of impurities and effect of carrier freeze-out. This simulator is able to attach lumped resistors, capacitors, and inductors to contacts as well as distributed contact resistances. Both voltage and current boundary conditions during a simulation can be specified. In addition, a light source can be specified in MEDICI for examining the photoelectrical response of photodiodes and phototransistors.

Advanced application modules (AAM) are optionally available for MEDICI. These modules include programmable device AAM, circuit analysis AAM, lattice temperature AAM, and heterojunction device AAM. Programmable device AAM provides the ability to simulate the programming characteristics of nonvolatile memory devices. Circuit analysis AAM allows circuit simulation in conjunction with MEDICI device simulation. Lattice temperature AAM is able to solve the lattice heating equation self-consistently with the other device equations. Heterojunction device AAM simulates heterojunction device behavior. Programmable device AAM is important for nonvolatile memory device design, such as EPROMs, EEPROMs, and flash EEPROMs. Circuit analysis AAM is useful for mixed-mode device and circuit simulation. The impact of device phenomena on circuit behavior can be examined. This module is critical for advanced BiCMOS circuit design since the bipolar transistor of the BiCMOS logic is operated at a high-current regime where high injection and base pushout effects become very significant. The lattice temperature advanced application module is useful for analyzing the effects of lattice heating on a device's electrical performance. A typical example is bipolar transistor thermal runaway. Heterojunction device AAM is useful for the study of a variety of heterojunction devices such as heterojunction bipolar transistors (HBTs) and high-electron-mobility transistors (HEMTs).

ATLAS provides a comprehensive set of physical models such as drift–diffusion transport models, energy balance transport models, and lattice heating and heatsinks. It is able to simulate graded and abrupt heterojunctions, hot carrier injection, band-to-band and Fowler–Nordheim tunneling, nonlocal impact ionization, thermionic currents, optoelectronic interactions, and stimulated emission and radiation. ATLAS has a modular architecture that includes other simulation tools such as MIXEDMODE, which offers circuit simulation capabilities that employ numerical physically based devices as well as compact analytical models, and INTERCONNECT3D, which provides capabilities for three dimensional parasitic (interconnect) extraction.

In general, device simulation provides the ability to perform computer simulation experiments without using semiconductor wafers. The simulation is transparent to semiconductor technologies and independent of expensive fabrication equipment. In addition, the turn-around design time of advanced or prototype devices is reduced. The efficient use of device simulators, however, requires knowledge of semiconductor device physics for the insightful interpretation of device simulation results. The accuracy of device simulation relies heavily on the physical parameters used in the simulator. The simulation results are sensitive to the design of mesh points, especially for nonplanar device structures. The precise doping profile for small-geometry devices is not well defined, which makes the comparison between the simulation result and experimental data difficult. Furthermore, thermionic emission and tunneling mechanisms across the heterojunction are presently not modeled in many simulators. Nevertheless, the device simulation provides a relatively accurate model of device response with respect to boundary conditions. It also provides insight into device operation and doping profile optimization.

1.5. APPLICATIONS OF DEVICE SIMULATION

The use of device simulation provides physical insight into device operation for model development and reduces the design cycle for new transistor design. Figure 1.1

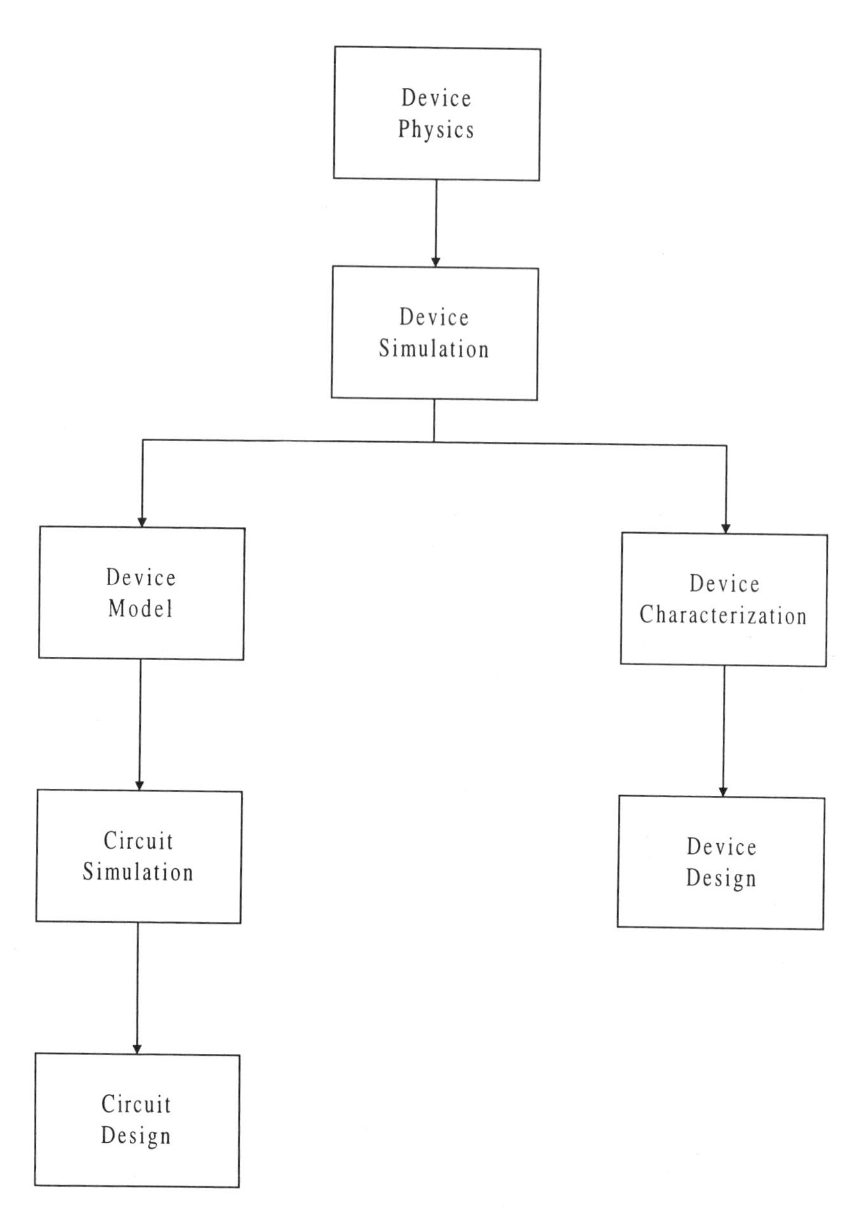

FIGURE 1.1. The role of device simulation in device and circuit design.

shows the flowchart of device and circuit design using device simulation. For example, device physics provides analytical equations and physical parameters for device simulation. The device simulator solves the fundamental semiconductor equations numerically. The dc, ac, and transient simulation results are used for device design and device model development. The device model is implemented in the circuit simulator. The circuit simulation results are then used for the design of integrated circuits. A more detailed flowchart in Fig. 1.2 demonstrates the relationship of device, process, and circuit simulation in Technology CAD. The semiconductor process time, temperature, dopants,

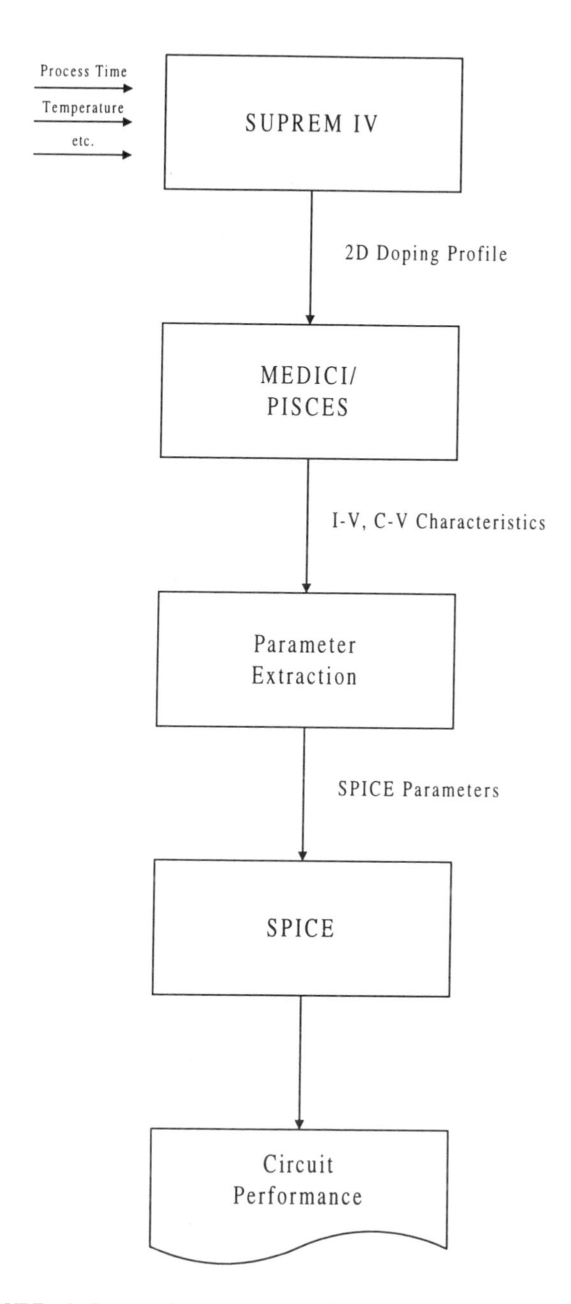

FIGURE 1.2. Process, device, and circuit simulation in Technology CAD.

etc., are specified in the two-dimensional process simulator such as SUPREM IV (1933). The process simulation produces the doping profile in two dimensions. This doping profile is loaded into the two-dimensional device simulator such as MEDICI. The device simulation generates I–V and C–V data of semiconductor devices. With a parameter extraction program, SPICE (Nagel, 1975) circuit parameters are obtained. Based on these circuit parameters, SPICE simulation predicts the circuit performance for IC design.

REFERENCES

Arora, N. D., J. R. Hauser, and D. J. Roulston (1982), *IEEE Trans. Electron Devices* **ED-29**, 292.

Bardeen, J. and W. H. Brattain (1948), *Phys. Rev.* **74**, 230.

Bergren, K.-F. and B. E. Sernelius (1985), *Solid St. Electron.* **28**, 12.

Bube, R. H. (1974), *Electronic Properties of Crystalline Solids: An Introduction to Fundamentals* (Academic Press, New York).

Buturla, E. M., P. E. Cottrell, B. M. Grossman, and K. A. Salsburg (1988), *IBM J. Res. Develop.* **25**, 218.

del Alamo, J. and R. M. Swanson (1984), *IEEE Trans. Electron Devices* **ED-31**, 1878.

del Alamo, J., S. Swirhun, and R. M. Swanson (1985), *IEDM Tech. Dig.* 290.

Dorkel, J. M. and Ph. Leturcq (1981), *Solid-St. Electron.* **24**, 821.

Dumke, W. P. (1983), *Appl. Phys. Lett.* **42**, 196.

Dziewior, J. and W. Schmid (1977), *Appl. Phys. Lett.* **31**, 346.

Dziewior, J. and D. Silber (1979), *Appl. Phys. Lett.* **35(2)**, 170.

Fischer, C., P. Habas, O. Heinreichsberger, H. Kosina, Ph. Lindorfer, P. Pichler, H. Potzl, C. Sala, A. Schutz, S. Selberherr, M. Stiftinger, and M. Thurner (1994), *MINIMOS 6.0*, User's Guide (Institute for Microelectronics, Technical University Vienna).

Franz, A. F., G. A. Franz, W. Kausel, G. Nanx, P. Dickinger, and C. Fischer (1989), *BAMBI 2.1 User's Guide* (Institute for Microelectronics, Technical University Vienna).

Green, M. A. (1990), *J. Appl. Phys.* **67(6)**, 2944.

Kennedy, D. P. and P. C. Murley (1973), *IBM J. Res. Dev.* **17**, 2.

Kennedy, D. P. and R. R. O'Brien (1969), *IBM J. Res. Dev.* **13**, 662.

Klaasen, D. B. M. (1992), *Solid-St. Electron.* **35**, 953.

Laux, S. E. (1985), *IEEE Trans. Electron Devices* **ED-32**, 2028.

Littlejohn, M. A., R. J. Trew, J. R. Hauser, and J. M. Golio (1983), *J. Vac. Sci. Technol. B* **1**, 449.

Misiakos, K., C. H. Wang, and A. Neugroschel, (1989), *IEEE Electron Device Lett.* **EDL-10**, 111.

Mock, M. S. (1973), *Solid-St. Electron.* **24**, 959.

Nagel, L. W. (1975), SPICE2: A computer program to simulate semiconductor circuits, Rep. No. ERL-M520 (Electronics Research Laboratory, University of California, Berkeley).

Roulston, D. J., N. D. Arora, and S. G. Chamberlain (1982), *IEEE Trans. Electron Devices* **ED-29**, 284.

Silvaco International Inc. (1996), ATLAS User Manual, Device Simulation Software.

Slotboom, J. W. (1969), *Electron. Lett.* **5**, 677.

Slotboom, J. W. and H. C. de Graaff (1976), *IEEE Trans. Electron Devices* **ED-19**, 857.

Swirhun, J. S. E., J. del Alamo, and R. M. Swanson (1986), *IEEE Electron. Device Lett.* **EDL-7**, 168.

Sze, S. M. (1981), *Physics of Semiconductor Devices*, 2nd ed. (Wiley Interscience, New York).

Technology Modeling Associates, Inc. (1984), *TMA SEDAN-2: One-Dimensional Device Analysis Program*, User Manual.

Technology Modeling Associates, Inc. (1989), *PISCES: Two-Dimensional Device Analysis Program*, User's Manual.

Technology Modeling Associates, Inc. (1993), *DAVINCI, Three-Dimensional Semiconductor Device Simulation.*

Technology Modeling Associates, Inc. (1993), SUPREM IV, *Two-Dimensional Semiconductor Process Simulation.*

Technology Modeling Associates, Inc. (1993), *MEDICI: Two-Dimensional Semiconductor Device Simulation*, User's Manual.

Technology Modeling Associates in conjunction with Electrical and Computer Engineering Department, University of Waterloo (1993), *BIPOLE3: Bipolar Semiconductor Device Simulation*, User's Manual.

Vandorpe, D. and N. H. Xuong (1971), *Electron. Lett.* **7**, 47.

2

P–N Junction

Most semiconductor devices contain at least one junction between p-type and n-type material. P–n junction theory serves as the foundation of the physics of semiconductor devices. The basic theory of current–voltage characteristics of p–n junctions was established by Shockley (1949). This theory was then expanded by Sah *et al.* (1957) and Moll (1958).

2.1. DEVICE PHYSICS OF p–n JUNCTION

The electric field E and potential V ($E = -dV/dx$) in the transition (or space-charge) region of a p–n junction can be obtained by solving Poisson's equation:

$$\frac{\partial E}{\partial x} = \frac{\rho(x)}{\varepsilon} = \frac{q}{\varepsilon}\left[p(x) - n(x) + N_D^+(x) - N_A^-(x)\right] \tag{2.1.1}$$

In the transition region, electron and hole concentrations are much less than the acceptor or donor concentration. Using the depletion approximation and assuming the complete ionization (i.e., $N_D^+ = N_D$ and $N_A^- = N_A$), Eq. (2.1.1) reduces to

$$\frac{\partial E}{\partial x} \approx \frac{q}{\varepsilon}\left[N_D(x) - N_A(x)\right] \tag{2.1.2}$$

If the doping concentrations in the neutral regions are uniform, and impurity concentration at the metallurgical junction changes abruptly (a step junction), one obtains

$$\frac{-\partial V^2}{\partial x^2} \approx \frac{-qN_A}{\varepsilon} \qquad \text{for } -x_p < x < 0 \tag{2.1.3}$$

$$\frac{-\partial V^2}{\partial x^2} \approx \frac{qN_D}{\varepsilon} \qquad \text{for } 0 < x < x_n \tag{2.1.4}$$

where x_p and x_n are the edges of the space-charge layer in the quasi-neutral p and n regions, respectively. The electric field in the transition region is found by integrating (2.1.3) and (2.1.4)

$$E(x) = \frac{-qN_A(x + x_p)}{\varepsilon} \qquad \text{for } -x_p \le x < 0 \tag{2.1.5}$$

$$E(x) = \frac{qN_D(x - x_n)}{\varepsilon} \qquad \text{for } 0 < x \le x_n \tag{2.1.6}$$

Integrating (2.1.5) and (2.1.6) once again provides the potential distribution $V(x)$:

$$V(x) = \frac{qN_A x_n}{\varepsilon}\left(x - \frac{x^2}{W}\right) \tag{2.1.7}$$

where W is the depletion region width.

As the p–n junction is more positively biased, the depletion width reduces. The effective built-in potential is $V_{bi} - V_A$, where V_{bi} is the built-in potential at thermal equilibrium ($V_{bi} = kT/q \ln(N_A N_D/n_i^2)$) and V_A is the applied voltage. Since $x_n = N_A W/(N_A + N_D)$ and $V(x = W) = qN_A x_n W/2\varepsilon = V_{bi} - V_A$, the depletion width for a step junction is obtained as follows

$$W = \sqrt{\frac{2\varepsilon(N_A + N_D)(V_{bi} - V_A)}{qN_A N_D}} \tag{2.1.8}$$

If the impurity distribution in the transition region is linear, Poisson's equation with the depletion approximation yields

$$\frac{\partial E}{\partial x} = \frac{q}{\varepsilon}[p(x) - n(x) + ax] \approx \frac{qax}{\varepsilon} \tag{2.1.9}$$

where a is the impurity gradient in cm^{-4}.

Integrating Eq. (2.1.9) once gives the field distribution

$$E(x) = -qa\frac{(W/2)^2 - x^2}{2\varepsilon} \tag{2.1.10}$$

Integrating (2.1.10) once again yields the effective potential

$$V_{bi} - V_A = \frac{qaW^3}{12\varepsilon} \tag{2.1.11}$$

The depletion width for a linear junction is thus

$$W = \left[\frac{12\varepsilon(V_{\text{bi}} - V_A)}{qa}\right]^{1/3} \tag{2.1.12}$$

For the above case, free carriers in the space-charge region are neglected when calculating field, potential, and depletion width. To account for free electrons and holes in the space-charge region, we use

$$n(x) = n_i e^{qV(x)/kT} \tag{2.1.13}$$

$$p(x) = n_i e^{-qV(x)/kT} \tag{2.1.14}$$

Inserting (2.1.13) and (2.1.14) into (2.1.1) and (2.1.9), respectively, gives

$$\frac{\partial^2 V}{\partial x^2} = \frac{q}{\varepsilon}\,[n_i e^{-qV(x)/kT} - n_i e^{qV(x)/kT} + N_D^+(x) - N_A^-(x)] \qquad \text{for a step junction} \tag{2.1.15}$$

$$\frac{\partial^2 V}{\partial x^2} = \frac{q}{\varepsilon}\,[n_i e^{-qV(x)/kT} - n_i e^{qV(x)/kT} - ax] \qquad \text{for a linear junction} \tag{2.1.16}$$

Equations (2.1.15) and (2.1.16) can be resolved numerically to determine $V(x)$, $E(x) = -\int V(x)\,dx$, and W (at $\rho(x) = 0$) for a step or linear p–n junction.

In reality, the fabrication of semiconductor junctions that involves the diffusion or the ion implantation process produces a Gaussian profile. The developed analytical equations are not rigorously applicable to the real p–n junction. For this diffused structure, the one-sided step junction solution to Poisson's equation works well for reverse bias,

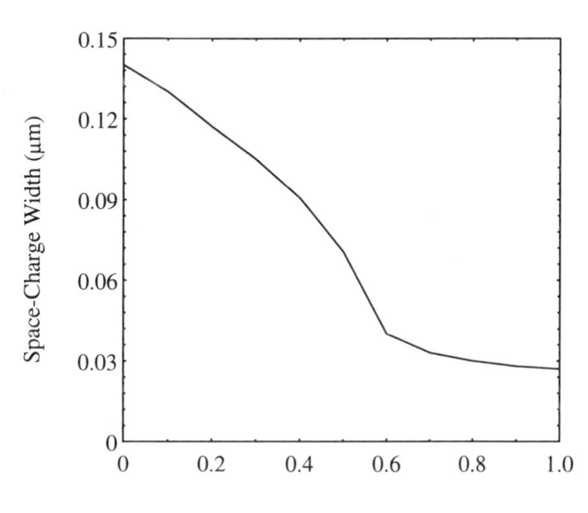

FIGURE 2.1. Space-charge region width versus applied bias.

since the depletion layer extends mainly into the lightly doped n-region. In forward bias, the junction properties tend toward the linearly graded case. An excellent engineering solution to the space-charge layer thickness can be obtained by using device simulation for any shape of impurity profile.

Figure 2.1 shows the MEDICI simulation result of space-charge region width versus the applied voltage. The space-charge region width decreases with increasing junction voltage as expected. Accurate presentation of the space-charge region width under forward bias for all types of doping profiles is one of the advantages of device simulation over analytical results obtained by assuming simplified junction profiles.

2.2. DC CHARACTERISTICS OF A p–n JUNCTION DIODE

2.2.1. *Forward-Bias Current–Voltage Characteristics*

Consider first a p^+–n long-base diode where its n-region width is much longer than the hole diffusion length L_p. The hole concentration in the quasi-neutral n region is determined by the continuity equation for holes:

$$\frac{d^2p(x)}{dx^2} = \frac{p(x)}{L_p^2} \tag{2.2.1}$$

The solution to the differential equation is

$$p(x) = p(0)e^{-x/L_p} \tag{2.2.2}$$

where $p(0)$ is the hole concentration at $x = 0$.

The hole diffusion current density at $x = 0$ is found by taking the derivative and is

$$J_p = -qD_p\frac{dp}{dx} = \frac{qD_p p_{n0}}{L_p}\left(e^{qV_j/kT} - 1\right) \tag{2.2.3}$$

where p_{n0} is the hole concentration at thermal equilibrium and V_j is the junction voltage.

If the width of the quasi-neutral n region W_n is much less than L_p (the definition of narrow-base diode), then the recombination throughout the neutral region is negligible. On the assumption that drift current can be neglected, we obtain

$$J_p = -qD_p\frac{dp}{dx} = \text{constant} \tag{2.2.4}$$

This implies that

$$-\frac{dp}{dx} = \frac{p(0) - p(W_n)}{W_n} \approx \frac{p(0)}{W_n} \tag{2.2.5}$$

The diffusion current density at $x = 0$ for a narrow-base diode is thus

$$J_p = \frac{qD_p p_{n0}}{W_n}(e^{qV_j/kT} - 1) \tag{2.2.6}$$

The only difference between this result and that for the long-base diode lies in the substitution of L_p by W_n. A similar expression of J_n for the p^+ side of a narrow-base diode is

$$J_n = \frac{qD_n n_{p0}}{W_p}(e^{qV_j/kT} - 1) \tag{2.2.7}$$

where n_{p0} is the electron concentration at equilibrium at the edge of space-charge layer in the p^+ side and W_p is the width of the quasi-neutral p region.

The terminal current density of a narrow-base diode is

$$J_t = q\left(\frac{D_p p_{n0}}{W_n} + \frac{D_n n_{p0}}{W_p}\right)(e^{qV_j/kT} - 1) \tag{2.2.8}$$

The terminal current includes the diffusion current as well as the drift current. With assumptions the drift current may be derived analytically. For instance, for a p^+–n junction with a uniformly doped narrow base, the terminal current is predominately hole current. The electron current is assumed to be zero with respect to the hole current:

$$J_n = q\mu_n nE + qD_n \frac{dn}{dx} \approx 0 \tag{2.2.9}$$

The electric field from (2.2.9) is

$$E = -\frac{D_n}{\mu_n}\frac{1}{n}\frac{dn}{dx} = -\frac{kT}{q}\frac{1}{n}\frac{dn}{dx} \tag{2.2.10}$$

Charge neutrality gives the electron concentration from hole concentration distribution $p(x)$:

$$n(x) = p(x) + N_D \tag{2.2.11}$$

Therefore the field becomes

$$E = -\frac{kT}{q}\frac{1}{N_D + p(x)}\frac{dp(x)}{dx} \tag{2.2.12}$$

The total hole current is

$$J_p = -qD_p \frac{p(x)}{N_D + p(x)}\frac{dp(x)}{dx} - qD_p \frac{dp(x)}{dx} \tag{2.2.13}$$

When the minority carrier concentration is much larger than the doping concentration, high-level injection occurs. The conductivity of the quasi-neutral region is increased. The condition of high-level injection can be written as

$$p(x) >> N_D \tag{2.2.14}$$

Equation (2.2.13) reduces to

$$J_p = -2qD_p \frac{dp(x)}{dx} \tag{2.2.15}$$

Regarding the transport of carriers under high injection, it can be concluded that the drift current is equal to the diffusion current. Therefore, its total current can be written as diffusion current with a diffusion constant twice as large. Under the assumption of linear distribution of the carriers due to negligible recombination,

$$\frac{dp(x)}{dx} = \frac{-p}{W_n} \tag{2.2.16}$$

and

$$p \approx n = n_i e^{qV_j/2kT} \tag{2.2.17}$$

Taking into account Eqs. (2.2.16) and (2.2.17), the hole current density becomes

$$J_p = \frac{2qD_p n_i}{W_n} e^{qV_j/2kT} \tag{2.2.18}$$

In Eq. (2.2.18) the current depends again exponentially on the junction voltage, but with the coefficient $q/2kT$ instead of q/kT.

When the p^+–n junction is under high forward bias, the semiconductor experiences a significant ohmic drop in the neutral regions. The actual junction voltage can be much less than the applied voltage V_A and is

$$V_j = V_A - (I_n + I_p)(R_n + R_p) \tag{2.2.19}$$

where R_n is the series resistance in the n region and R_p is the series resistance in the p region.

Figure 2.2 displays the I–V characteristics of the p^+–n junction for different n-type dopings. The p^+–n junction with lower n-type doping has higher terminal current. Also, the high injection and series resistance effects are significant for the p–n junction with lightly doped n region.

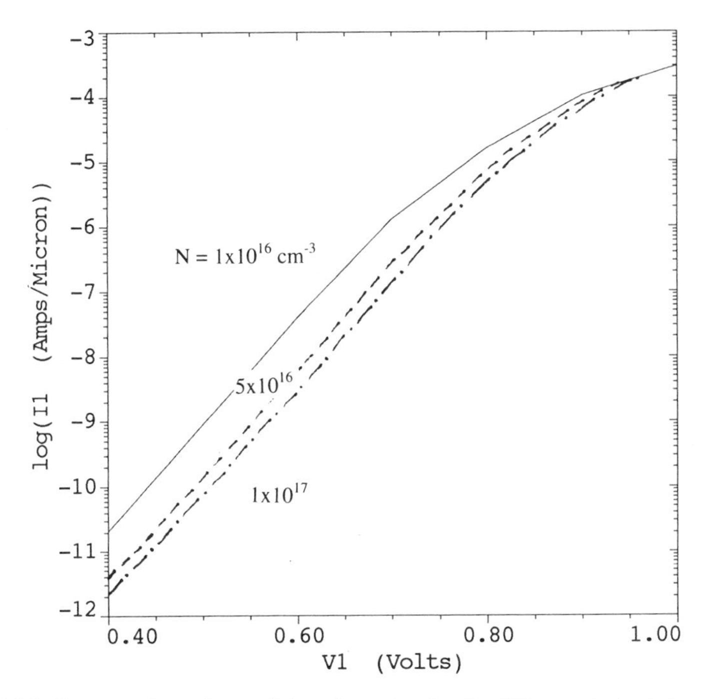

FIGURE 2.2. Current–voltage characteristics of a p–n junction for different n-type substrate dopings.

2.2.2. *Reverse-Bias and Low-Forward-Bias Current–Voltage Characteristics*

When the p–n junction is reverse biased, the carrier concentration is less than n_i inside the depletion layer. Excess carriers created by a forward-biased junction are unlikely to be found in the depletion region or in the quasi-neutral regions. The reverse voltage increases the electric field in the depletion region, as shown in the energy-band diagram in Fig. 2.3. The electric field then sweeps electrons and holes in the depletion region across the junction. Thermal agitation tends to reestablish the equilibrium condition by generating electrons and holes in the depletion region. Thermally generated holes are

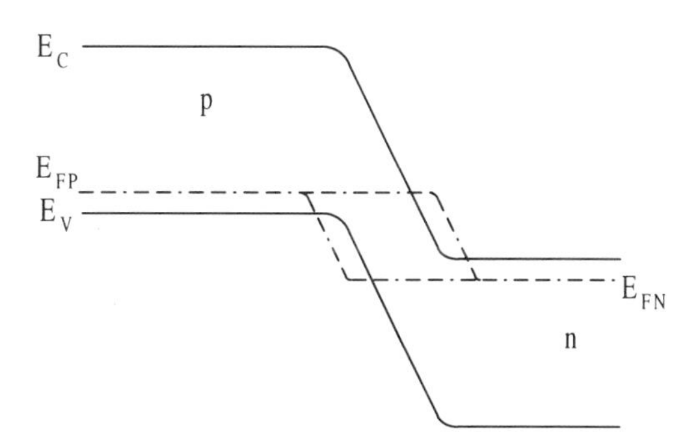

FIGURE 2.3. Energy-band diagram of a reversed-biased p–n junction.

swept across the depletion region by the field to the p-type quasi-neutral region, and thermally generated electrons are swept across the depletion region to the quasi-neutral n region. These carriers flow to the metal contacts and give rise to the terminal current. The terminal current is negative because the directions of carrier flow are opposite to those for a forward-biased junction. The current due to generation in the depletion region is

$$J_{\text{gen}} = -q \int_0^W U \, dx \tag{2.2.20}$$

where the generation–recombination rate in the depletion region is governed by the Shockley–Read–Hall (SRH) recombination

$$U = \frac{\sigma_n \sigma_p \nu_{\text{th}} N_t n_i^2 (e^{qV_j/kT} - 1)}{\sigma_n (n + n_i e^{(E_T - E_i)/kT}) + \sigma_p (p + n_i e^{(E_i - E_T)/kT})} \tag{2.2.21}$$

where σ_p is the hole capture cross section, σ_n is the electron capture cross section, ν_{th} is the thermal velocity, and N_t is the trap density.

Since at the reverse bias $n << n_i$ and $p << n_i$, Eq. (2.2.21) reduces to

$$U = \frac{\sigma_n \sigma_p \nu_{\text{th}} N_t n_i}{\sigma_n e^{(E_T - E_i)/kT} + \sigma_p e^{(E_i - E_T)/kT}} \tag{2.2.22}$$

The current due to the generation in the depletion region is thus

$$J_{\text{gen}} = -q \int_0^W U \, dx \approx -qUW = -\frac{qn_i W}{\tau_{\text{eff}}} \tag{2.2.23}$$

where τ_{eff} is the effective lifetime.

Under low-forward-bias conditions, space-charge recombination current in the p–n junction dominates the diffusion current in the quasi-neutral regions. The major recombination–generation processes in the depletion region are the capture processes. Under the assumptions $E_t = E_i$ and $\sigma_n = \sigma_p = \sigma$, Eq. (2.2.22) reduces to

$$U = \frac{\sigma \nu_{\text{th}} N_t n_i^2 (e^{qV_j/kT})}{n + p + 2n_i} = \frac{\sigma \nu_{\text{th}} N_t n_i (e^{qV_j/kT} - 1)}{e^{q(\psi - \phi_n)/kT} + e^{q(\phi_p - \psi)/kT} + 2} \tag{2.2.24}$$

The maximum value of U exists in the depletion region where the potential ψ is halfway between ϕ_p and ϕ_n, and (2.2.24) becomes

$$U \approx \frac{1}{2} \sigma \nu_{\text{th}} N_t n_i e^{qV/2kT} \tag{2.2.25}$$

and

$$J_{\text{rec}} = q \int_0^W U\, dx \approx \frac{qWn_i}{2\tau_{\text{eff}}} e^{qV/2kT} \tag{2.2.26}$$

Similar to the generation current in reverse bias, the recombination current in the forward bias is also proportional to n_i. The total forward current density at low forward bias is the sum of Eqs. (2.2.8) and (2.2.26):

$$J_F = \left(\frac{qD_p p_{n0}}{W_p} + \frac{qD_n n_{p0}}{W_n} \right) (e^{qV/kT} - 1) + \frac{1}{2} qW\sigma v_{\text{th}} N_t n_i e^{qV/kT} \tag{2.2.27}$$

2.2.3. *Junction Avalanche Breakdown*

When the p–n junction is under substantial reverse bias, the electric field in the depletion region can be fairly high. Free carriers are accelerated by the field to a point where they acquire sufficient energy to create additional carriers when collisions occur. If the distance is large enough for the carriers to be accelerated to a high enough velocity, generation of additional carriers during collisions continues to sustain the avalanche multiplication process. For example, assume that a current I_{p0} is incident at the left-hand side of the depletion region with width W. If the electric field in the depletion region is high enough that electron–hole pairs are generated by impact ionization process, the hole current I_p will increase with distance through the depletion region and will reach a value of $M_p I_{p0}$ at W. Similarly, the electron current I_n will increase from $x = W$ to $x = 0$. The total current $I\,(= I_p + I_n)$ is constant at steady state. The incremental hole current at x equals the number of electron–hole pairs generated per second in the distance dx:

$$\frac{dI_p}{dx} = \alpha_n I_n + \alpha_p I_p \tag{2.2.28}$$

Solving (2.2.28) and using the boundary condition $I = I_p(W) = M_p I_{p0}$ gives (Sze, 1981)

$$I_p(x) = I_p(0) + I_p(W) \frac{\int_0^x \alpha_n \exp\left[-\int_0^x (\alpha_p - \alpha_n)\, dx' \right] dx}{\exp\left[-\int_0^x (\alpha_p - \alpha_n)\, dx' \right]} \tag{2.2.29}$$

Equation (2.2.29) can be rewritten as

$$1 - \frac{1}{M_p} = \int_0^W \alpha_p \exp\left[-\int_0^x (\alpha_p - \alpha_n)\, dx' \right] dx \tag{2.2.30}$$

The avalanche breakdown is defined as the state at which M_p approaches infinity. Hence, the breakdown condition is

$$\int_0^W \alpha_p \exp\left[-\int_0^x (\alpha_p - \alpha_n)\, dx'\right] dx = 1 \tag{2.2.31}$$

or

$$\int_0^W \alpha_n \exp\left[-\int_x^W (\alpha_n - \alpha_p)\, dx'\right] dx = 1 \tag{2.2.32}$$

if the avalanche breakdown is initiated by electrons.

Once the electric field versus distance within the depletion layer is known, the ionization integral can be evaluated numerically. Several sets of ionization coefficient data have been published that enable the multiplication factor to be computed from a solution of Poisson's equation, giving the field E versus distance x (Eltoukhy and Roulston, 1982; Morgan and Smits, 1960).

An empirical expression is often used to relate the avalanche multiplication with the breakdown voltage V_{br} as follows:

$$M = \frac{1}{1 - (V/V_{\text{br}})^m} \tag{2.2.33}$$

where m is an empirical parameter ranging from 2 to 6.

A physical model is also available to describe the avalanche multiplication factor (Poon and Meckwood, 1972):

$$M = 1 + \int_0^W \alpha_n \exp\left[-\frac{E_{\text{crit}}}{E(x)}\right] dx \tag{2.2.34}$$

where E_{crit} is the critical electric field.

Impact ionization can be simulated in MEDICI. The generation rate for electron–hole pairs due to impact ionization can be expressed by

$$G^{\text{II}} = \frac{\alpha_{n,ii} \mid \vec{J}_n \mid}{q} + \frac{\alpha_{p,ii} \mid \vec{J}_p \mid}{q} \tag{2.2.35}$$

where $\alpha_{n,ii}$ and $\alpha_{p,ii}$ are the electron and hole ionization coefficients used in MEDICI. The ionization coefficients are expressed in terms of the local electric field.

2.2.4. *Temperature Dependence of Steady-State Current*

The temperature dependence of the diode steady-state current is important in many applications. Although the diode equation contains the absolute temperature explicitly in the exponent $qV/n_d kT$, where $n_d = 1$ for diffusion current and $n_d = 2$ for recombination current, the principal temperature dependence of the characteristics results from the extremely strong implicit temperature dependence of the preexponential factor. For example, the electron and hole diffusion coefficients in the diffusion current are a function of temperature given by Einstein relation and temperature-dependent mobility. The strongest temperature-dependent parameter in the preexponential factor is the intrinsic carrier concentration n_i. Its temperature-dependent equation is given by Eq. (1.2.10) in Chapter 1. Since the space-charge recombination current is proportional to n_i and the diffusion current is proportional to n_i^2, it is believed that the diffusion current is more sensitive to temperature variation.

The change of temperature affects the avalanche current also. At lower temperatures, the avalanche breakdown voltage decreases. This is because the impact ionization rate increases with decreasing temperature due to an increase of mean free path or a decrease of optical phonon scattering.

2.2.5. *Two-Dimensional Effect*

The actual *p–n* junction diode is three dimensional. If the length of a diffused junction is much larger than its width, a two-dimensional analysis is sufficient. Examining the two-dimensional cross section of a *p–n* junction, the sidewall of a diffused junction contributes the junction capacitance and sidewall injection. This two-dimensional effect

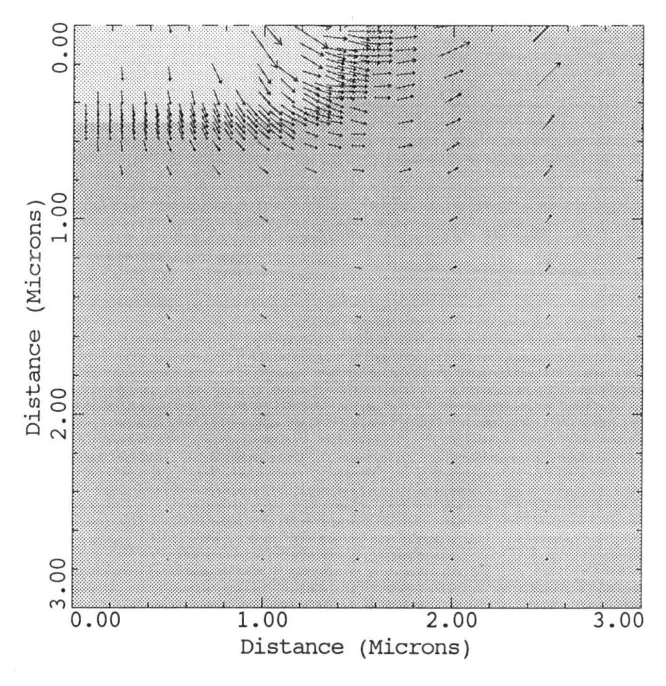

FIGURE 2.4. Current vectors of a *p–n* junction.

becomes significant when the width of the junction is comparable with the junction depth. Figure 2.4 shows the two-dimensional current vectors of a *p–n* junction. The current flow is from the top of the ohmic contacts. Significant sidewall injection is observed as well.

The two-dimensional junction curvature effect must be considered when calculating impact ionization multiplication. Since the top corner of a *p–n* junction has highest doping concentration and field intensity, its impact ionization rate is larger than those at other regions. Figure 2.5 shows the distribution of impact ionization rates from MEDICI. In this plot, the *p–n* junction is biased at −14 V. It is clear from this plot that the impact ionization rate is larger at the surface of the *p–n* junction.

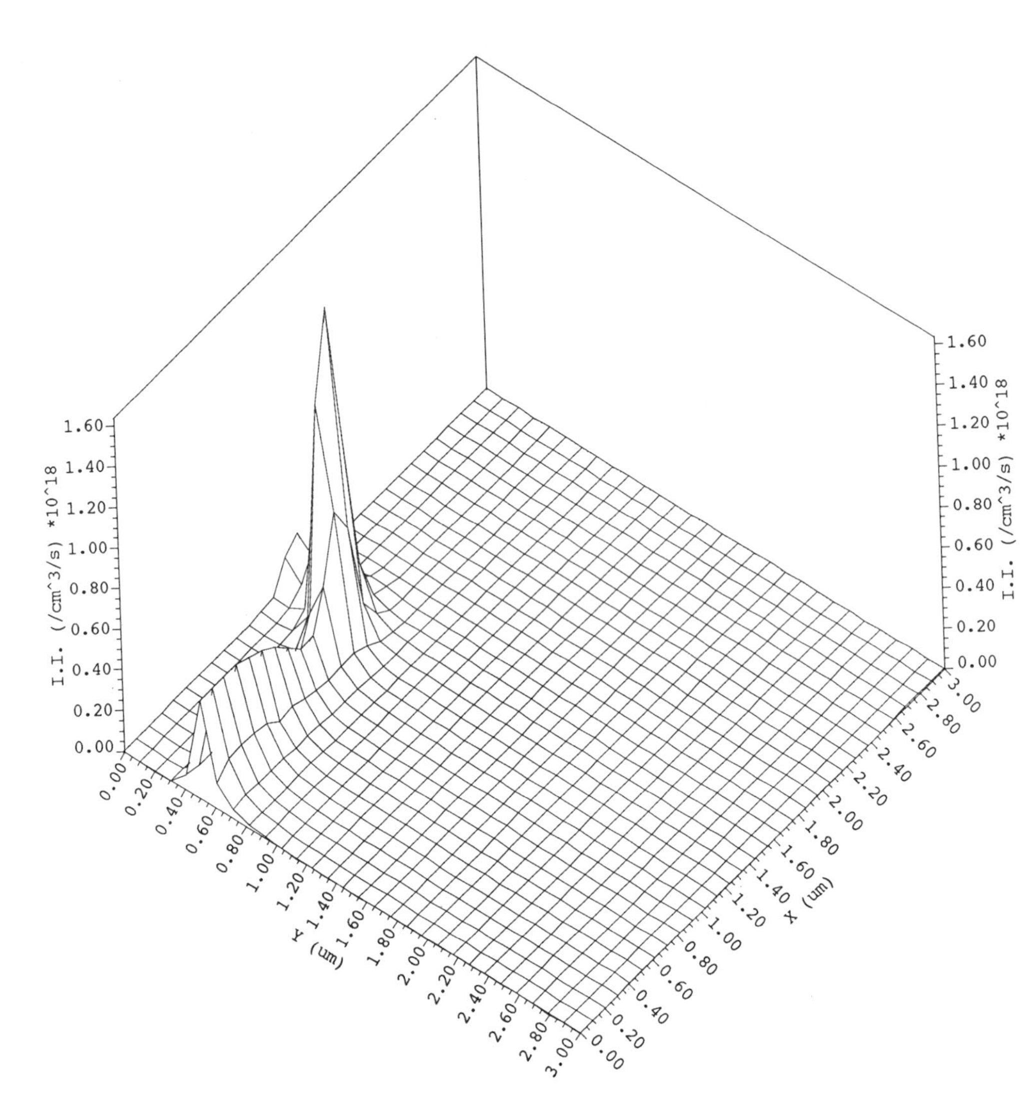

FIGURE 2.5. Impact ionization rates of a *p–n* junction from MEDICI.

2.3. AC PARAMETERS OF A p–n JUNCTION DIODE

The small-signal parameters of a p–n junction diode include the junction capacitance, diffusion capacitance, and conductance. The junction capacitance represents the modulation of the space-charge in the transition region with respect to the change of the junction voltage. The diffusion capacitance represents the modulation of minority carrier charges in the quasi-neutral regions with respect to the change of the junction voltage. The modulation of both types of charges introduces frequency response degradation at high frequency.

2.3.1. *Junction Capacitance*

The change in applied voltage modulates the space-charge layer thickness. As a result, the charge Q in the space-charge region changes with respect to the junction voltage. The small-signal capacitance associated with the dipole of charge in the depletion layer of a p–n junction is defined by

$$C_j = \left| \frac{dQ}{dV} \right| \tag{2.3.1}$$

With the depletion approximation, the space-charge can be easily found by neglecting free electrons and holes in that region. The depletion capacitance per unit area is

$$\begin{aligned} C_j &= \frac{\varepsilon}{W} \\ &= \sqrt{\frac{q\varepsilon N_A N_D}{2(N_A + N_D)(V_{\mathrm{bi}} - V_j)}} && \text{for a step junction} \\ &= \left[\frac{qa\varepsilon^2}{12(V_{\mathrm{bi}} - V_j)} \right]^{1/3} && \text{for a linearly graded junction} \end{aligned} \tag{2.3.2}$$

The depletion approximation only accounts for the change of charge due to shifting the edges of the space-charge region and neglects the change of the free-carrier charge in the volume of the space-charge region associated with the carrier injection from one side of the junction to the other. Thus, the depletion capacitance in (2.3.2) is only accurate when the p–n junction is reverse biased.

Chawla and Gummel (1971) introduced a simple analytical model that is valid for reverse, zero, and small-forward voltages. The analytical equation is valid for the exponential-constant junction and its two extremes, the step and linear-graded junctions. The model retains the same form as that of Shockley's depletion-capacitance model but replaces V_{bi} with an effective built-in voltage so that the effect of the free-carrier charges in the space-charge region is taken into account. Chawla and Gummel developed the capacitance model by first converting the Morgan and Smits model (Morgan and Smits, 1960) into an asymptotic form for large reverse bias. Then they compared the asymptotic form with the result of their contact-to-contact numerical simulation. The numerical results agree with the asymptotic form provided the junction gradient $a > 10^{14}$ cm^{-4}.

Lindholm (1982) derived a capacitance model for high forward voltages for a general junction based on phenomenological reasoning. He decomposed the capacitance into the dielectric capacitance C_ε and the free-carrier capacitance C_F:

$$C_\varepsilon = \frac{\varepsilon}{W} \tag{2.3.3}$$

$$C_F = q\int_0^W \frac{\partial p}{\partial V}\,dx \tag{2.3.4}$$

The separation stems from the physical reasoning that C_ε is the capacitance corresponding to the change of the free carriers at the edges of the space-charge region, whereas C_F is the capacitance corresponding to the change of the free carriers within the volume of the space-charge region. For reverse bias, $p = 0$, C_j reduces to C_ε, which is the depletion capacitance in the classical treatment. By reasoning that the derivative of C_j with respect to the thickness of the space-charge region equals zero, Lindholm obtained C_j at very high forward bias. At moderate forward voltages, the Hermite polynomial can be used to bridge the gap between the low forward bias and the very high forward bias (Liou *et al.*, 1985).

2.3.2. *Diffusion Capacitance and Conductance*

When a small ac signal is applied to a junction that is forward biased by a voltage V_0 and the current density is J_0, the total voltage and current density are defined by

$$V(t) = V_0 + \tilde{V}_1 e^{j\omega t} \tag{2.3.5}$$

$$J(t) = J_0 + \tilde{J}_1 e^{j\omega t} \tag{2.3.6}$$

where $\tilde{V}_1$ ($<< V_0$) and $\tilde{J}_1$ are the small-signal amplitude of the voltage and current density, respectively. The small-signal ac component of the hole density is

$$\begin{aligned} p_n(x,t) &= p_{n0} e^{q(V_0 + \tilde{V}_1 e^{j\omega t})/kT} \\ &= p_{n0} e^{qV_0/kT} + p_{n0}\frac{q\tilde{V}_1}{kT} e^{qV_0/kT} e^{j\omega t} \end{aligned} \tag{2.3.7}$$

The first term in Eq. (2.3.7) is the dc component, and the second term is the time-dependent small-signal ac component at the depletion boundary. Substituting p_n into the continuity equation for holes yields

$$j\omega p_n = -\frac{p_n}{\tau_n} + D_p \frac{\partial^2 p_n}{\partial x^2} \tag{2.3.8}$$

or

$$\frac{\partial^2 p_n}{\partial x^2} - \frac{p_n(1 + j\omega\tau_p)}{D_p \tau_n} = 0 \tag{2.3.9}$$

Solving the differential equation and making the appropriate substitutions, we have

$$J_1 = \frac{q\tilde{V}_1}{kT}\left(\frac{qD_p p_{n0}\sqrt{1 + j\omega\tau_p}}{L_p} + \frac{qD_n n_{p0}\sqrt{1 + j\omega\tau_n}}{L_n}\right) \tag{2.3.10}$$

From Eq. (2.3.10) the ac admittance per unit area is obtained:

$$Y = \frac{\tilde{J}_1}{\tilde{V}_1} = G + j\omega C_d \tag{2.3.11}$$

For low frequencies ($\omega_p << 1/\tau_p$), the conductance is

$$G = \frac{q}{kT}\left(\frac{qD_p p_{n0}}{L_p} + \frac{qD_n n_{p0}}{L_n}\right) e^{qV_0/kT} \tag{2.3.12}$$

and the diffusion capacitance is

$$C_d = \frac{q}{kT}\left(\frac{qL_p p_{n0}}{2} + \frac{qL_n n_{p0}}{2}\right) e^{qV_0/kT} \tag{2.3.13}$$

Note that the conductance is identical to that obtained by differentiating the diffusion current with respect to voltage. The diffusion capacitance is the same as that obtained by differentiating the minority carrier charges with respect to the quasi-neutral regions. For high frequencies, the diffusion capacitance decreases with increasing frequency and is approximately proportional to $\omega^{-1/2}$. For this reason and because C_d is proportional to $\exp(qV_0/kT)$, the diffusion capacitance is important at low frequencies and under forward-bias conditions.

2.4. TRANSIENT BEHAVIOR OF A p–n JUNCTION DIODE

When the bias condition across a p–n junction diode is suddenly changed, the minority carriers stored in the quasi-neutral regions cannot respond instantaneously. The diode exhibits transient characteristics from the time the boundary condition is abruptly changed to the time the stored charges return to the steady-state condition. The distribution of the minority carrier concentration in the base with time can be described based on constant voltage or constant current at the terminal. When the device is suddenly turned

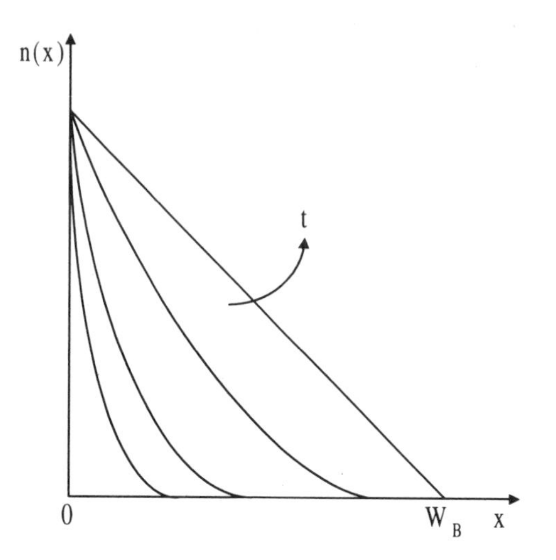

FIGURE 2.6. Minority carrier concentration during turn-on transient under a constant voltage drive.

on and its junction voltage is assumed finite instantaneously, the minority carrier concentration builds up gradually, as shown in Fig. 2.6. If the *p–n* junction is under a constant current drive, its minority concentration as a function of time changes and is displayed in Fig. 2.7. The slope of minority carrier concentration is constant under a constant current drive. In the circuit environment, the diode is used in conjunction with an external resistor. The minority carrier concentration distribution as a function of time in fact has combined effects of constant current and constant voltage drives. This is evidenced by the MEDICI simulation. The MEDICI input file is shown in Example 2.1. In the example X and Y MESH points are designed first. The ELECTRIC contacts are on the top of the silicon diode. The *p–n* junction doping profiles are specified using PROFILE card. REGRID is used to refine the mesh. A lumped resistor is attached to the p^+ contact with a value of $1 \times 10^5\ \Omega$. Physical models in the simulation include SRH

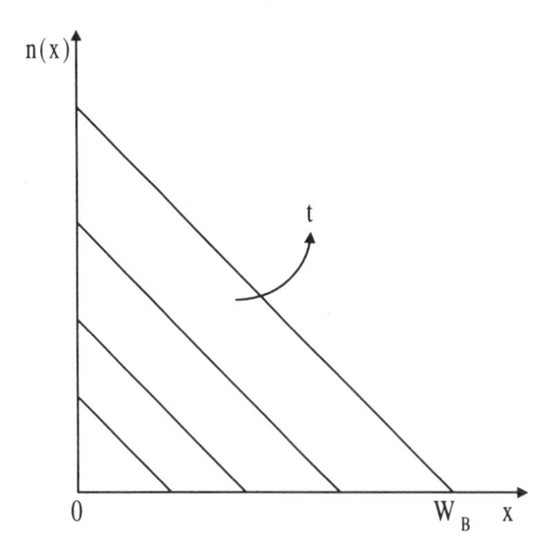

FIGURE 2.7. Minority carrier concentration during turn-on transient under a constant current drive.

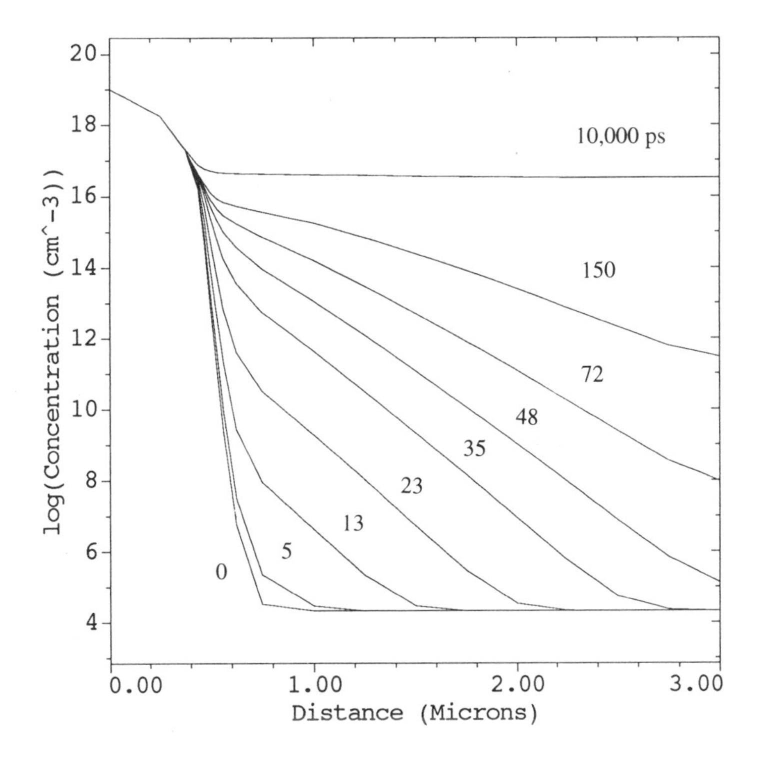

FIGURE 2.8. Minority carrier concentration during turn-on transient of a diode with an external resistor.

recombination, Auger recombination, concentration-dependent mobility, and field-dependent mobility. The simulation uses the Newton method. Both electrons and holes are simulated. The input is switched from 0 to 2 V with 1-ps rise time and 10-ns duration. The diode is under forward-bias transient conduction. The hole concentration from top (p^+ contact) to bottom (substrate) at 0, 5, 13, 23, 35, 48, 72, 150, and 10,000 ps is displayed in Fig. 2.8. As seen in this figure, the hole concentration at the edge of the emitter–base space-charge region (at about 0.6 μm) and the slope of the minority carrier concentration in the lightly doped n region change during the switching-on transient. This indicates that the analysis using either a constant voltage or a constant current drive is inadequate to describe the real switching response. The turn-off transient of a p–n junction can be analyzed similarly. The turn-off transient behavior can be used to characterize the minority carrier lifetime and diffusion length of the diode. These methods include the open-circuit voltage decay (OCVD), the short-circuit current decay (SCCD), and the reverse recovery transient (RRT).

2.4.1. Open-Circuit Voltage Decay

The open-circuit voltage decay method sets the initial conditions by exciting the diode with forward voltage or with incident light before $t = 0$. At time $t = 0$, one switches

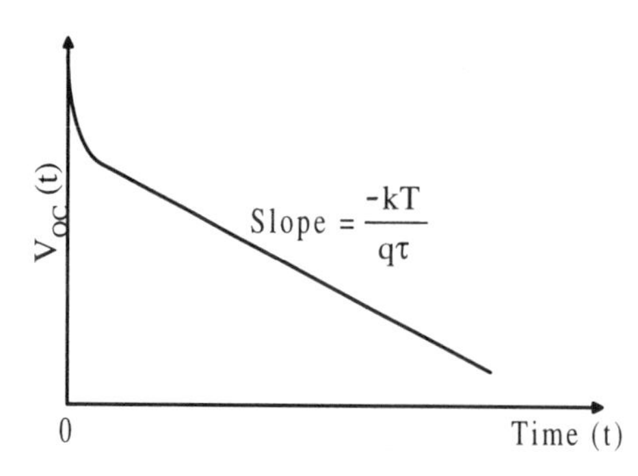

FIGURE 2.9. Transient response of open-circuit-voltage decay.

off the excitation and observes the voltage decay. In the theoretical OCVD described in (Lederhandler and Giacoletto, 1955), $v(t)$ shows an initial phase of rapid decay corresponding to an infinite number of large natural frequencies or short relaxation times characteristic of a distributed system. Then $v(t)$ follows the response from the dominant natural frequency associated with the pole on the real axis nearest the origin. This response, in terms of the minority carrier density, is proportional to $\exp(-t/\tau)$, where τ is the recombination lifetime of the quasi-neutral base. Because of the exponential relation between the minority carrier density at the edge of this region and the nonequilibrium component of the junction voltage, the voltage at the terminals falls off in a straight line having slope = $-(kT/q)/\tau$. This simple relation applies for a long-base diode, for which the thickness of the quasi-neutral base greatly exceeds the minority carrier diffusion length. Figure 2.9 illustrates the response, from which one infers the value of τ.

Under low level of injection, open-circuit voltage decays linearly with time to a close approximation and measurement of the slope of this straight-line decay suffices to determine τ. This result applies if the minority carriers in the base dominate in determining the response, as in the case of Ge diodes. But if the carriers associated with junction SCR or with QNE contribute appreciably to the response, then the OCVD method is not applicable. As indicated in Lederhandler and Giacoletto (1955), this response is observed in Si diodes, which enables determination of τ. It is not observed in Si *p–n* junction solar cells, despite the widespread assumption to the contrary both for Si photovoltaics and Si power devices. Apparently, the first work to recognize the invalidity of the direct use of OCVD for Si was that of Neugroschel *et al.* (1978), in which results from OCVD were compared with those obtained by admittance-bridge techniques. Later authors, including Mahan and Barnes (1981) and Green (1983), also recognized the invalidity of the direct use of OCVD for Si devices and presented modified versions of the method designed to extract the value of τ.

That OCVD works for Ge devices but not for Si devices stems from the differences in the energy gap and hence in the intrinsic carrier density n_i. The intrinsic density of Ge exceeds that of Si by about three orders of magnitudes. From the theory of *p–n* junctions, the excess hole and electron densities in the junction space-charge region are proportional to n_i; these mobile carrier densities contribute to the quasi-static capacitance of this region. This capacitance is neglected in the ideal or conventional OCVD interpretation. In contrast, the excess hole and electron densities in the quasi-neutral base (whose recom-

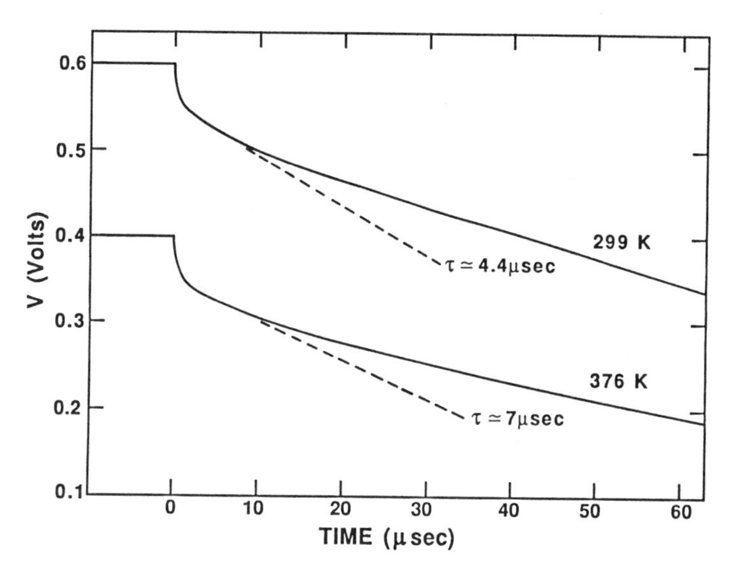

FIGURE 2.10. Experimental data of open-circuit-voltage decay.

bination is the central feature of conventional OCVD) are proportional to n_i^2. The pertinent ratio is n_i. For Ge, the ratio is so large that the quasi-static capacitance can be neglected. For Si and even more so for GaAs, the capacitance slows the relaxation of the excess holes and electrons, producing a curving derivation from the ideal falling straight line. The experimental data for Si p–n junction diodes in Fig. 2.10 illustrates this phenomenon. In Fig. 2.10 dashed lines correspond to the lifetimes τ determined by the capacitance method (Neugroschel *et al.*, 1978).

2.4.2. Short-Circuit Current Decay

The initial conditions of short-circuit current decay can be established electrically or optically. In this method, the very small external resistance used to monitor the current $i(t)$ provides a nearly perfect short-circuit across the junction space-charge region (Fig. 2.11). Thus, the junction region attains a nearly perfect equilibrium value in which the excess hole and electron concentrations practically vanish. Hence, in principle, only the relaxation of excess charge in the quasi-neutral regions determines the observed $i(t)$. From this response, illustrated in Fig. 2.12, one can determine both the recombination lifetime τ and back surface recombination velocity S_B of the quasi-neutral base. The response is described by an infinite series of exponential decays:

$$i(t) = \sum_{i=1}^{\infty} i_i(0)e^{-t/\tau_{di}} = i(0)e^{-t/\tau_{di}} = i(0)e^{-t/\tau_d} + \sum_{i=2}^{\infty} i_i(0)e^{-t/\tau_{di}} \tag{2.4.1}$$

where τ_{di} is the decay time constant of the ith mode, τ_d is the decay time constant of the first order mode and $i_i(0)$ is the corresponding initial value at $t = 0$. As shown in Jung *et*

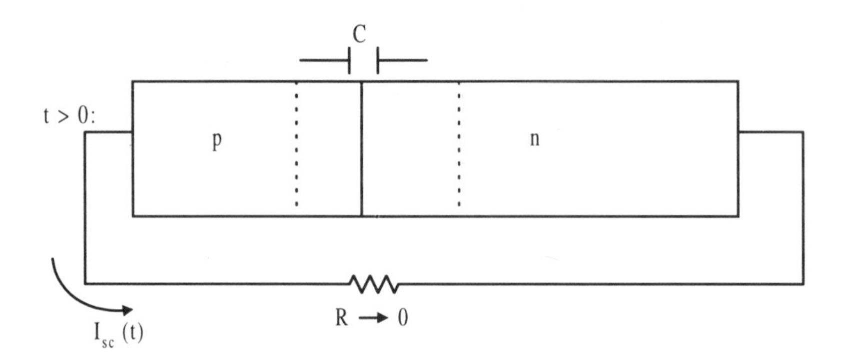

FIGURE 2.11. Schematic of short-circuit-current decay.

al. (1984), the first-order mode dominates and the higher decay modes can be neglected. One measures the slope of the straight-line part of ln[i(0)], the intercept of its extrapolation, and the forward voltage, $v(t<0)$, which sets the initial condition. These observations suffice to determine S_B and τ. The component of $i(t)$ derived from conduction and recombination in the quasi-neutral emitter contributes only to the initial part of the transient.

From a mathematical standpoint, the poles corresponding to relaxation in the quasi-neutral emitter mix with the poles corresponding to the fast relaxation of the quasi-neutral base. Following the near vanishing of the resultant fast part of the response,

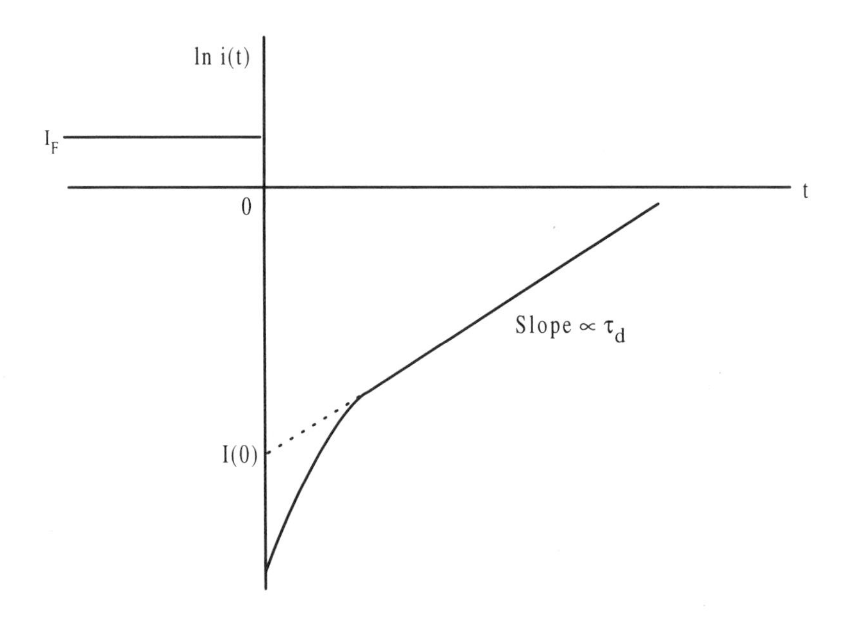

FIGURE 2.12. $i(t)$ of short-circuit-current decay.

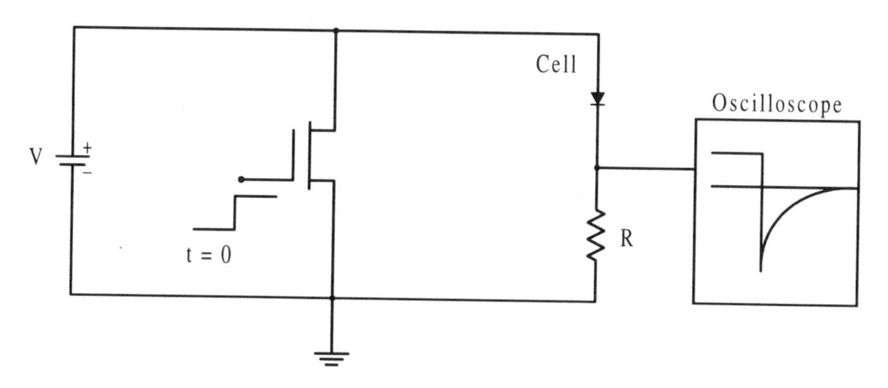

FIGURE 2.13. MOS transistor of short-circuit-current decay circuit.

only the dominant-pole response remains. This contributes to the straight-line portion of the response in Fig. 2.12, from which S_B and τ are determined. The circuit used for electrical short-circuit current decay (Fig. 2.13) utilizes an MOS transistor switch, which provides faster switching ($\approx$1 ns).

The switch must have a characteristic switching time one order of magnitude smaller than the observed decay time. This decay time may be as much as one order of magnitude smaller than the recombination lifetime. This is because the excess holes and electrons vanish from the quasi-neutral base not only by volume recombination but also by conduction through the junction region on one side of the base and through the base surface on the other. Further, the external resistance R(ext) interacts with the equilibrium or zero-bias value of the junction capacitance, and the resultant time constant R(ext)C must be an order of magnitude or more smaller than the decay time of the straight-line portion of log[$i(t)$]. In this regard, note that R(ext) is the sum of the resistance used in the external circuit for monitoring the current decay, the resistance of the contacts to the device being measured, and the series resistance of the device. Thus, attaining the inequality $R(\text{ext})C \ll \tau_d$ demands the exercise of some precautions. If the series resistance of the solar cell or diode is large, other methods exist to overcome this difficulty.

In electrical SCCD, the short circuit across the junction depresses the excess hole and electron densities to zero at both the quasi-neutral emitter and the quasi-neutral base edges of the junction space-charge region. Thus, the poles in the complex frequency plane corresponding to the quasi-neutral emitter simply add to those corresponding to the quasi-neutral base. Because the poles associated with the emitter produce a shorter relaxation time relative to the dominant-pole relaxation time of the base, which is the observable used in determining τ and S_B, the transient response from the emitter has no effect on the SCCD method. In this sense, the response from the heavily doped emitter is decoupled from that of the base. This observation holds despite the presence of so-called heavy doping effects, which may yield a large steady-state recombination current of the quasi-neutral emitter.

The discharging of excess holes and electrons within the junction SCR in the SCCD method occurs within a time of about 10^{-11} s, which is much less than any time associated with discharge of the quasi-neutral regions. This absence in effect of excess holes and electrons within the SCR greatly simplifies interpretation of the observed transient. It is one of the main advantages of this method of measurement.

2.4.3. *Reverse Recovery Transient*

In most switching applications, a diode is switched from forward conduction to a reverse-biased state, and vice versa. The turn-off condition for the reverse recovery transient is depicted in Fig. 2.14. Here, one forces a reverse current through the diode by reversing the source voltage to some large negative value and then extracts the excess stored charge. The slope of the carrier concentration at the space-charge layer boundary is then reversed in direction due to the current reversal. However, since it takes a finite time t_s for $p(0)$ to become zero, the current will remain approximately constant for a finite time. When $p(0) = 0$ the total terminal voltage across the diode is zero.

When the stored charge is depleted and the carrier concentration at $x = 0$ becomes negative, the junction exhibits a negative voltage. Since the reverse-bias voltage of a junction can be large, the source voltage begins to divide between R and the junction. As time proceeds, the magnitude of the reverse current becomes smaller. The junction capacitance and circuit resistance form an RC time constant, which also affects the transient response. The current finally reaches its reverse saturation current level, which is a characteristic specific to the diode being studied.

The reverse-recovery method differs from the OCVD only in that an external current source, I(reverse), consisting of an ideal voltage source in series with a relatively large resistance, speeds the removal of the excess holes and electrons stored within the device. As the main observation, the reverse-recovery method employs the duration of time that the current stays constant, which is approximately the duration that the voltage across the junction region stays in the forward direction. In OCVD, I(reverse) = 0, and the voltage $v(t)$ is observed. Because the incremental current or time variation in current is zero, that is, I(reverse) = 0, the two methods, OCVD and RRT, are alike. The device is connected to an incremental open circuit. Thus, mathematically, the initial–boundary-value problems for the two methods have the same natural frequencies (or relaxation times or poles in the complex frequency plane).

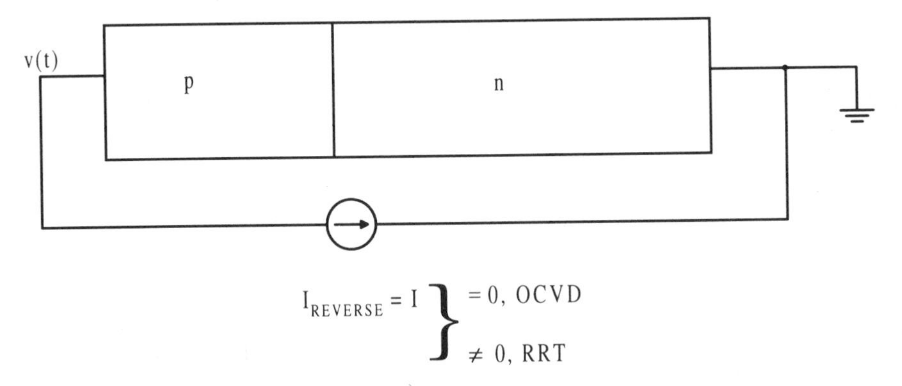

FIGURE 2.14. Schematic of reverse-recovery transient.

2.5. SCHOTTKY DIODES

The Schottky diode is electrically similar to an abrupt one-side p^+–n junction, but several important differences make Schottky diodes attractive and useful. First, the diode operates as a majority carrier device under low-level injection conditions. As a result, storage time due to the storage of minority carriers is eliminated and a fast response is obtained. Second, for a fixed voltage bias, the current through the device is typically more than two orders of magnitude greater than it is for the junction device. When the Schottky diode is connected between the base and collector of the bipolar transistor, the saturation effect and turn-off storage time of the bipolar transistor are significantly reduced, due to the contribution of the Schottky diode; that is, the Schottky diode provides a limiting voltage between the base–collector junction of the bipolar transistor that prevents hard saturation of the BJT and improves the switching speed of bipolar logic.

The current transport in metal–semiconductor contacts is mainly due to majority carriers. Four basic transport processes are usually presented:

1. Transport of electrons from the semiconductor over the potential barrier into the metal
2. Quantum-mechanical tunneling of electrons through the barrier
3. Recombination in the space-charge region
4. Hole injection from the metal to the semiconductor

In addition, the Schottky diode may have edge leakage current due to a high electric field at the contact periphery or interface current due to traps at the metal–semiconductor interface.

The transport of electrons over the potential barrier can be described by the thermionic emission. The thermionic emission theory assumes that

1. The barrier height is much larger than kT.
2. Thermal equilibrium is established at the plane that determines emission.
3. The existence of a net current flow does not affect this equilibrium so that one can superimpose the current flux from metal to semiconductor and that from semiconductor to metal.

The current density of a Schottky diode has a similar expression as that of a p–n junction:

$$J_n = A^* T^2 e^{-q\phi_B/kT}(e^{qV_j/n_d kT} - 1) \tag{2.5.1}$$

where A^* is the effective Richardson constant, ϕ_B is the barrier height, and n_d is the diode ideality factor.

2.6. HETEROJUNCTION

The use of energy gap variations beside electric fields to control the forces acting on electrons and holes results in a greater design freedom and permits an optimization of doping levels and geometries, leading to higher heterostructure performance for microwave and high-speed circuit applications (Kromer, 1982). Three types of heterojunctions are frequently used. These are the abrupt heterojunction, linearly graded heterojunction, and heterojunction with a setback layer. For an abrupt *N–p* heterojunction (capital *N* represents a wide-band-gap semiconductor), the conduction energy spike necessitates the thermionic emission and tunneling for electron transport from the *N* to *p* quasi-neutral regions and the valence-band discontinuity suppresses hole injection from the *p* to *N* regions. The energy-band diagram of a heterojunction is shown in Fig. 2.15. Note that in the energy-band diagram, the electron quasi-Fermi level is discontinuous at the emitter–base heterojunction. This is because in an abrupt *N–p* heterojunction a forward bias reduces the barrier to electron flow into the base yet increases the barrier to electron flow in the opposite direction. This leads to a situation where the net electron flow may not be small compared to each of the counterdirected flows. In other words, unlike in a homojunction, the flow of electrons cannot be considered as a perturbation from equilibrium. Therefore, the electron quasi-Fermi level is discontinuous at the metallurgical boundary of an abrupt heterojunction (Pulfrey, 1993).

The carrier injection efficiency can be improved if a thin region adjacent to the heterojunction is graded. Such a layer, normally having a thickness between 100 and 300 Å, can effectively remove the spike and thus make the therminoic and tunneling mechanisms less important. A linearly graded heterojunction with the complete removal of energy-band spike and discontinuity has similar transport behavior to that of a homojunction.

Another approach to improve the injection efficiency is to insert a thin intrinsic GaAs spacer or setback layer at the heterojunction. The setback layer reduces the impurity out-diffusion from the heavily doped region and alters the potential barrier of the heterojunction for improving the injection efficiency. The addition of the spacer layer increases not only the injected current across the heterojunction but also the recombination at the heterojunction space-charge region.

A linear-graded heterojunction with a setback layer is shown in Fig. 2.16. Region I ($-X_1 < x < -X_p$) has a constant permittivity. The graded layer ($-X_p < x < 0$) has a

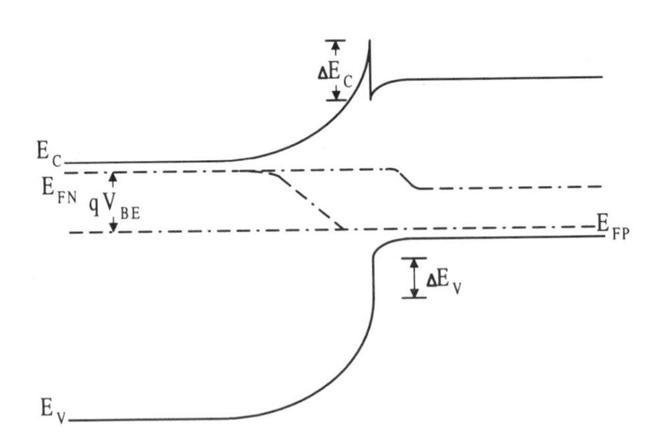

FIGURE 2.15. Heterojunction energy-band diagram.

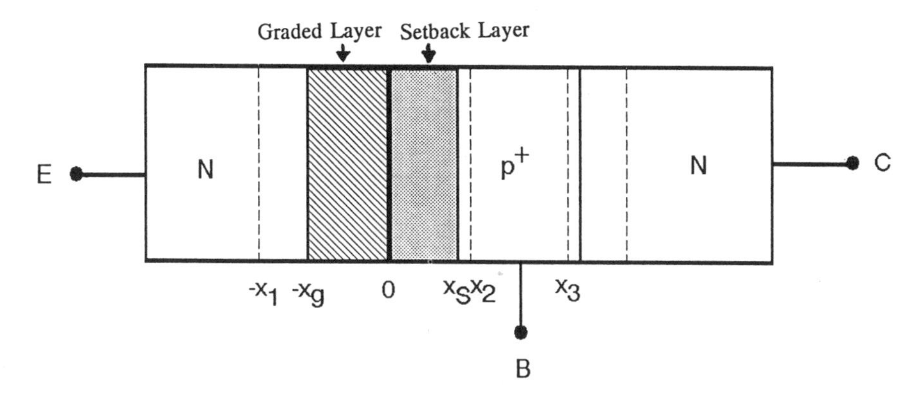

FIGURE 2.16. Schematic of a heterojunction.

position-dependent permittivity as $\varepsilon(x) = \varepsilon_2 - (\varepsilon_1 - \varepsilon_2)x/X_0$. The permittivity in the setback layer ($0 < x < X_S$) and region II ($X_S < x < X_2$) is independent of position. Assuming free-carrier concentrations are much smaller than donor and acceptor concentrations (i.e., the depletion approximation), one finds the electric field in each region using the following equations:

$$E_1(x) \approx \frac{qN_D(x + X_1)}{\varepsilon_1} \qquad \text{for } -X_1 < x < -X_g \tag{2.6.1}$$

$$E_g(x) = \frac{\rho x}{\varepsilon(x)} + \frac{\varepsilon_2 E_g(0)}{\varepsilon(x)} \qquad \text{for } -X_g < x < 0 \tag{2.6.2}$$

$$E_s(x) \approx \frac{qN_A(X_2 - X_s)}{\varepsilon_2} \qquad \text{for } 0 < x < X_s \tag{2.6.3}$$

$$E_2(x) \approx \frac{qN_A(X_2 - x)}{\varepsilon_2} \qquad \text{for } X_s < x < X_2 \tag{2.6.4}$$

The electric displacement ($D = \varepsilon E$), $D_1(-X_g) = D_g(-X_g)$, provides the boundary condition to solve for X_2:

$$X_2 = X_S + X_1 \frac{N_D}{N_A} \tag{2.6.5}$$

Furthermore, using $V_{\text{bi}} - V = -V_2(X_2)$ yields

$$X_2 = \frac{-b + \sqrt{b^2 - 4ac}}{2a} \tag{2.6.6}$$

where

$$a = \frac{qN_D}{2\varepsilon_1}\left(\frac{N_A}{N_D}\right)^2 + \frac{qN_A}{2\varepsilon_2}$$

$$b = -\frac{qN_D}{\varepsilon_1}\left(\frac{N_A}{N_D}\right)^2 X_S - \frac{qX_g N_A}{\varepsilon_1} + \frac{qN_A X_g}{\varepsilon_2 - \varepsilon_1}\ln\left(\frac{\varepsilon_2}{\varepsilon_1}\right)$$

$$c = \frac{qN_D}{2\varepsilon_1}\left(\frac{N_A}{N_D}X_S + X_g\right)^2 - \frac{qN_A X_S^2}{2\varepsilon_2} + \frac{qN_D X_g^2 - qN_A X_g X_S \ln(\varepsilon_2/\varepsilon_1)}{\varepsilon_2 - \varepsilon_1}$$

$$- \frac{qN_D X_g^2 \varepsilon_2}{(\varepsilon_2 - \varepsilon_1)^2}\ln\left(\frac{\varepsilon_2}{\varepsilon_1}\right) - (V_{bi} - V)$$

The modulation of space charge in the depletion region with respect to the applied bias results in a depletion capacitance. The depletion capacitance of a heterojunction is expressed as

$$C_J = \left|\frac{dQ_n}{d(V_{bi} - V)}\right| = qN_A\left|\frac{dX_2}{d(V_{bi} - V)}\right| \tag{2.6.7}$$

Using (2.6.6) in deriving $dX_2/d(V_{bi} - V)$ and inserting the resulting equation into (2.6.7) gives the depletion capacitance per unit area

$$C_J = \frac{qN_A}{\sqrt{b^2 - 4ac}} \tag{2.6.8}$$

Equation (2.6.8) is a general solution for the heterojunction with a linearly graded layer and a setback layer. For a linearly graded heterojunction without a setback layer ($X_S = 0$), (2.6.8) reduces to

$$C_J = \frac{qN_A}{\sqrt{\Delta}}$$

where

$$\Delta = \left(\frac{qN_A X_g}{\varepsilon_1} + \frac{qN_A X_g \ln(\varepsilon_2/\varepsilon_1)}{\varepsilon_2 - \varepsilon_1}\right)^2 - 4\left[\frac{qN_D}{2\varepsilon_1}\left(\frac{N_A}{N_D}\right)^2 + \frac{qN_A}{2\varepsilon_2}\right]$$

$$\times\left[\frac{qN_DX_g^2}{2\varepsilon_1}+\frac{qN_DX_g^2}{\varepsilon_2-\varepsilon_1}-\frac{qN_DX_g^2\varepsilon_2}{(\varepsilon_2-\varepsilon_1)^2}\ln\left(\frac{\varepsilon_2}{\varepsilon_1}\right)-(V_{bi}-V)\right]$$

For an abrupt heterojunction with a setback layer ($X_g = 0$), (2.6.8) reduces to

$$C_J=\left[2(V_{bi}-V)\frac{\varepsilon_1N_D+\varepsilon_2N_A}{q\varepsilon_1\varepsilon_2N_AN_D}+\frac{X_S^2}{\varepsilon_2^2}\right]^{-1/2}$$

For an abrupt heterojunction without a setback layer ($X_g = X_S = 0$), (2.6.8) reduces to the well-known formula (Shur, 1987) as follows:

$$C_J=\sqrt{\frac{q\varepsilon_1\varepsilon_2N_AN_D}{2(V_{bi}-V)(\varepsilon_1N_D+\varepsilon_2N_A)}}$$

Figure 2.17 plots the effect of setback layer thickness on the heterojunction capacitance at $V = -1.0, -0.5, 0, 0.5$, and 1.0 V. In this figure the squares, lines, and triangles represent the analytical predictions, using (2.6.8), and the circles represent the MEDICI simulation. In the MEDICI simulation, the heterojunction module is used with the standard capabilities of MEDICI to provide with heterojunction properties. The parameters available for describing the properties of the heterojunction are the usual parameters available for describing the properties of the materials to meet at the heterojunction. Some of these include the energy-band-gap parameters E_g, electron affinity χ, densities of states

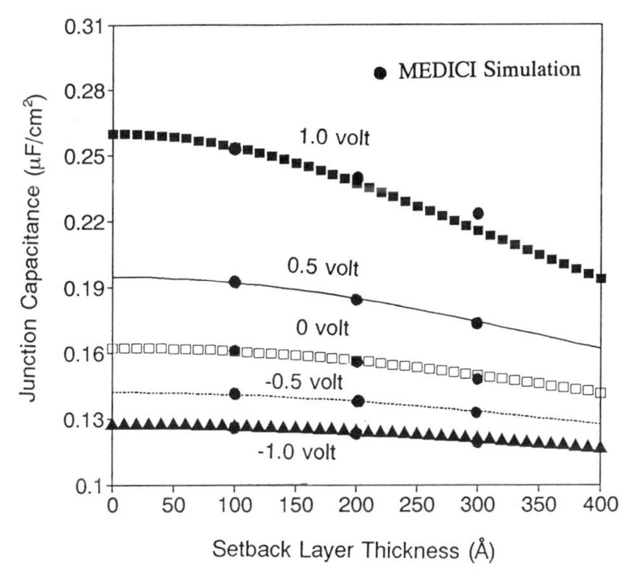

FIGURE 2.17. Heterojunction capacitance versus bias.

for the conduction band N_C, and densities of states for the valence band N_V, as well as the various parameters for describing recombination, mobility, etc. The intrinsic Fermi potential and the vacuum level are related by

$$\psi_{\text{vacuum}} = \psi - \theta \tag{2.6.9}$$

where ψ is the intrinsic Fermi potential and θ is the band structure given by

$$\theta = \chi + \frac{E_g}{2q}\frac{kT}{q}\ln\left(\frac{N_C}{N_V}\right) \tag{2.6.10}$$

Due to differences in band gap, electron affinity, and densities of states in adjacent materials, Poisson's equation is written in the form

$$\vec{\nabla}\varepsilon\vec{\nabla}(\psi - \theta) = -q(p - n + N_D^+ - N_A^-) - \rho_F \tag{2.6.10}$$

The form of the continuity equations remains unchanged for heterojunctions except that the electric field terms $\vec{E}_n$ and $\vec{E}_p$ in the transport equations account for gradients in conduction and valence-band edges:

$$\vec{E}_n = -\nabla\left[\psi + \frac{kT}{q}\ln(n_{ie})\right] = -\nabla\left[\frac{-E_C}{q} + \frac{kT}{q}\ln(N_C)\right] \tag{2.6.12}$$

$$\vec{E}_p = -\nabla\left[\psi - \frac{kT}{q}\ln(n_{ie})\right] = -\nabla\left[\frac{-E_V}{q} + \frac{kT}{q}\ln(N_V)\right] \tag{2.6.13}$$

In Fig. 2.17 the agreement between the analytical results and MEDICI heterojunction simulation is very good. The junction capacitance decreases with setback layer thickness for all junction voltages owing to the increase of space-charge layer thickness. The sensitivity of the heterojunction capacitance with respect to the setback layer thickness is more pronounced at higher forward biases.

EXAMPLE 2.1

```
COMMENT  PN Diode Turn-On Transient Simulation
COMMENT  Define mesh points
   MESH  ^DIAG.FLI
 X.MESH  Y.MAX=3.0  H1=0.50
 Y.MESH  Y.MAX=3.0  H1=0.25
 REGION  NUM=1  SILICON
COMMENT  Define electrical contacts
 ELECTR  NUM=1  TOP  X.MIN=0.0  X.MAX=1.0
 ELECTR  NUM=2  TOP  X.MIN=2.5  X.MAX=3.0
COMMENT  Define doping profiles
PROFILE  N-TYPE  N.PEAK=1E17  UNIF
```

```
        PROFILE  P-TYPE N.PEAK=1E19 X.MIN=0.0 WIDTH=1.0
              +  X.CHAR=0.2 Y.MIN=0.0 Y.JUNC=0.5
        COMMENT  Refine the mesh points
         REGRID  DOPING LOG RATIO=3 SMOOTH=1
         REGRID  DOPING LOG RATIO=3 SMOOTH=1
         REGRID  DOPING LOG RATIO=3 SMOOTH=1
        COMMENT  Add a resistor in the contact
        CONTACT  NUM=1 RESIST=1E5
        COMMENT  Physical models used in device simulation
         MODELS  SRH AUGER CONMOB FLDMOB
        COMMENT  Use Newton method to solve both minority and majority carriers
           SYMB  NEWTON CARRIERS=2
          SOLVE  INI OUT.FILE=PN00
        COMMENT  Switching transient simulation
          SOLVE  V1=2 TSTEP=1E-12 TSTOP=10E-9 OUTF=PN01
        COMMENT  plot the hole concentration at different time points
           LOAD  IN.FILE=PN00
        PLOT.1D  HOLES Y.LOG X.ST=0 X.EN=0 Y.ST=0 Y.EN=3
+device=p/postscript x.length=11 y.length=11
           LOAD  IN.FILE=PN04
        PLOT.1D  HOLES Y.LOG X.ST=0 X.EN=0 Y.ST=0 Y.EN=3 UNCH
+device=p/postscript x.length=11 y.length=11
           LOAD  IN.FILE=PN07
        PLOT.1D  HOLES Y.LOG X.ST=0 X.EN=0 Y.ST=0 Y.EN=3 UNCH
+device=p/postscript x.length=11 y.length=11
           LOAD  IN.FILE=PN10
        PLOT.1D  HOLES Y.LOG X.ST=0 X.EN=0 Y.ST=0 Y.EN=3 UNCH
+device=p/postscript x.length=11 y.length=11
           LOAD  IN.FILE=PN13
        PLOT.1D  HOLES Y.LOG X.ST=0 X.EN=0 Y.ST=0 Y.EN=3 UNCH
+device=p/postscript x.length=11 y.length=11
           LOAD  IN.FILE=PN15
        PLOT.1D  HOLES Y.LOG X.ST=0 X.EN=0 Y.ST=0 Y.EN=3 UNCH
+device=p/postscript x.length=11 y.length=11
           LOAD  IN.FILE=PN17
        PLOT.1D  HOLES Y.LOG X.ST=0 X.EN=0 Y.ST=0 Y.EN=3 UNCH
+device=p/postscript x.length=11 y.length=11
           LOAD  IN.FILE=PN20
        PLOT.1D  HOLES Y.LOG X.ST=0 X.EN=0 Y.ST=0 Y.EN=3 UNCH
+device=p/postscript x.length=11 y.length=11
           LOAD  IN.FILE=PN30
        PLOT.1D  HOLES Y.LOG X.ST=0 X.EN=0 Y.ST=0 Y.EN=3 UNCH
+device=p/postscript x.length=11 y.length=11
```

REFERENCES

Chawla, B. R. and H. K. Gummel (1971), *IEEE Trans. Electron Devices* **ED-18**, 178.

Eltoukhy, A. A. and D. J. Roulston (1982), *Solid-St. Electron.* **25**, 829–831.

Green, M. A. (1983), *Solid-St. Electron.* **26**, 1117–1122.
Jung, T. W., F. A. Lindholm, and A. Neugroschel (1984), *IEEE Trans. Electron Devices* **ED-31**, 588–595.
Kromer, H. (1982), *IEEE Pro.* **70**, 13.
Lederhandler, S. R. and L. J. Giacoletto (1955), *Proc. IRE* **3**, 477–483.
Lindholm, F. A. (1982), *J. Appl. Phys.* **53**, 7606–7608.
Liou, J. J., F. A. Lindholm, and J. S. Park (1985), *IEEE Trans. Electron Devices* **ED-32**, 2415–2419.
Mahan, J. E. and D. L. Barnes (1983), *Solid-St. Electron.* **24**, 989–994.
Moll, J. L. (1958), *Proc. IRE* **46**, 1076.
Morgan, S. P. and F. M. Smits (1960), *Bell Syst. Tech. J.* **39**, 1573–1602.
Neugroschel, A., P. J. Chen, S. C. Pao, and F. A. Lindholm (1978), *IEEE Trans. Electron Devices* **ED-25**, 485–490.
Poon, H. C. and J. C. Meckwood (1972), *IEEE Trans. Electron Devices* **ED-19**, 90.
Pulfrey, D. L. (1993), *IEEE Electron Devices* **ED-40**, 1183–1185.
Sah, C. T., R. N. Noyce, and W. Shockley (1957), *Proc. IRE* **45**, 1228.
Shockley, W. (1949), *Bell Syst. Tech. J.* **28**, 435.
Shur, M. (1987), *GaAs Devices and Circuits* (Plenum, New York).
Sze, S. M. (1981), *Physics of Semiconductor Devices*, 2nd ed. (Wiley Interscience, New York).

3

Bipolar Junction Transistors

The bipolar junction transistor (BJT) was invented by a research team at Bell Laboratories in 1947. In 1948 John Bardeen and Walter Brattain announced the development of the point-contact transistor (Bardeen and Brattain, 1948). In the following year William Shockley's paper on junction diodes and transistors was published (Shockley, 1949). The development of conventional bipolar theory and fabrication was very active from the 1950s to 1960s. Because of the need for low power dissipation and high packaging density for VLSI applications, major R & D efforts focused on the study of complementary metal–oxide semiconductor (CMOS) field-effect transistors in the 1970s. R & D in BJT technology became dormant for approximately a decade. The activity of bipolar development was revitalized when the polysilicon emitter was introduced in the BJT at IBM's Watson lab in the early 1980s (Ning *et al.*, 1981a). Conventional bipolar theory was reexamined and improved for advanced polysilicon bipolar transistor design and fabrication.

3.1. DEVICE PHYSICS

Consider the energy-band diagram of the n–p–n bipolar transistor in the forward-active mode in Fig. 3.1 at which the emitter–base junction is forward biased and the collector–base junction is reverse biased. The forward-biased junction voltage reduces the potential barrier at the emitter–base junction and triggers the injection of electrons from the emitter to the base. The reduced potential barrier also helps holes diffuse through the forward-biased emitter–base junction into the emitter region. The supply of electrons at the emitter contact provides the emitter current, and the supply of holes at the base contact results in the base current. Since the collector junction is reverse biased, the energy band bends down and prevents hole injection from the base to the collector. The electrons injected into the base, however, are assisted by the field in the collector–base depletion region and are swept to the collector. The collection of electrons at the collector contact gives rise to the collector current.

The magnitude of collector current depends on the electron injection from the emitter and is proportional to $\exp(qV_{BE}/kT)$ (Sze, 1981). For the first-order approximation,

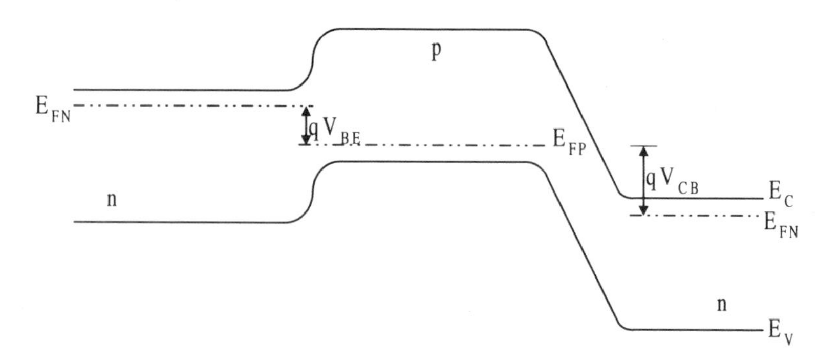

FIGURE 3.1. Energy-band diagram of an *n–p–n* bipolar transistor.

$I_C \approx I_S \exp(qV_{BE}/kT)$, where I_S is the saturation current at the base–emitter voltage $V_{BE} =$ 0 V. To improve the accuracy of the model, physical mechanisms such as base-width modulation (Early, 1952) and base conductivity modulation (Webster, 1954) must be accounted for. Base-width modulation affects the effective base width, and base conductivity modulation changes the effective base concentration. A larger effective base width and concentration product increases the base Gummel number and decreases the collector current. To examine the detail of base-width and conductivity modulation effects on the collector current, we derive the Gummel–Poon collector current equation (Gummel and Poon, 1970).

3.1.1. Collector Current

The derivation of the collector current for the BJT with nonuniform base doping has been given by Getrev (1976). The electron and hole current densities of an *n–p–n* BJT are (Streetman, 1990; Sze, 1981; Getreu, 1976)

$$J_n = q\mu_n(x)n(x)\,E(x) + qD_n(x)\frac{dn(x)}{dx} \tag{3.1.1}$$

$$J_p = q\mu_p(x)p(x)\,E(x) - qD_p(x)\frac{dp(x)}{dx} \tag{3.1.2}$$

Assume the hole current density is much larger than the electron current density ($J_n >> J_p \approx 0$). Rearranging (3.1.2) and solving for the field gives

$$E = \frac{D_p}{\mu_p}\frac{1}{p(x)}\frac{dp(x)}{dx} \tag{3.1.3}$$

Substitution of E into J_n results in

$$J_n = qD_n(x)\,\frac{n(x)}{p(x)}\,\frac{dp(x)}{dx} + qD_n(x)\,\frac{dn(x)}{dx} \tag{3.1.4}$$

Using the Einstein relationship (Streetman, 1990; Sze, 1981)

$$\frac{D_n}{\mu_n} = \frac{D_p}{\mu_p} = \frac{kT}{q}$$

we obtain

$$p(x)\,J_n = qD_n\,\frac{d[n(x)\,p(x)]}{dx} \tag{3.1.5}$$

If the base width of the BJT is much less than the diffusion length of the minority carriers, the base recombination is negligible. J_n is constant for dc and independent of x. Integrating both sides of (3.1.5) and removing J_n from the integral yields

$$J_n = \frac{qD_n[n(X_c)p(X_c) - n(X_E)p(X_E)]}{\int_0^{X_B} p(x)\,dx} \tag{3.1.6}$$

where X_B is the base width.

Using the boundary conditions of the p–n product at the emitter and collector junctions (Bardeen and Brattain, 1948)

$$n(X_E)\,p(X_E) = n_i^2 e^{qV_{BE}/kT}$$

$$n(X_C)\,p(X_C) = n_i^2 e^{qV_{BC}/kT}$$

the collector current density ($J_C = -J_n$) is therefore

$$J_n = \frac{-qD_n n_i^2\,(e^{qV_{BE}/kT} - e^{qV_{BC}/kT})}{\int_0^{X_B} p(x)\,dx} \tag{3.1.7}$$

where V_{BE} is the base–emitter junction voltage and V_{BC} is the base–collector junction voltage. The collector current in (3.1.7) accounts for base-width modulation and base conductivity modulation through the integration of the hole concentration from $x = X_E$ to $x = X_C$.

The base charge Q_B ($= q\int p(x)\,dx$) is divided into five components (Getreu, 1976), as shown in Fig. 3.2.

$$Q_B = Q_{B0} + Q_E + Q_C + Q_F + Q_R \tag{3.1.8}$$

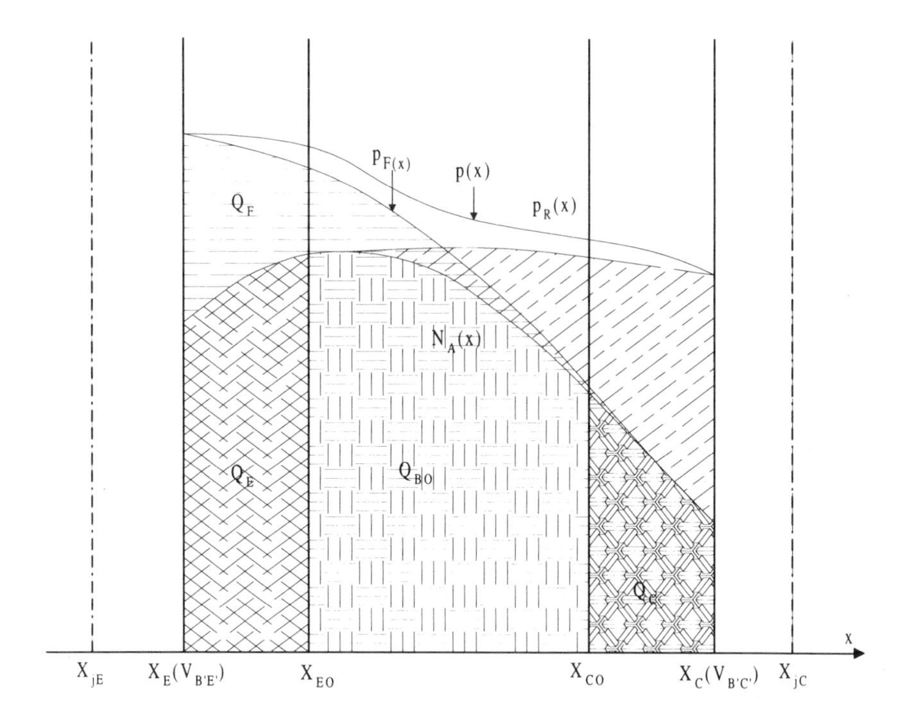

FIGURE 3.2. Base charge of the bipolar transistor.

The base charge at thermal equilibrium is

$$Q_{B0} = q \int_{X_{E0}}^{X_{C0}} N_A(x)\, dx \tag{3.1.9}$$

where N_A is the base doping concentration.

The base charge Q_E due to base-width modulation at the emitter–base junction is

$$Q_{BE} = q \int_{X_E}^{X_{E0}} N_A(x)\, dx \tag{3.1.10}$$

The base charge Q_C due to base-width modulation at the collector–base junction is

$$Q_C = q \int_{X_{C0}}^{X_C} N_A(x)\, dx \tag{3.1.11}$$

The base charge Q_F due to base conductivity modulation in the forward-active mode is

$$Q_F = q \int_{X_E}^{X_C} [p_F(x) - N_A(x)]\, dx \tag{3.1.12}$$

where p_F is the hole concentration in the base in the forward-active mode.

The base charge Q_R due to base conductivity modulation at the inverse-active mode is

$$Q_R = q \int_{X_E}^{X_C} [p_R(x) - N_A(x)]\, dx \tag{3.1.13}$$

where p_R is the hole concentration in the base at the inverse-active mode.

The base-width modulation at the collector–base junction caused by changing the collector–emitter voltage V_{CE} for a given V_{BE} is called the Early effect (Early, 1952). The base-width modulation at the emitter–base junction caused by changing the emitter–collector voltage for a given V_{BC} is called the late effect. The Early and late effects become very important when the base thickness X_B is small and base concentration is low. This implies that Q_E and Q_C are comparable to Q_{B0} when base-width modulation is significant. The base conductivity modulation is caused by high injection in the base. It is also called the Webster effect (Webster, 1954). Base conductivity modulation becomes very important when the base doping is low and the applied bias is high. To reduce base-width modulation and base conductivity modulation effects for modern bipolar transistors, higher base doping is needed. A higher base doping concentration also reduces the base resistance and improves the switching speed of the BJT.

3.1.2. Base Current

The base current consists of the base bulk recombination, emitter bulk recombination, emitter–base space-charge layer recombination, collector–base space-charge layer recombination, and emitter–base surface recombination. The emitter bulk recombination current is due to hole back injection into the emitter, and base bulk recombination is caused by electron–hole recombination in the base. The bulk recombination current is governed by the diffusion process (Sze, 1981), and its magnitude is proportional to $\exp(qV_{BE}/kT)$. The space-charge recombination current is a result of the SRH recombination process (Sah *et al.*, 1957) in the depletion or space-charge region. This recombination current is proportional to $\exp(qV_{BE}/2kT)$. Because of high surface states at the semiconductor surface, the recombination current at the silicon and silicon dioxide interface between the base and emitter contacts can be significant. The surface recombination current, which is process dependent, is proportional to $\exp(qV_{BE}/n_F kT)$, where $1 \le n_F \le 2$. The space-charge and surface recombination reduce the current gain of the BJT at low V_{BE} and result in $1/f$ noise in the base current of the BJT (Pawlikiewicz and van der Ziel, 1987). The bulk recombination current dominates the space-charge recombination current when V_{BE} is sufficiently high. For modern bipolar transistors, the recombination current in the emitter is much larger than that in the base because the emitter doping is higher.

3.1.3. Current Gain

The current gain β generally varies with collector current and base–emitter voltage. A typical β plot is shown in Fig. 3.3. At low base–emitter voltages, the contribution of the recombination current in the depletion region and at the surface is large compared with the diffusion current of minority carriers across the base. The current gain increases with the collector current as follows:

$$\beta \propto \frac{e^{qV_{BE}/kT}}{e^{qV_{BE}/2kT}} \propto \sqrt{I_C} \tag{3.1.14}$$

As the base current reaches the mid-current region, the current gain increases to a high plateau. In this region, both collector current and base current are determined by diffusion current. The slight decrease of β at moderate V_{BE}'s is due to base-width modulation, which decreases the collector current. For even higher V_{BE}'s, the injected carrier density in the base approaches the majority carrier density, and the injected carriers effectively increase the base doping. Since the base current during injection is determined by diffusion current in the emitter, high injection in the base, which decreases the collector current, does not change the base current. Consequently, the current gain at higher collector currents decreases with collector current as follows:

$$\beta \propto \frac{e^{qV_{BE}/2kT}}{e^{qV_{BE}/kT}} \propto \frac{1}{\sqrt{I_C}} \tag{3.1.15}$$

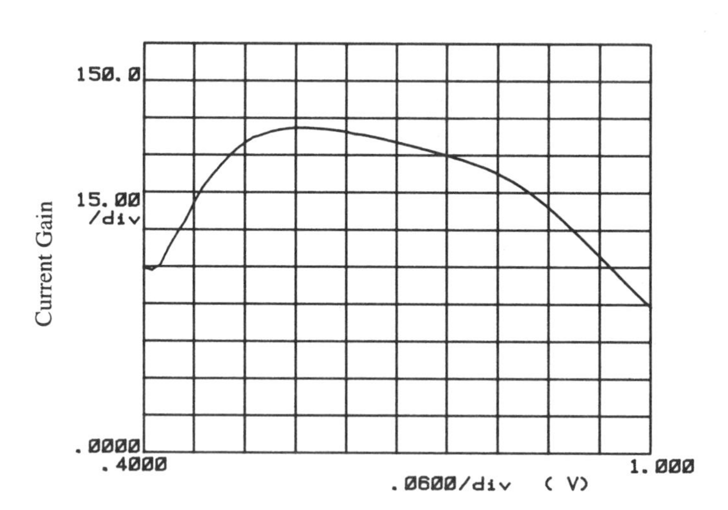

FIGURE 3.3. Current gain versus base–emitter voltage.

Another degradation mechanism of the current gain at higher collector currents is base pushout or Kirk effect (Kirk, 1962). Base pushout increases the effective base width. This increases the base Gummel number and decreases the collector current and current gain. In addition to base widening, physical effects such as emitter crowding (Yuan and Liou, 1989), quasi-saturation (Jeong and Fossum, 1985), and collector current spreading (Yuan and Eisenstadt, 1988a), are important at higher collector currents. These mechanisms, which are not accounted for in the Gummel–Poon equations, will be discussed in detail in Sections 3.2.4 and 3.2.5.

3.2. DC CHARACTERISTICS AND SIMULATION

The current–voltage characteristics during forward-active, inverse-active, saturation, and inverse saturation modes of operation and for different temperatures are presented in this section. Physical mechanisms of emitter current crowding (dc, ac, and transient crowding), base pushout, collector current spreading, current-dependent base resistance (Yuan *et al.*, 1988), avalanche multiplication (Sze and Gibbons, 1966), and base current reversal (Lu and Chen, 1989) will be illustrated in detail. Experimental data and device simulation results are given in conjunction with qualitative analysis. Experimental results are shown to guide our presentation. Device simulations are conducted to provide physical insight into device operation, not to duplicate experimental data. The use of device simulation is also useful for model development, process diagnosis, and device design in a Technology CAD environment.

3.2.1. Current–Voltage Characteristics for Forward-Active and Saturation Operation

When the bipolar transistor is in the forward-active mode, electrons diffuse through the forward-biased emitter–base junction into the base, and holes diffuse through the forward-biased junction into the emitter. The injected electrons in the base are swept by the field at the collector–base junction to the collector. When the bipolar transistor is in the saturation mode, electrons are injected from both the emitter and collector regions and holes are injected from the base contact to supply recombination currents in the emitter, base, and collector. In this mode of operation, the collector current is decreased and the base current is increased, resulting in a significant reduction of current gain in saturation. This result is consistent with physical insight into minority carrier concentration distribution in the base. For example, when the bipolar transistor is in saturation, the minority carrier concentration slope is smaller. This decreases the collector current since J_C is proportional to dn/dx. The redistribution of minority carrier concentration in the base increases the base charge. The recombination current in the base is also increased. In addition, recombination in the collector begins when the collector–base junction becomes forward biased.

The saturation effect has never been correctly accounted for in the bipolar current equations (Gummel and Poon, 1970; Getrev, 1976). For example, in the derivation of collector current for the Gummel–Poon model, $J_p \approx 0$ is assumed. The assumption states that the current gain of the BJT is high. For the bipolar transistor operated in saturation, however, the current gain is small. The assumption of negligible hole current density is invalid for the BJT in saturation. Therefore, Eq. (3.1.7) is valid only for the BJT in the forward-active mode, not in the saturation mode, of operation.

The use of a 2-D device simulator provides physical insight into device operation in any mode. The internal physical responses such as electron and hole concentrations, current vectors, and potential contours of the BJT can be obtained from device simulation. Physical simulation is useful for analyzing the impact of different modes of operation on a device's dc, ac, and transient behaviors as well as for designing high-performance transistors.

We will now simulate a bipolar transistor. A generic advanced BJT cross section was constructed by incorporating common features of many state-of-the-art transistors reported in the literature (Konaka *et al.*, 1986; Tang *et al.*, 1987). Only a half cross section and doping profile is needed due to the cross section's symmetry. The cross section of the BJT and its vertical doping profile are shown in Figs. 3.4 and 3.5, respectively. The BJT has a 1.2-μm emitter width ($W_E/2 = 0.6$ μm), a 0.15-μm emitter junction depth, a 0.3-μm base junction depth, and a 0.7-μm epilayer depth. The doping profiles are assumed Gaussian for the emitter, Gaussian for the base, and uniform for the epilayer collector. The peak dopings are 2×10^{20} cm^{-3}, 9×10^{17} cm^{-3}, and 2×10^{16} cm^{-3} in the emitter, intrinsic base, and epilayer, respectively. The lateral emitter–base spacing was designed

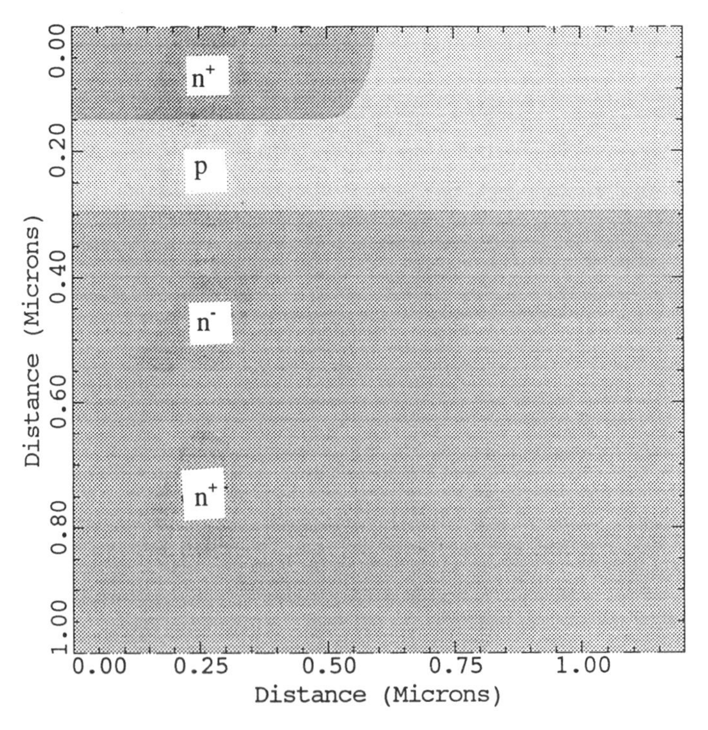

FIGURE 3.4. Cross section of the BJT used in device simulation.

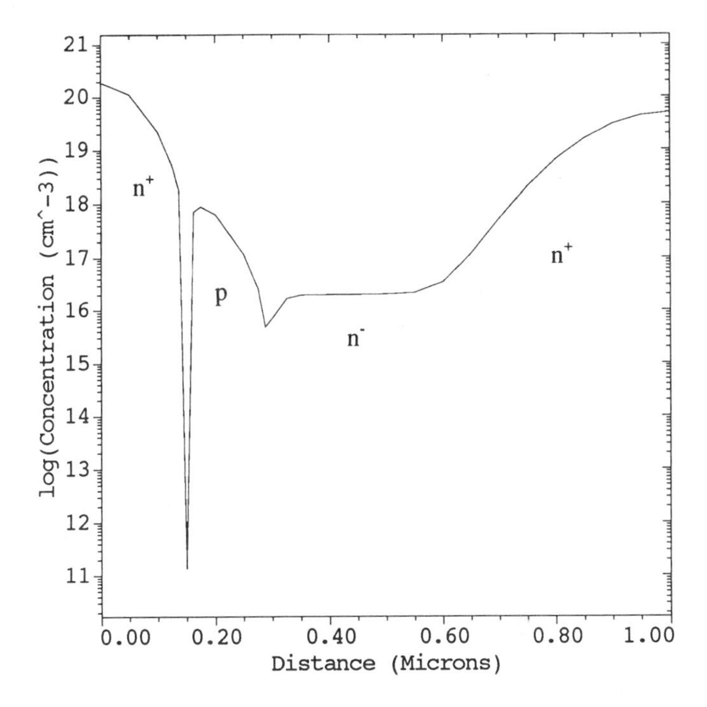

FIGURE 3.5. Vertical doping profile of the BJT.

to avoid sidewall tunneling and perimeter punch-through in the BJT (Li *et al.*, 1988). The emitter dopant lateral straggles are assumed to be 75% of the vertical straggles for the emitter–sidewall lateral diffusion (Chuang *et al.*, 1987). Polysilicon contacts are simulated for the emitter and base by using an effective surface recombination velocity of 3×10^4 cm/s (Neugroschel, 1987a). The physical models used in device simulation include Shockley–Read–Hall recombination, Auger recombination, concentration- and field-dependent mobilities, concentration-dependent lifetimes, and band-gap narrowing. The device parameters used include $\tau_{n0} = 1 \times 10^{-7}$ s, $\tau_{p0} = 1 \times 10^{-7}$ s, $N_C = 2.8 \times 10^{18}$ cm^{-3}, $N_V = 1.04 \times 10^{18}$ cm^{-3}, $N_{\mathrm{SRHN}} = 5 \times 10^{16}$ cm^{-3}, $N_{\mathrm{SRHH}} = 5 \times 10^{16}$ cm^{-3}, $c_n = 2.8 \times 10^{-31}$, $c_p = 9.9 \times 10^{-32}$, $\varepsilon_{Si} = 11.8$, and $E_G = 1.08$ eV. Note that the standard drift–diffusion transport model is used in the simulation because the BJT has a base thickness about 0.1 μm. Physical mechanisms such as hot electron, ballistic transport, and quantum-mechanical behavior in the base of the bipolar transistor are not considered.

Figure 3.6 shows a typical Gummel plot of an *n–p–n* bipolar transistor using the MEDICI simulator. The MEDICI input file is shown in Example 3.1. In Fig. 3.6 the collector and base currents increase with base–emitter voltage. At low base–emitter voltages, the base current is relatively high due to contribution of space-charge recombination. At high base–emitter voltages, significant ohmic resistance effects ($I_B R_B + I_C R_C$) are observed. Figure 3.7 shows the current gain versus base–collector voltage from the MEDICI simulation. Significant saturation effect is observed at $V_{BC} = 0.6$ V. Figure

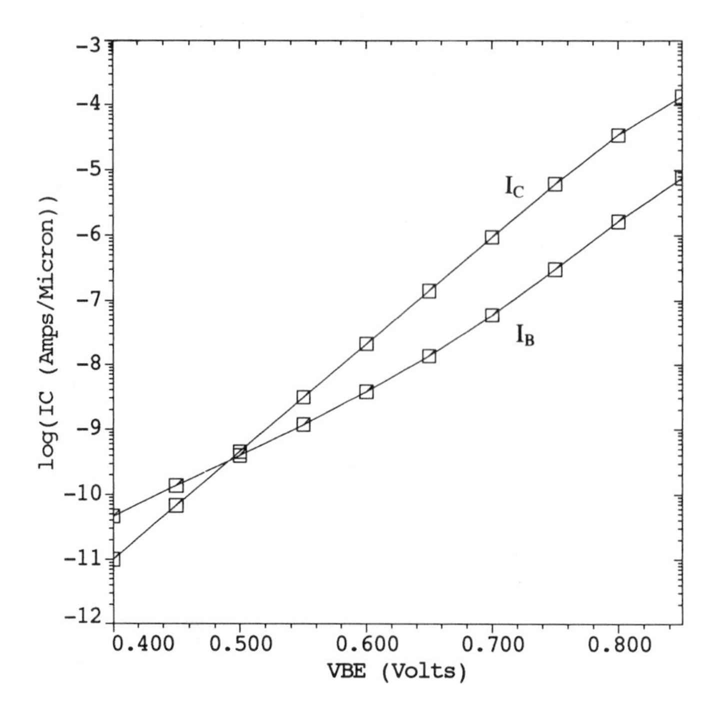

FIGURE 3.6. Collector and base currents versus base–emitter bias simulated by MEDICI.

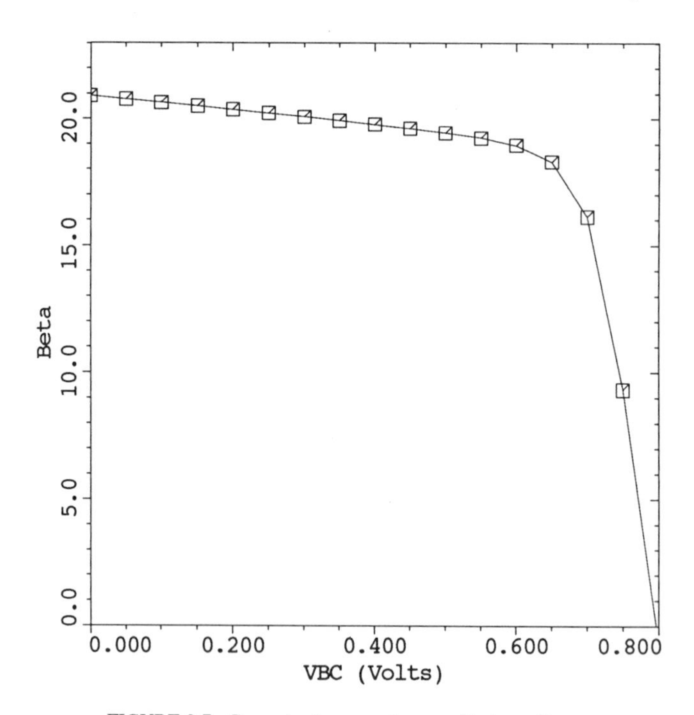

FIGURE 3.7. Current gain versus base–collector voltage.

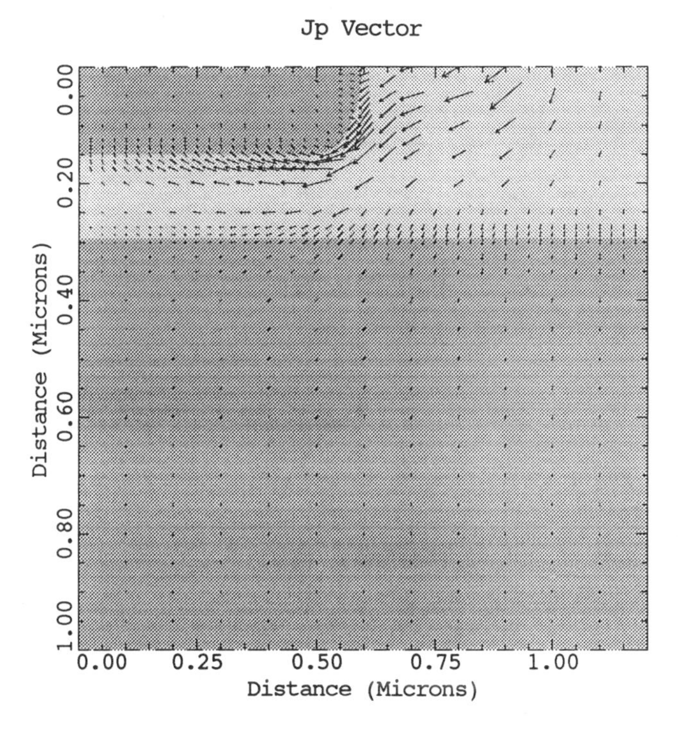

FIGURE 3.8. Current vectors of the BJT in saturation.

3.8 shows hole current vectors in saturation. In this mode of operation, holes are injected into the emitter and collector. The base current is increased and the current gain is decreased.

3.2.2. Current–Voltage Characteristics for Reverse-Active Operation

If we now interchange the collector and emitter terminals, then the two junctions effectively interchange roles. When the collector–base junction is forward biased and the emitter–base junction is reverse biased, this is known as reverse-active operation of the BJT. Furthermore, when both collector–base and emitter–base junctions are forward biased and the forward-biased collector–base junction has a greater value than that of the emitter–base junction, this condition is reverse saturation. For modern bipolar transistors, the emitter and collector doping profiles are not balanced or interchangeable. The emitter, which has a Gaussian profile, is heavily doped to increase the emitter injection efficiency and reduce the emitter resistance. The base has a doping profile with an aiding and a retarding field toward electron movement (Roulston, 1990). In general, the aiding field is stronger than the retarding field for advanced bipolar transistors. A lightly doped epilayer is used in the collector to reduce the collector–base junction capacitance and to increase the collector breakdown voltage (Sze, 1988). A heavily doped buried layer is

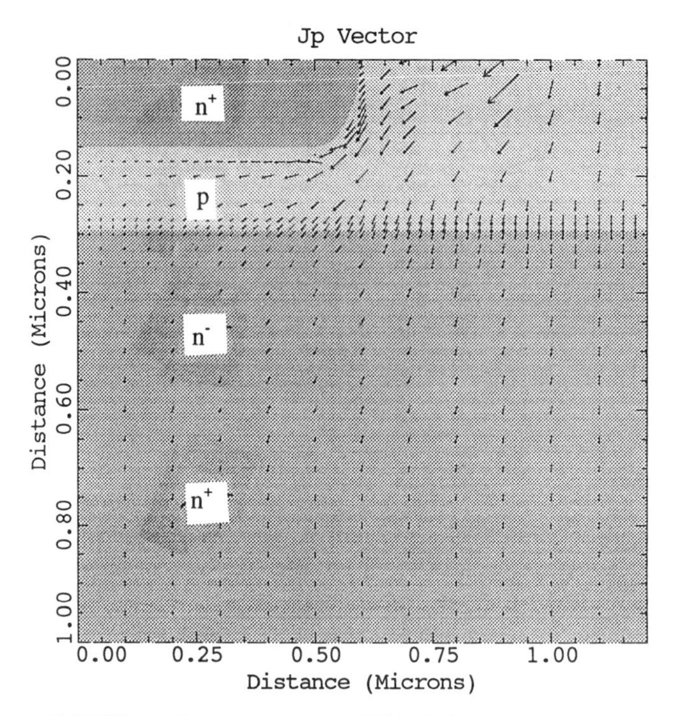

FIGURE 3.9. Current vectors of the BJT in the inverse-active mode.

designed in the collector to reduce the collector resistance. When the BJT emitter and collector are interchanged, the injection efficiency from the emitter (which is the collector in the forward-active mode) is reduced. The epilayer is flooded with excess carriers and the resistivity is reduced due to a high level of injection. The base has a stronger retarding field than an aiding field. The collector saturation current is decreased. The reverse operation also decreases the collector–base breakdown voltage due to higher collector and base concentrations and the collection of electrons in the collector due to a smaller collector area. For the inverse-active mode, the emitter injection efficiency, saturation current, and current gain of the BJT are totally different from those in the forward-active mode. The Gummel–Poon bipolar transistor model used in circuit simulation, however, assumes a constant saturation current density and emitter injection efficiency in the collector current. This current equation can lead to circuit simulation errors when the BJT is operating in the inverse-active mode.

The MEDICI device simulator is used to study the current characteristics of the BJT in the inverse-active mode. In this mode of operation, the collector current is reduced due to reduction of injection efficiency and a retarding field in the base. The hole current vectors in the inverse-active mode are shown in Fig. 3.9.

3.2.3. Current–Voltage Characteristics at Different Temperatures

Low-temperature operation is being applied and contemplated for electronic systems ranging from single-transistor circuits for basic research to VLSI integrated circuits for ultrafast computers. It has also been recognized that semiconductor devices possess unique characteristics when they are operated at a low temperature (Rosenberg, 1963). This stems mainly from the fact that impurity atoms are not completely ionized at low temperatures, which results in a decrease in free carriers and a doping-dependent dielectric permittivity due to the polarization of the de-ionized impurity atoms (Castellan and Seitz, 1951). The low-temperature device behavior is influenced heavily by the temperature-dependent mobilities and intrinsic concentration as well as carrier freeze-out at low temperatures.

The collector and base currents increase with temperature due to an increase of intrinsic concentration at higher temperature (Thurmond, 1975). The increase of collector current at higher temperatures indicates that the turn-on voltage of the BJT decreases with temperature (–2 mV/°C).

The maximum forward current gain for a moderate collector current is determined by the ratio of the electron diffusion current in the base to the recombination current in the emitter:

$$\beta \approx \frac{qD_n n_{iB}^2 \int_0^{X_E} n(x)\,dx}{qD_p n_{iE}^2 \int_0^{X_B} p(x)\,dx} \tag{3.2.1}$$

where n_{iB} is the intrinsic carrier concentration in the base and n_{iE} is the intrinsic carrier concentration in the emitter. If n_{iB} and n_{iE} as well as D_n and D_p have identical temperature dependence, the current gain of the BJT is independent of temperature. Experimental data on advanced bipolar transistors, however, show a very sensitive current gain to temperature. The current gain decreases from $\beta = 100$ at $T = 300$ K to $\beta = 10$ at $T = 77$ K (Buhanan, 1969).

Figure 3.10 shows the current gain versus inverse temperature (1000/T). The temperature dependence of the current gain is due to band-gap narrowing in the emitter:

$$n_{iE}^2 = n_i^2\, e^{\Delta E_G/kT} \tag{3.2.2}$$

where $\Delta E_G = 9\{\log(N_E/10^{17}) + [0.5 + \log(N_E/10^{17})]^{1/2}\}$ meV (Slotboom and de Graaff, 1976). Inserting (3.2.2) into (3.2.3) and using $D/\mu = kT/q$ gives the temperature-dependent β:

$$\ln[\beta(T)] = \frac{\Delta E_G(T_0)}{kT_0} - \frac{\Delta E_G(T)}{kT} + \frac{1}{T}\ln\left[\frac{\mu_{nB}(T)\,\mu_{pE}(T_0)}{\mu_{pE}(T)\,\mu_{nB}(T_0)}\right] + \frac{1}{T}\ln(T_0) \tag{3.2.3}$$

where T_0 is the room temperature.

Equation (3.2.3) is approximately a linear relationship of log(β) versus 1/T. Such dependence has been confirmed by the experimental data in Fig. 3.10.

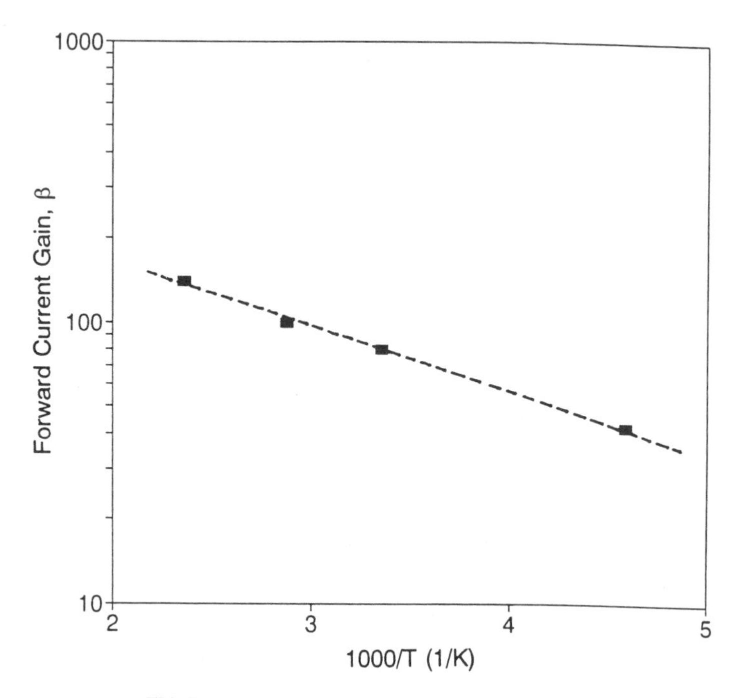

FIGURE 3.10. Current gain versus inverse temperature.

When designing bipolar transistors for low-temperature operation where the current gain must be comparable with that of room temperature, a lower emitter doping (Woo and Plummer, 1987), a higher base doping (Stork *et al.*, 1989), or a lower emitter doping and a higher base doping (Cressler *et al.*, 1990) could be used. These doping profiles reduce the band-gap narrowing between the emitter and the base and the sensitivity of β at low temperatures.

3.2.4. Emitter Crowding and Sidewall Injection

Submicron bipolar transistors resulting from double-polysilicon technology exhibit multidimensional current flows, especially when the BJT operates at high collector currents. These second-order effects include sidewall injection (Kurkx, 1987), emitter crowding, base pushout, and collector current spreading. Understanding these physical mechanisms is important for making high-performance transistors achieve high current-driving capability, larger current gains, and higher cutoff frequencies. We now proceed to examine emitter crowding and sidewall injection of the BJT.

As the base current flows through the active base region, a potential drop in the horizontal direction causes a progressive lateral reduction of dc bias along the emitter–base junction. Consequently, emitter current crowding occurs at the peripheral emitter edges (dc crowding), as evidenced by the electron current density plot in Fig. 3.11. This

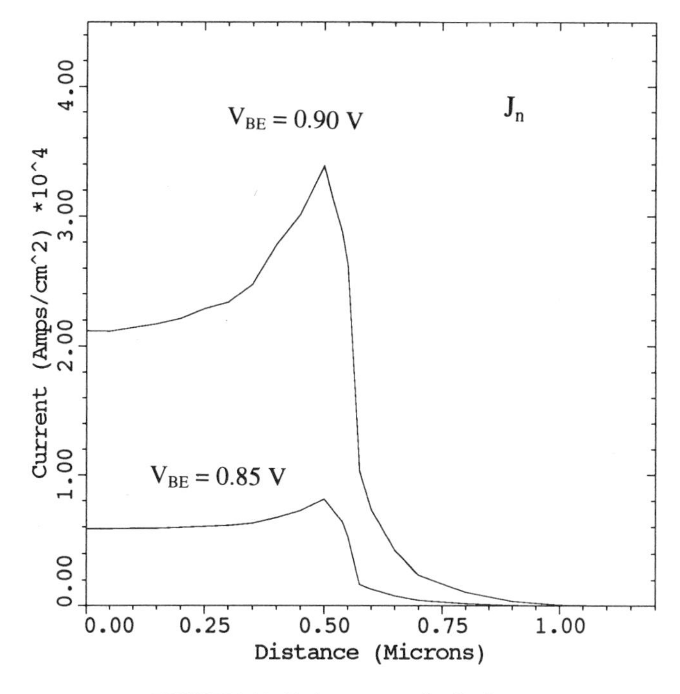

FIGURE 3.11. Emitter current distribution.

nonuniform current distribution effect is enhanced during transient operation in which the base resistance and junction capacitance contribute finite *RC* time constant (delay) in the base region (Tang, 1985). Thus, the emitter edge region of a BJT turns on earlier than the emitter center region during a switch-on transient. Also, the charge at the emitter edge region is larger than at the center during switching (transient crowding).

In order to analytically represent the emitter crowding effect, a variable, emitter crowding factor f_{CR}, is defined as the ratio of the emitter current with emitter crowding to the emitter current without emitter crowding (Yuan and Liou, 1989):

$$f_{\mathrm{CR}} = \frac{L_E \int_0^{W_E} J_E[V_{BE}(x,t)]\,dx}{L_E \int_0^{W_E} J_E[V_{BE}(0,t)]\,dx} \tag{3.2.4}$$

where J_E is the position- and time-dependent emitter current density and W_E is the emitter width. In general, $J_E(x,t)$, the nonuniform transient emitter current density, cannot be integrated analytically. Equation (3.2.4) is solved numerically by using Simpson's integration method as

$$f_{CR} = \frac{J_E(0,t) + J_E(W_E,t) + 4\sum_{j=1}^{m/2} J_E\left[\frac{(2j-1)W_E}{m,t}\right] + 2\sum_{j=1}^{m/2-1} J_E\left(\frac{2jW_E}{m,t}\right)}{3mJ_E(0,t)} \tag{3.2.5}$$

where m is the number of the partitioned base regions.

When the bipolar transistor operates at high frequency, ac emitter crowding occurs (Ghosh, 1965). Emitter area utilization is thus reduced. The ac crowding is not necessarily dependent on dc crowding and can be quite significant during high-frequency operation. For high-frequency circuit design, multiemitter fingers are used to reduce ac crowding while maintaining the same dc current driving capability.

The multidimensional effects are internal physical responses that cannot be measured by experiment. Device simulation, however, provides physical insight into transistor operation for examining second-order effects. An analysis of the PISCES simulation (Cressler *et al.*, 1990) of an advanced BJT was performed to identify emitter crowding and multidimensional currents (Yuan and Liou, 1989).

Device simulation is advantageous when analyzing the two-dimensional effects of lateral injection from the emitter–base sidewall on the base current and current gain. When the lateral dimensions of the emitter area are on the same order of magnitude as the emitter width, the emitter–base sidewall plays an important role in the performance

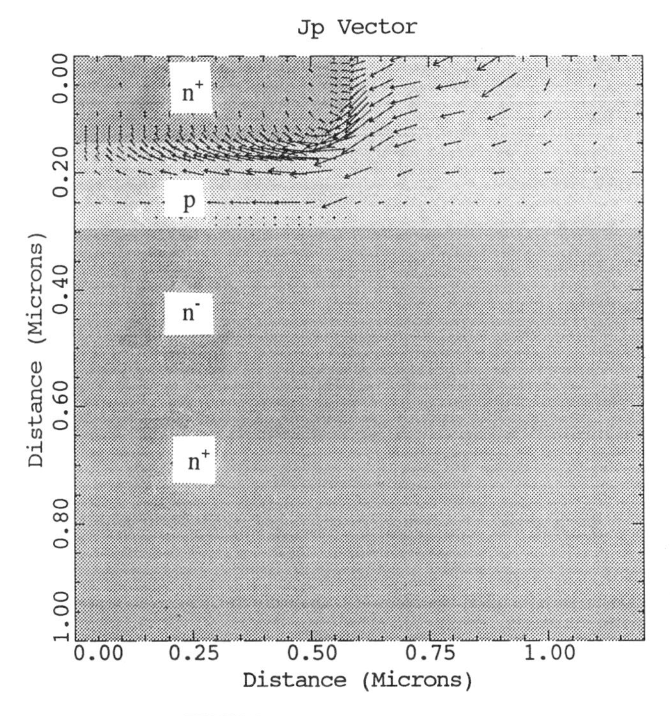

FIGURE 3.12. Sidewall injection.

of the bipolar transistors (Verret and Brighton, 1987). Using the 2-D simulation the base current paths in the forward-active mode are illustrated in Fig. 3.12. The figure displays the hole current vectors of the BJT biased at $V_{BE} = 0.85$ V and $V_{CE} = 3.0$ V. The current vectors suggest that sidewall injection contributes an important component of the base current. A quantitative measure of the sidewall current is given by integrating the electron current density and hole current density along the emitter–base junction sidewall. Simulation of the current gain versus emitter width suggests that the impact of sidewall injection is proportional to the ratio of emitter area and emitter perimeter.

Figure 3.13 shows the electron current density versus position at different times. The emitter crowding is significant during the initial turn-on transient. The emitter crowding effect is enhanced in transient operation because the base resistance and junction capacitance contribute finite RC time constant (delay) in the base region. Thus, the emitter edge of a BJT turns on earlier than the emitter center during a switch-on transient. From the above analysis and simulation, one finds that emitter current crowding depends on emitter width size, base–emitter voltage magnitude, base resistivity, rise time of the input waveform, and frequency. To reduce emitter current crowding, the emitter width must be narrow. To maintain dc current driving capability while reducing ac and transient crowding, multiple emitter and base fingers must be designed. Multiple emitter fingers provide uniform current distribution under the emitter fingers up to very high frequencies.

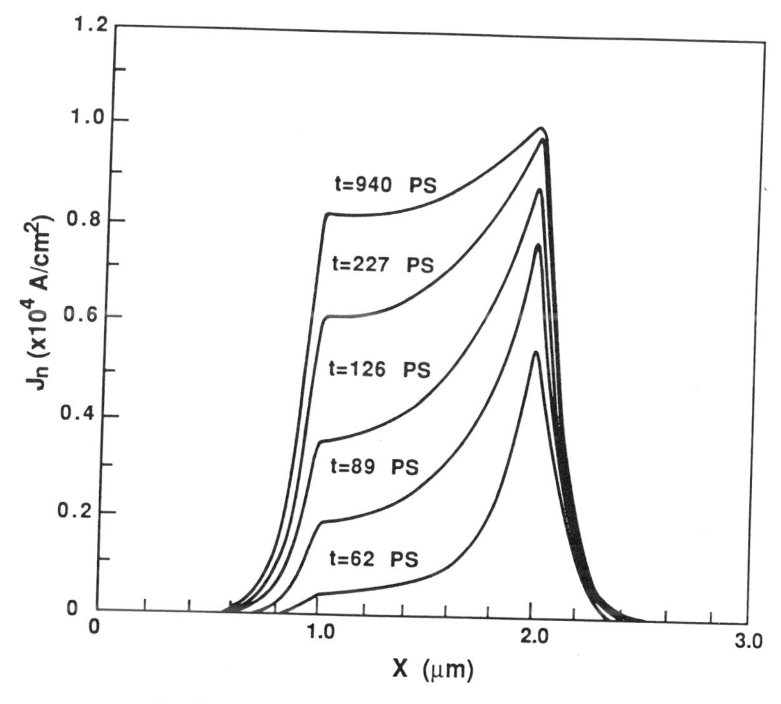

FIGURE 3.13. Emitter current as a function of time.

3.2.5. Base Pushout and Collector Spreading

When the bipolar transistor enters the high-collector-current regime, cutoff frequency and large-signal current gain begin to decrease rapidly with increasing collector current density at the onset of base pushout (Kirk, 1962). This high-current effect has been investigated by numerous authors for many years (Whitter and Tremere, 1969; van der Ziel and Agouridis, 1966; de Graaff, 1973). In general, two distinct physical models were proposed. In the 1-D model (Whitter and Tremere, 1969), the current flow is strictly one dimensional. Because the area over which current flows is fixed at the emitter area A_E, it is necessary to increase the current density above the critical current density J_0 to realize currents greater than I_0. The space-charge limitation allows larger numbers of holes to enter the epitaxial collector of an n^+–p–n–n^+ BJT. Increasing electron current will thus cause an increase in the number of holes and electrons in the collector. That is, part of the collector region is electrically equivalent to the base region. The region of the epitaxial layer that is neutral due to electron storage will be referred to as the "current-induced base." The current-induced base thickness is (Sze, 1981)

$$\Delta X_B = W_C \left[1 - \sqrt{\frac{J_0 - q v_s N_C}{J_C - q v_s N_C}} \right] \tag{3.2.6}$$

where W_C is the epilayer thickness and $J_0 = q v_s (N_C + 2\varepsilon V_{CB}/qW_C^2)$. As J_C becomes larger than J_0, ΔX_B increases; and when J_C becomes much larger than J_0, ΔX_B approaches W_C.

The distribution of electric field as a function of collector current in Fig. 3.14 illustrates the change of space-charge layer and E-field pattern before and after base

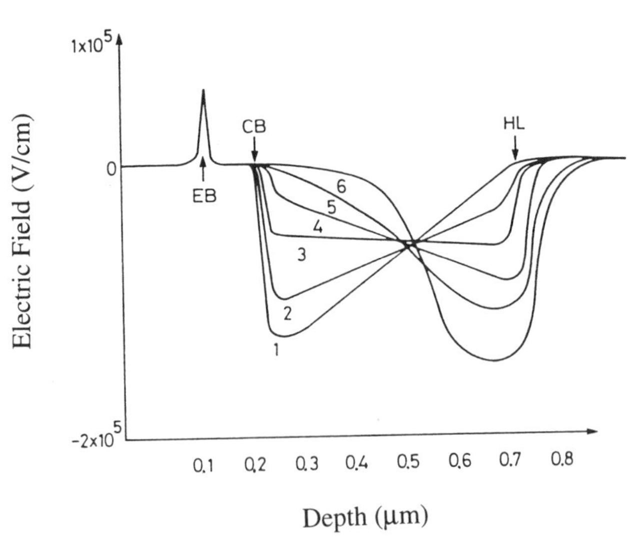

FIGURE 3.14. Electric field pattern at different levels of injection.

pushout. When the collector–base voltage is increased, the electric field in the collector–base space-charge region increases and the space-charge layer expands. The peak electric field is located at the collector–base metallurgical junction. When the base–emitter voltage increases, the peak electric field at the collector–base junction gradually decreases due to electron injection from the emitter (curve 2 in Fig. 3.14). The peak electric field decreases until a uniform distribution is attained (curve 3 in Fig. 3.14). As injection increases further, the peak electric field moves toward the collector region (curve 4 in Fig. 3.14). The peak field starts to rise again at higher collector current densities. When the collector current density approaches the current density for Kirk effect, base pushout occurs. The effective base region is widened, and the peak electric field is pushed into the collector high–low junction (curve 5 in Fig. 3.14).

In the 2-D model (van der Ziel and Agouridis, 1966) the assumption is that the current density cannot exceed J_0 anywhere in the collector. This is equivalent to the assumption that holes cannot be stored in the collector. Thus, in order to exceed the space-charge limitation existing at the critical current I_0, it is necessary for two-dimensional effects to take place (i.e., lateral injection in the base). At currents below I_0, the one-dimensional picture is still invoked, while at currents above I_0 the area over which carriers are collected increases according to $A_E I_C / I_0$ for $I_C > I_0$. For advanced bipolar transistors, however, the heavily doped extrinsic base resulting from double-polysilicon technology makes 2-D base spreading negligible. In contrast, the base pushout in the collector is two dimensional and results in lateral collector current spreading in the epitaxial collector (Yuan and Eisenstadt, 1988). The bipolar transistor for a wide range of operation is shown in the I_C versus V_{CE} plot in Fig. 3.15. Currents $I_1 = qA_E v_s N_C$ and $I_2 = qA_E \mu_n N_C V_{CB}/W_C$ are used to define operating regions; I_1 is the collector current that would flow if all carriers in the collector–base junction region moved with saturated velocity, and I_2 is the current below which the transistor is operating in the active region (Bouler and Lindholm, 1973). For collector current $I_C \leq I_1 \leq I_2$ (region I) and $I_C \leq I_2 \leq I_1$ (region II), the transistor is operated in the active mode and the transition region of the base–collector junction is reverse biased for small-to-medium collector–emitter voltage. For a base–collector junction under reverse bias, two tendencies exist if V_{CB} is held constant and I_C is increased. One tendency causes the space-charge region to expand, resulting from a reduction in the net charge in the space-charge region caused by the injection of free carriers in the region. The other tendency causes the space-charge region to shrink because an increasing collector current gives rise to an increasing voltage drop across the undepleted portion of the collector region. For $I_C \leq I_2 \leq I_1$ (region II), because of the relatively low V_{CB}, the contraction tendency dominates. For $I_C \leq I_1 \leq I_2$ (region I), large V_{CB} favors the expansion tendency. Note that regions I and II are separated by a critical voltage V_c, where V_c corresponds to V_{CE} at which $I_1 = I_2$. For $I_1 < I_C < I_2$ (region III), the transistor is also operating in the active mode, but because $I_C > I_1$ the collector region is fully depleted and space-charge-limited flow occurs. For $I_2 \leq I_C \leq I_1$ (region IV), $I_2 \leq I_1 \leq I_C$ (region V), and $I_1 \leq I_2 \leq I_C$ (region VI), the transistor, depending on I_C, is in the quasi-saturation or saturation region and the transition region is under forward bias due to the domination of the voltage drop ($I_C R_C$) in the undepleted collector region. For regions V and VI, in which $I_C > I_1$, the

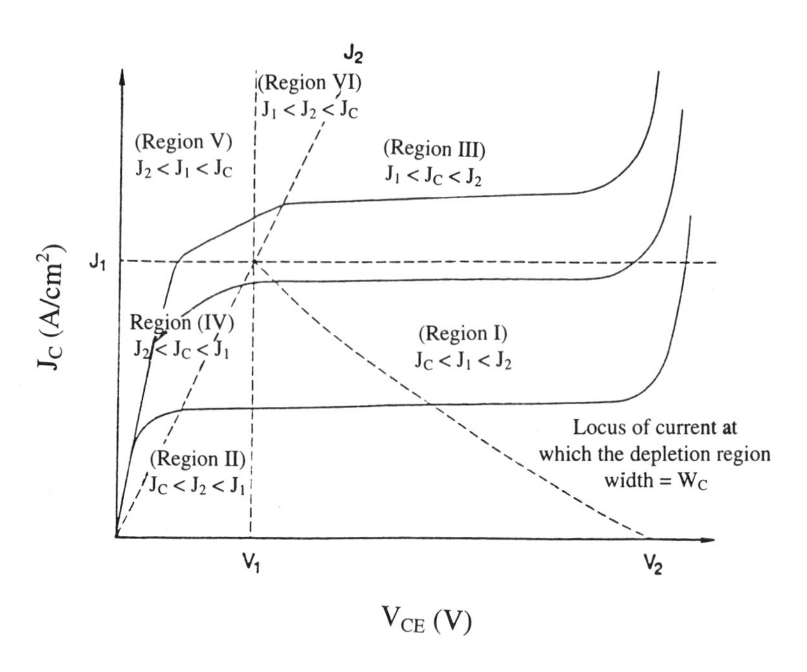

FIGURE 3.15. Illustration of the six operating regions I_C versus V_{CE} characteristics.

mobile electrons overcompensate the donors and the sign of the space-charge is reversed, which causes the space-charge region to behave differently from that of region IV.

For the bipolar transistor operated with high collector current, base pushout occurs at high collector current and high V_{CE} (region III) and quasi-saturation occurs at high collector current and low V_{CE} (regions V and VI). Both ohmic and nonohmic quasi-saturation occur when the internal base–collector junction is forward biased and the external base–collector terminal is reverse biased, while the excess carrier concentration in the epilayer is less than the epilayer concentration for ohmic quasi-saturation and the excess carrier concentration in the epilayer is larger than the epilayer concentration for nonohmic quasi-saturation (Jeong and Fossum, 1985). To examine the quasi-saturation effect, an analysis of PISCES simulations of the advanced bipolar transistor was undertaken to identify the physical origin of the collector current spreading in the epitaxial collector (Yuan and Eisenstadt, 1988). The substrate contact is situated at the bottom of the buried collector. It was found that only a negligible perturbation on the lateral current flow in the lightly doped epitaxial collector results when a side-collector contact is used. This is discussed later in this section.

A PISCES simulation predicted the multidimensional current flow (Fig. 3.16). This diagram displays the PISCES simulation of the electron current density–vector plot of the advanced BJT biased at $V_{CE} = 2.0$ V and $V_{BE} = 0.9$ V ($I_C \approx 0.5$ mA/μm). In the high-current mode, excess carriers are injected into the epitaxial collector (base pushout). The excess electrons diffuse laterally in the epilayer collector under the extrinsic base due to high carrier concentration gradients in the horizontal direction. Collector current

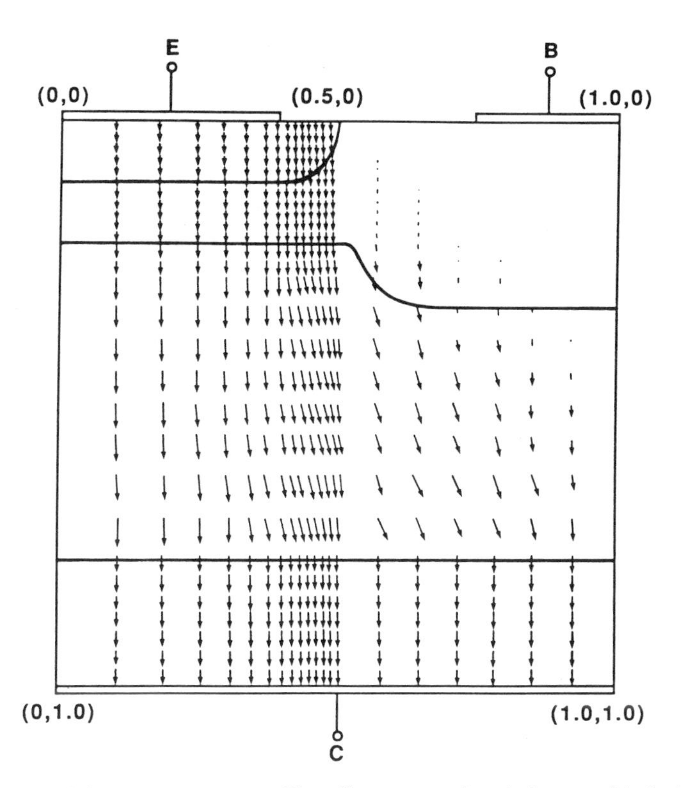

FIGURE 3.16. Plot of electron current vectors. The collector contact is at the bottom of the buried collector.

spreading is indicated by the horizontal component of current density vectors underneath the extrinsic base.

Note that Fig. 3.16 qualitatively shows where the multidimensional collector currents occur, but it does not lend itself to a quantitative estimation of the magnitude of these currents. The grid in Fig. 3.16 is nonuniform (for better simulation accuracy and convergence) and is dense at emitter, base and emitter–base junctions because of the position-dependent doping density at these regions. The magnitudes of the current density vectors at the grid points which are sparsely located are enhanced when compared to those of dense grid points.

The effect of electric field on the current distribution in the epitaxial collector can be seen by putting the collector contact on the right side of the BJT. Figure 3.17 shows a BJT simulation from PISCES of the current density vectors for a right-side-collector contact, with $V_{CE} = 2.0$ V and $V_{BE} = 0.9$ V, the same bias as that of Fig. 3.16. Although there are great differences in the current vectors in the buried collector (due to lateral ohmic drop), all the current in the intrinsic BJT remains virtually the same. The magnitudes of the vertical and lateral current density vectors in Fig. 3.16 and Fig. 3.17 are typically within 0.5% of each other. This indicates that the electric field from the right-side collector contact only controls the current flow in the buried-collector region,

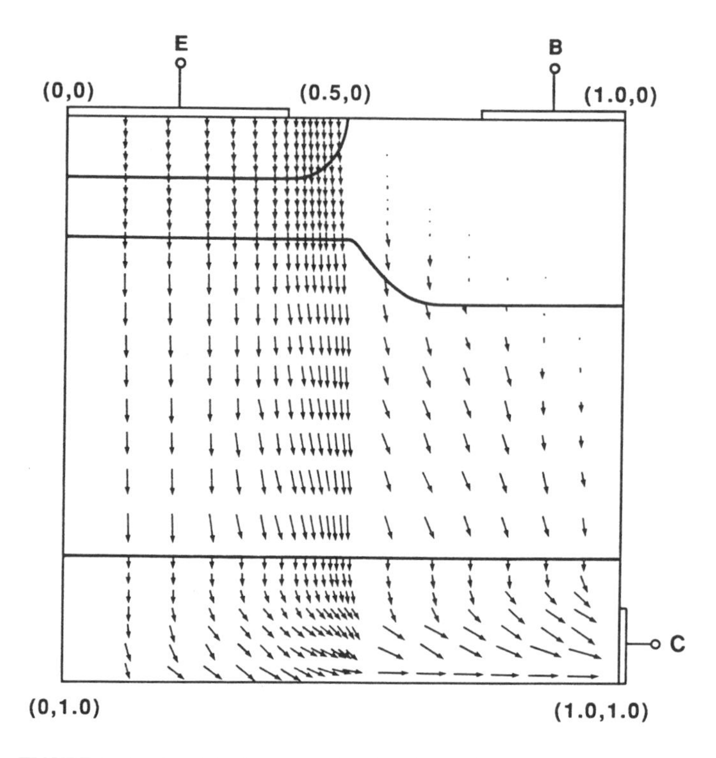

FIGURE 3.17. Plot of electron current vectors. The collector contact is at the side.

and that it does not significantly affect the lateral flow. Thus, there is no significant drift component in the collector spreading mechanism.

A 1-D collector transistor was simulated in order to isolate the effects of collector spreading in the advanced BJT operation. This 1-D collector BJT, shown in Fig. 3.18, has the same emitter and base regions as the 2-D BJT in Fig. 3.16. However, below the extrinsic base region of the 1-D collector transistor, the epitaxial and buried-collector regions are replaced with SiO_2. This forces the collector current to flow solely in the vertical direction below the intrinsic base, hence the name 1-D collector transistor.

Comparing the current gain, β, and cutoff frequency, f_T, of the 2-D BJT to those of the 1-D collector BJT at high currents is one of the keys to understanding the role of collector current spreading. The 2-D BJT exhibits a larger β and f_T than the 1-D collector BJT as both transistors are driven further into saturation (Yuan and Eisenstadt, 1988).

Figure 3.19 indicates how base pushout is improved by lateral diffusion in the collector. This figure displays a hole concentration plot for a vertical slice along the center of the emitter of both the 2-D BJT and 1-D collector BJT. This simulation is performed for BJTs with $W_E = 1$ μm, $V_{BE} = 0.9$ V, and $V_{CE} = 2.0$ V. The base in the 1-D BJT displays significantly more base widening than the 2-D collector BJT and results in degraded β and f_T at high current. The reduced quasi-saturation effects in 2-D BJT are due to the lateral diffusion current, which is due to a high concentration gradient in the horizontal

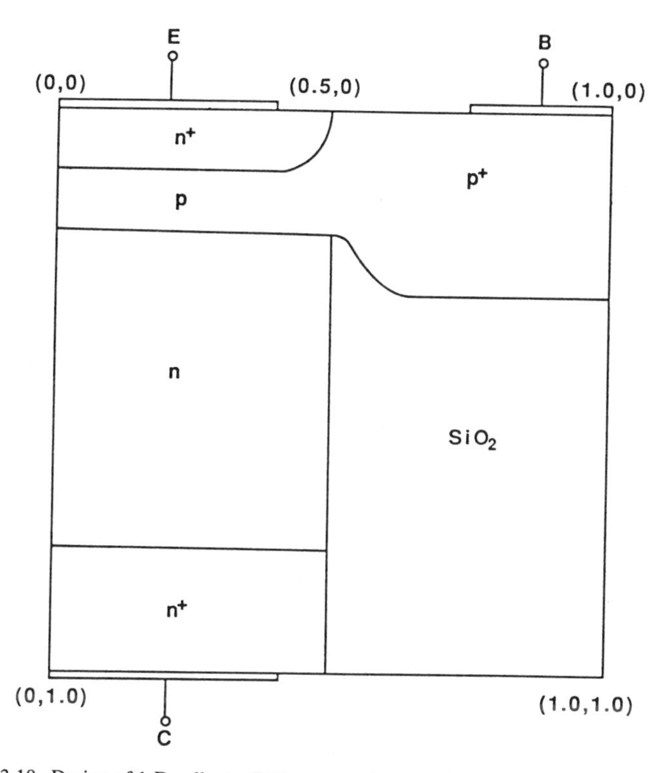

FIGURE 3.18. Design of 1-D collector BJT cross section to isolate the 2-D collector spreading.

direction. Further simulation of current spreading at different emitter widths, with the same doping profiles and boundary conditions as in Fig. 3.17, indicates that the lateral diffusion current is a function not of W_E but of the charge around the emitter periphery. This implies that the collector current spreading effect is more prominent for a narrow-emitter BJT at high currents.

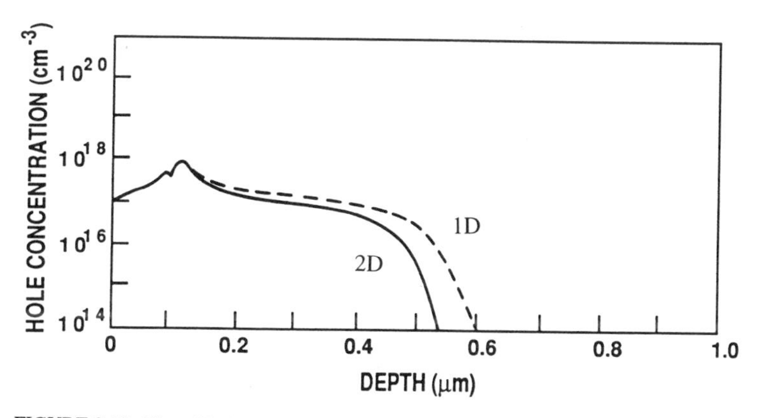

FIGURE 3.19. Plot of hole concentration from the center of the emitter to the collector.

The high-current effect is very important for power bipolar transistors and BiCMOS circuits. To suppress base widening and quasi-saturation effects, the collector width must be narrow and the collector doping concentration must be high. Narrow collector width and high collector doping increase the onset current density for base pushout. However, they also produce a lower collector–base breakdown voltage.

3.2.6. Current-Dependent Base Resistance

Base resistance plays a significant role in the switching speed of ECL logic and frequency response of bipolar amplifiers. Characterizing the BJT base resistance is a difficult task. For the past 20 years, various methods for deriving base resistance have been reported. Recently, Ning and Tang (1984) developed an elegant dc method for measuring base and emitter resistances. However, the accuracy of this method depends on having the intrinsic base resistance linearly proportional to the forward current gain at high currents. By varying the base–emitter voltages, the current-dependent base resistance was characterized at low currents (Neugroschel, 1987a). But this ac technique is sensitive to the parasitic capacitances associated with probed measurements on the integrated circuit.

The base resistance of the bipolar transistor consists of the intrinsic base resistance R_{B1}, extrinsic base resistance R_{BX}, and the base contact resistance $R_{B\text{con}}$. The base contact resistance is usually much smaller than R_{BI} and R_{BX}. For advanced bipolar transistors using double-polysilicon technology, the extrinsic base is heavily doped (Ning *et al.*, 1981b). Thus R_{BX} is current independent. The intrinsic base resistance, however, is subject to base-width modulation, base conductivity modulation, emitter current crowding, and base pushout effects. Therefore, R_{BI} is very sensitive to collector current. We now discuss how these physical mechanisms affect R_{BI}.

Figure 3.20 shows a simplified structure of an n^+–p–n–n^+ bipolar transistor. Using conventional terminology, the base width X_B is the vertical dimension between the emitter–base space-charge region and collector–base space-charge region. The emitter length L_E is defined as the dimension pointing into the figure. The cross-sectional area of the base (perpendicular to the base current path) is determined by the product of the quasi-neutral base width X_B and L_E. Since R_{BI} is inversely proportional to the cross-sectional area of the base current flow, the modulation of the emitter–base or the collector–base space-charge layers will change the magnitude of R_{BI}. For example, assume that the emitter–base junction is forward biased and there is a constant collector–base applied voltage. Then X_B is modulated by the moving edge of the emitter–base space-charge region when V_{BE} changes. As V_{BE} increases, the emitter–base space-charge region contracts, and X_B expands. This reduces R_{BI} because the base cross-sectional area X_BL_E increases, resulting in a larger base charge and a larger effective Gummel number.

When a bipolar transistor (n–p–n) is in high-current operation, the hole concentration (including the excess carrier concentration) in the base exceeds the acceptor (dopant) concentration to maintain charge neutrality. As a result, the base sheet resistance under the emitter decreases as the hole injection level increases. The intrinsic base resistance is then decreased as the level of injection increases.

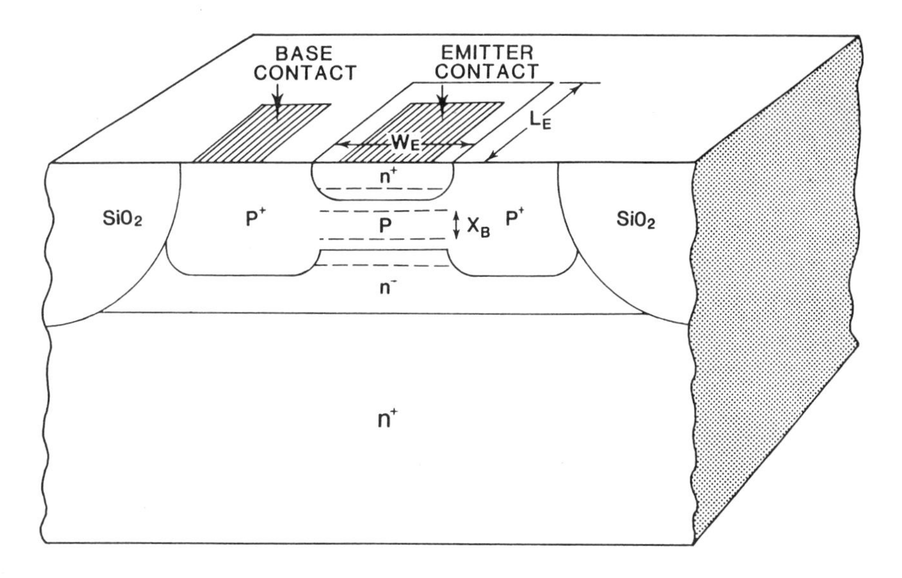

FIGURE 3.20. Three-dimensional BJT structure.

The nonuniform emitter current density distribution makes the effective emitter width smaller. Hence, emitter current crowding reduces the effective intrinsic base resistance. For the bipolar transistor with large emitter width at high base–emitter voltage, R_{BI} is more susceptible to emitter crowding effect.

For bipolar transistors operating at high currents, base pushout occurs. Base pushout increases the effective base width, which decreases intrinsic base resistance. For a bipolar transistor with low epilayer collector doping operating at high collector current density and low V_{CE}, the intrinsic base resistance is more vulnerable to the base pushout effect.

Figure 3.21 shows base resistance versus base–emitter voltage. The solid line represents the predictions using analytical expressions (Yuan *et al.*, 1988), the squares represent the PISCES simulation, and the circles represent the parameter extraction from experiment (Yuan and Liou, 1989). The agreement between the model prediction and experiment at high base–emitter voltage is fairly good. At low base–emitter voltage, it is virtually impossible to extract the base resistance using the series resistance drop method. Here, we use PISCES simulation to obtain the intrinsic base resistance. In the PISCES simulation, the intrinsic base resistance is extracted by

$$R_{BI} = \frac{L_E}{3q \int_0^{W_E} \int_0^{X_B} \mu_p(x, y)\, p(x, y)\, dx\, dy} \tag{3.2.7}$$

In Fig. 3.21 the agreement between the model and PISCES prediction is good. The small discrepancy between the model prediction and PISCES result at low V_{BE}'s is due to the nebulous boundary of the space-charge layer in device simulation.

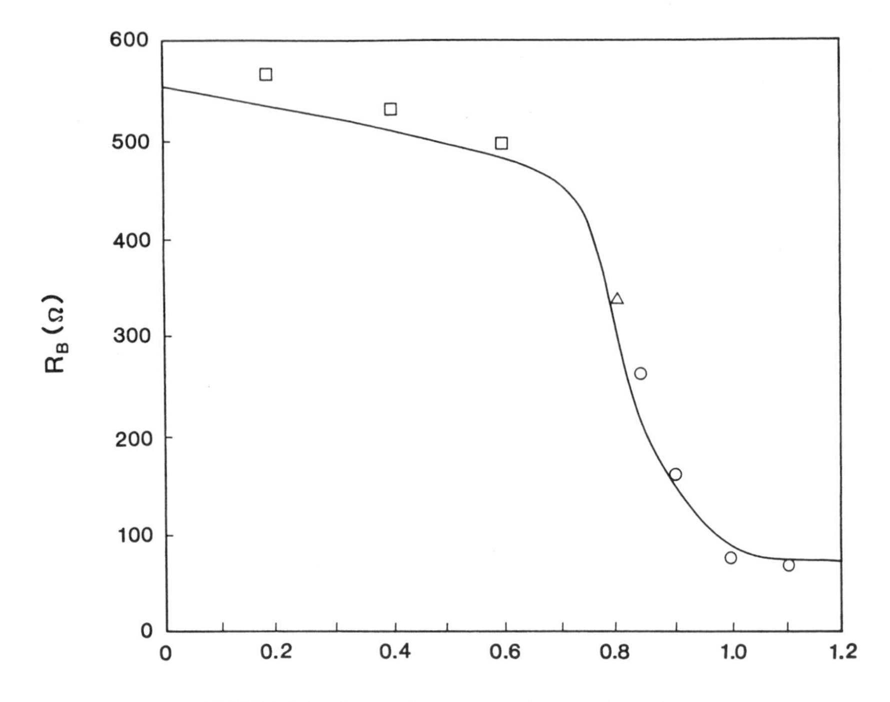

FIGURE 3.21. Base resistance versus base–emitter voltage.

3.2.7. Avalanche Multiplication

In the 1960s, avalanche breakdown of *p–n* junctions was studied as a function of semiconductor doping concentration (Overstaeten and De Man, 1970) and temperature (Crowell and Sze, 1966). For conventional bipolar junction transistors with collector doping concentrations on the order of 10^{15} cm^{-3}, the collector junction breakdown voltage is well above typical power supply voltages. Therefore, impact ionization was not an important parameter for circuit design. However, today's scaled bipolar transistors demand increased collector current density to maintain the same operating current level as conventional, nonscaled BJTs. Advanced bipolar transistors have base doping on the order of 10^{18} cm^{-3} and collector doping concentrations on the order of 10^{17} cm^{-3}. The base doping is raised to reduce the base resistance and to prevent the emitter-to-collector punch-through. The doping density in the collector is raised to support high switching-current density and to avoid high-current base pushout. As a result of higher collector doping, the electric field in the base–collector space-charge region increases accordingly. The resulting peak electric field can be as high as 10^5 V/cm, and impact ionization-induced current can become prominent even under normal operating conditions. Understanding the avalanche breakdown mechanism will help device and circuit designers fabricate reliable, high-performance transistors and integrated circuits. Figure 3.22 shows impact ionization rates of the BJT during avalanche multiplication.

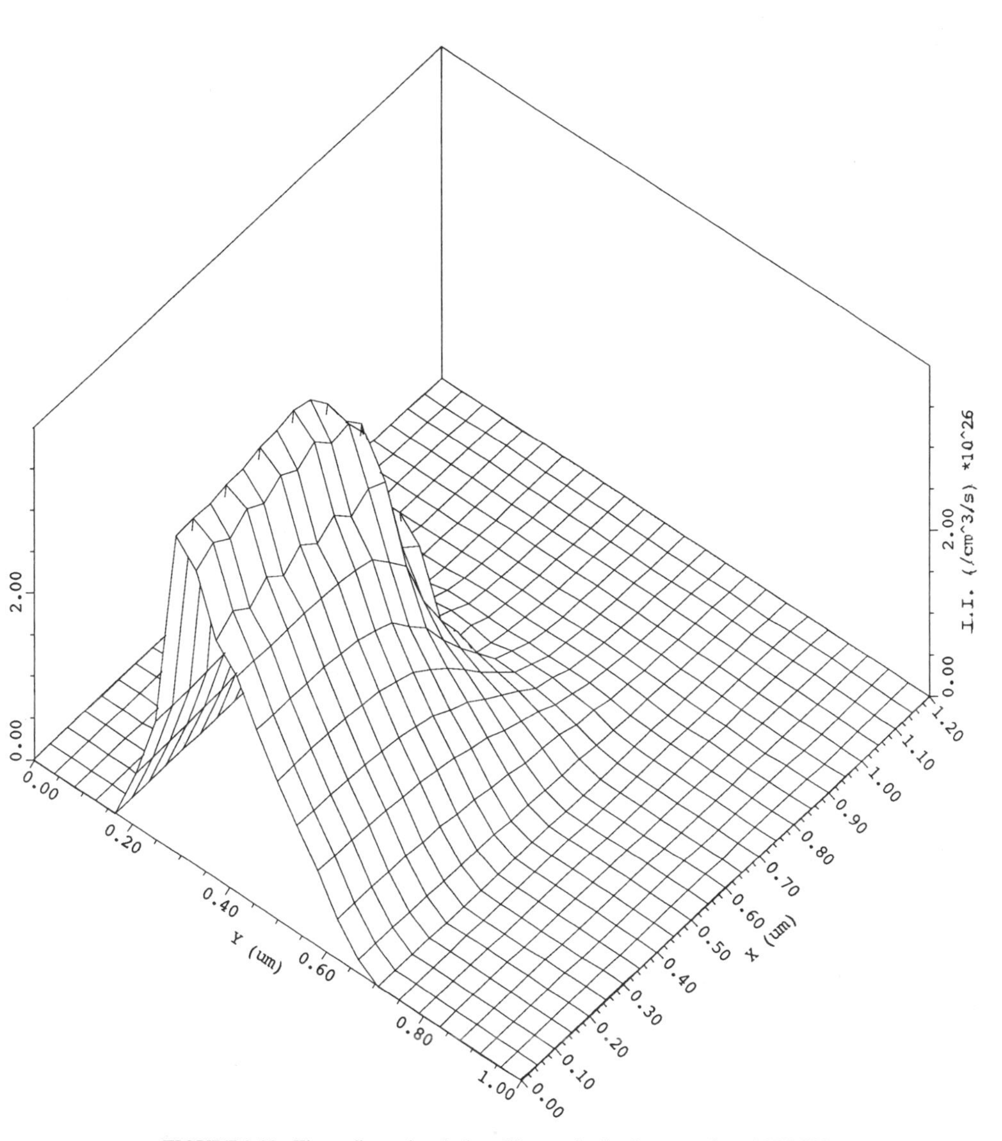

FIGURE 3.22. Three-dimensional plot of impact ionization rates from MEDICI.

3.2.7.1. Collector Current Including Avalanche Current In the collector–base depletion region for substantial reverse bias, the electric field intensity is fairly high. Free carriers can be accelerated by the field to a point where they acquire sufficient energy to create additional carriers when collisions occur. If there is a sufficient distance for the carriers to be accelerated to a high enough velocity, generation of additional carriers during collisions will continue to sustain the avalanche process. Electrons and holes generated in the base–collector space-charge region drift to the quasi-neutral regions under the influence of the electric field. For an n^+–p–n–n^+ bipolar transistor, impact-ionization-induced electrons are swept to the collector and increase the collector current according to the following equation:

$$I_C = (\xi + 1) I_C^0 \tag{3.2.8}$$

where ξ is the avalanche multiplication rate and I_C^0 is the collector current in the absence of avalanche multiplication.

3.2.7.2. Base Current Reversal Impact-ionization-induced holes also drift to the quasi-neutral base. Since the hole current from the base to the emitter is fixed by the base–emitter voltage, excess holes are forced to flow to the base terminal. The external base current decreases as

$$I_B = I_B^0 - \xi I_C \tag{3.2.9}$$

where I_B^0 is the base current in the absence of avalanche multiplication.

If the avalanche current ξI_C is larger than I_B^0, the terminal current will reverse its sign (Liou and Yuan, 1990). However, the base current can be made positive at higher base–emitter bias. This is because the hole diffusion current across the base–emitter junction increases at higher base–emitter bias so that I_B^0 is larger than ξI_C. The base–collector avalanche multiplication has strong dependence on the collector current density.

The avalanche multiplication rate decreases at higher base–emitter bias because it reduces the electric field in the collector–base space-charge region. This is explained as follows. When the collector–base voltage is increased, the electric field in the collector–base space-charge region increases and the space-charge layer expands. The electric field distribution in the base–collector space-charge region, as a function of base–emitter voltage for a given V_{CB}, is depicted in Fig. 3.14. The peak electric field is located at the collector–base metallurgical junction. From the local field model, the electron and hole impact ionization coefficients are highest at the collector–base metallurgical junction, as evidenced by the 3-D plot in Fig. 3.22. When the base–emitter voltage increases, the peak electric field at the collector–base junction gradually decreases due to electron injection from the emitter. The peak electric field decreases until a uniform distribution is attained. At this point, impact ionization at the collector–base junction reaches a minimum. As injection increases further, the peak electric field moves toward the collector region. The peak field starts to rise again at higher collector current density J_C. When the collector current density approaches J_1, base pushout occurs. The effective base region is widened, and the peak electric field is pushed into the collector high–low junction. When $J_C > J_1$, the impact ionization rate increases again due to the increased electric field at the collector high–low junction. This qualitative explanation of injection-modulated impact ionization is evidenced by experimental data (Liu *et al.*, 1991), as shown in Fig. 3.23, where the avalanche multiplication rate ξ is extracted by using the relation

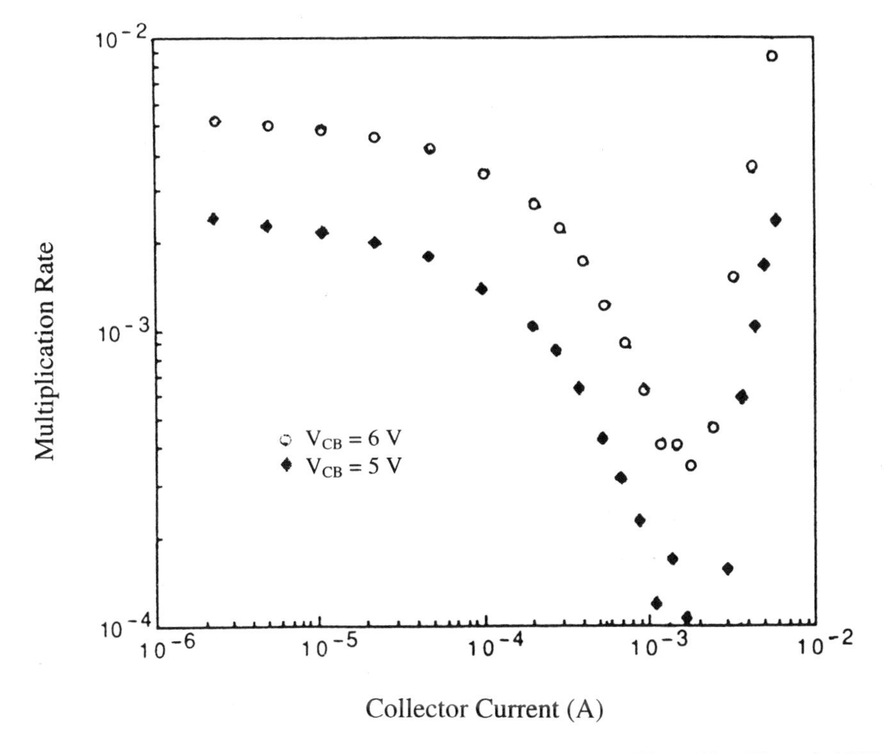

FIGURE 3.23. Avalanche multiplication rate versus collector current at different V_{CB} (Liu *et al.*, 1991).

$$\xi = \frac{I_B (V_{BE}, V_{BC} = 0) - I_B (V_{BE}, V_{BC})}{I_C (V_{BE}, V_{BC})} \tag{3.2.10}$$

Avalanche multiplication also has strong temperature dependence. At low temperature, the carrier mean free path is longer, and optical phonon scattering is less frequent (Sze, 1981). The avalanche multiplication factor increases exponentially with decreasing temperature, as evidenced by experimental data (Lu, 1990). Therefore, avalanche multiplication must be carefully addressed when designing bipolar circuits for low-temperature operation.

3.2.7.3. Nonlocal Impact Ionization Effect The popular drift–diffusion-type simulators assume locality of ionization events. This implies that the impact ionization coefficients at the base–collector metallurgical junction are highest. Since electrons must travel a non-negligible distance within the collector before attaining sufficient energy to initiate impact ionization, the electron energy usually lags the electric field. For an *n–p–n* bipolar transistor, the peak electron impact ionization coefficient shifts from the base–collector metallurgical junction into the collector, and the peak hole impact ionization coefficient shifts from the base–collector metallurgical junction into the base, as evidenced by Monte Carlo simulations (Zanoni *et al.*, 1992). The local electric field

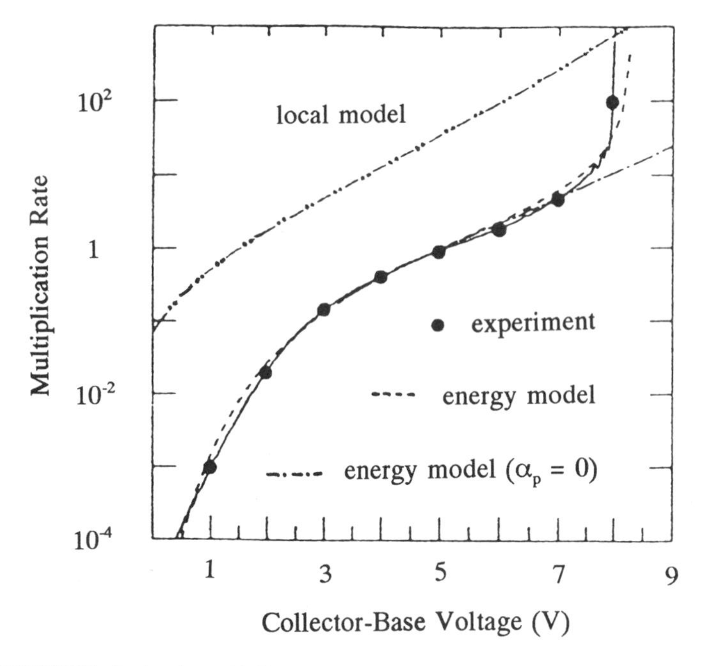

FIGURE 3.24. Avalanche multiplication rates collector–base voltage (Zanoni *et al.*, 1993).

model thus overestimates the avalanche multiplication rate compared to that obtained from the energy model, which uses mean carrier energy. Furthermore, as the collector–base voltage approaches the breakdown voltage, the contribution of secondary holes must be accounted for. Experimentally, the secondary holes cause a steep increase in the avalanche rate at a peak electric field of 9×10^5 V/cm (Zanoni *et al.*, 1993), as shown in Fig. 3.24. To improve the predictability of the avalanche multiplication model at very high electric field intensity, incorporation of a delay in the conventional current equation has been proposed (Di Carlo and Lugli, 1993). This simple analytical approach provides a steep increase of the avalanche multiplication rate for $\xi > 10$.

3.3. AC OPERATION AND DEVICE SIMULATION

High-frequency transistors and circuits are generally used for small-signal amplification applications such as ac voltage and current gain. The ac gain is basically determined by small-signal parameters. For instance, the ac voltage gain of the common-emitter small-signal bipolar amplifier is determined by the product of transconductance and output resistance. In this section, small-signal current gain, transconductance, output conductance, emitter–base junction capacitance, collector–base junction capacitance, cutoff frequency, maximum oscillation frequency, and *s*-parameters of the BJT are presented.

3.3.1. Small-Signal Current Gain

The small-signal current gain is defined as the change in collector current with respect to base current (i_C/i_B). The ac current gain is not necessarily equal to dc current gain (I_C/I_B). The small-signal current gain is a function of frequency. The current gain is constant at low frequency and decreases significantly at high frequency due to capacitive effects.

3.3.2. Transconductance

The transconductance g_m is defined as the change of collector current with respect to base–emitter voltage. The transconductance is simply equal to qI_C/kT when the Ebers–Moll collector current equation is used (Ebers and Moll, 1954). This simple equation, however, is not correct when the bipolar transistor is operating at high collector current. This is because emitter crowding, base pushout, and collector spreading, in addition to base-width and base conductivity modulation, make the collector current highly nonlinear. It is thus difficult to derive an analytical expression for transconductance that includes every physical effect. The use of device simulation, however, provides a better way to examine how high currents affect the transconductance of the BJT.

Figure 3.25 shows the transconductance versus the collector current of the bipolar transistor. At low collector current, the transconductance increases with I_C due to an

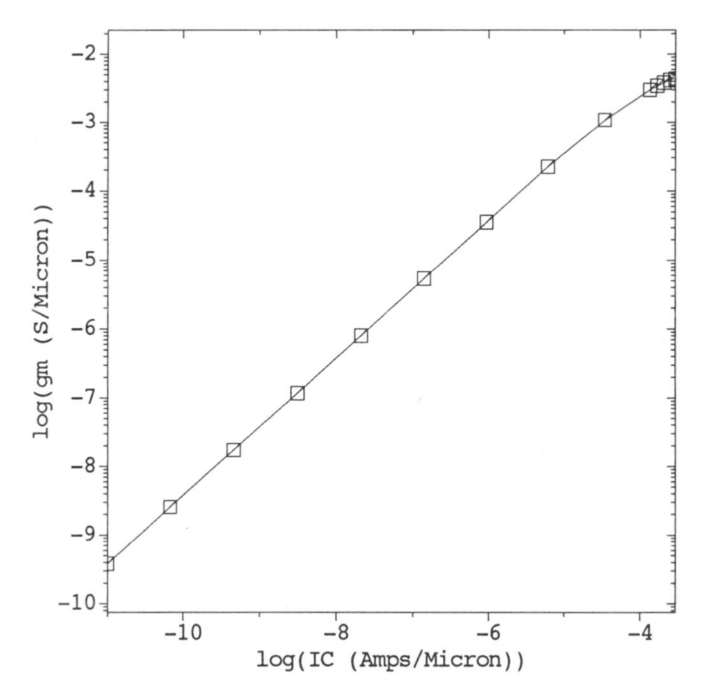

FIGURE 3.25. Transconductance versus collector current.

increase in collector current driving capability. At high collector current, the transconductance saturates due to high-current effects.

3.3.3. Output Conductance

The output conductance describes the base-width modulation or Early effect on the collector current in the small-signal equivalent circuit and is defined as the change of collector current with respect to collector–emitter voltage V_{CE}:

$$g_0 \equiv \frac{\partial I_C}{\partial V_{CE}} \approx \frac{\partial I_C}{\partial Q_B}\frac{\partial Q_B}{\partial V_{CB}} \tag{3.3.1}$$

Since the base charge Q_B is a function of base–emitter voltage as well as collector–base voltage, the output conductance can be nonlinear. For instance, the output conductance evaluated at V_{BE1} is not necessarily equal to that evaluated at V_{BE2}. This nonlinear characteristic is due to the change of position $x = 0$ in the base when determining Q_B as defined in (3.1.8).

The output conductance versus collector–emitter voltage at $V_{BE} = 0.95$ V is shown in Fig. 3.26. The output conductance decreases with collector–emitter voltage due to base-width modulation effects at both junctions.

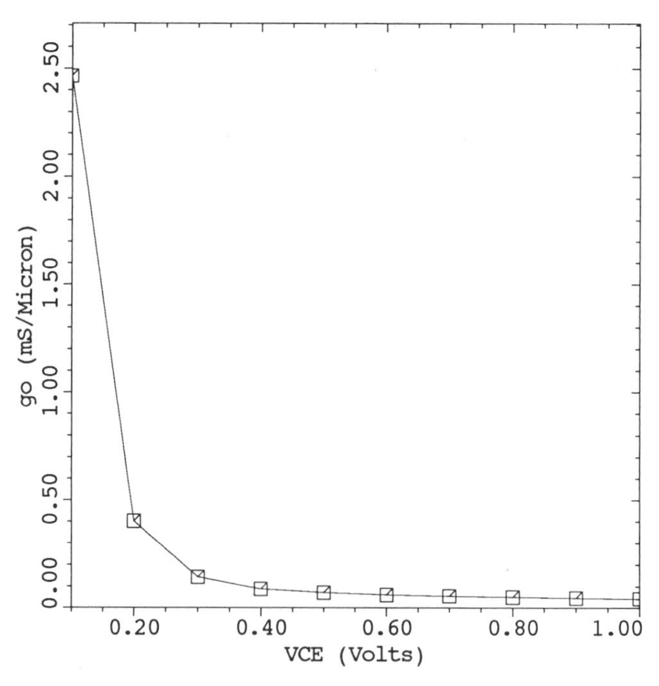

FIGURE 3.26. Output conductance versus collector current.

3.3.4. *Junction Capacitances*

The emitter–base junction capacitance of the bipolar transistor consists of the depletion capacitance and free-carrier capacitance (Liou *et al.*, 1987):

$$C_{JE} = \frac{\varepsilon A_E}{X_1} + qA_E \int_{X_1} \frac{\partial n}{\partial V_{BE}}\, dx \tag{3.3.2}$$

where X_1 is the emitter–base space-charge region thickness and A_E is the emitter area.

In (3.3.2) the depletion capacitance dominates during reverse bias and low forward bias, while the free-carrier capacitance becomes important at high forward bias. The junction capacitance increases with bias at low junction voltages and decreases with bias at high junction voltages. The decrease of C_{JE} at high junction biases is due to injection of mobile carriers into the space-charge region, which compensates the immobile carriers in the space-charge region.

The collector–base junction capacitance (C_{JC}) plays a significant role in determining the switching delay of the bipolar transistor (Liou, 1987). The collector–base junction capacitance is typically treated as the depletion capacitance on the basis that the collector–base junction is reverse biased. Due to free-carrier injection at high collector current densities, the collector–base space-charge region capacitance is determined by both the junction voltage and the collector current. As a result, C_{JC} can be considered as the depletion capacitance only when the collector current density is small enough that the injected free carriers are negligible in the space-charge region. Figure 3.27 shows the collector–base junction capacitance versus collector current. The collector–base junction capacitance begins to decrease when $I_C > 10^{-6}$ A/μm. This is due to injection of mobile carriers that increase the space-charge-layer width and decrease the collector–base junction capacitance. When $I_C > 10^{-4}$ A/μm, C_{BC} increases significantly as a result of the quasi-saturation effect, which forward-biases the collector–base junction and increases C_{BC} rapidly.

The base current reversal resulting from impact ionization also affects device ac parameters, such as collector–base junction capacitance. As impact ionization occurs at high collector–base voltage, free electrons and holes accumulate in the collector–base space-charge region before drifting to the collector and base terminals. Variation of the collector–base voltage results in a modulation of electron or hole charge in the space-charge region and the free-carrier capacitance. The collector–base junction capacitance is increased when impact ionization is significant (Yuan, 1992).

3.3.5. *Cutoff Frequency*

The most important parameter used to characterize the high-frequency behavior of a transistor is the common-emitter current gain–bandwidth product, sometimes called the transition frequency or cutoff frequency. The transition frequency, f_T, is defined as the frequency at which the short-circuit common-emitter current gain approaches unity. For transit time from the emitter to the collector, the cutoff frequency is

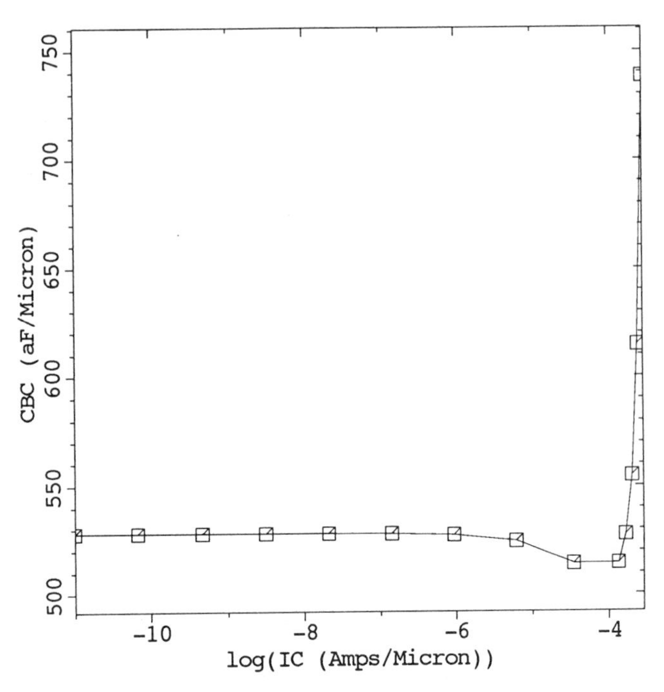

FIGURE 3.27. Collector–base junction capacitance as a function of collector current.

$$f_T = \frac{1}{2\pi\left(C_{JE}/g_m + W_B^2/\nu D_B + X_C/2v_s + R_C C_{JC}\right)} \tag{3.3.3}$$

where D_B is the base diffusion coefficient, v_s is the saturation velocity, and R_C is the collector series resistance. For a uniform base BJT, $\nu = 2$ for low injection and $\nu = 4$ for high injection. In (3.3.3) the first term represents the emitter charging time, the second term represents the base transit time, the third term represents the collector–base depletion layer transit time, and the fourth term represents the collector charging time.

The cutoff frequency increases with collector current at low I_C, reaches the maximum at moderate I_C, and decreases with collector current at high I_C. At low collector currents, the emitter charging time dominates. At moderate collector currents, the base transit time and base–collector depletion layer transit time are important. At high collector currents, the base transit time is increased significantly due to the base widening effect. To increase the peak cutoff frequency, the transistor must have a very narrow base thickness, a narrow collector region, and a higher collector concentration to suppress base widening effects at high collector current.

3.3.6. Maximum Oscillation Frequency

Another very important parameter that characterizes the high-frequency operation of the transistor is the maximum oscillation frequency, f_{max}. This term is defined as the frequency at which the power gain is unity and is given by

$$f_{\max} \approx \sqrt{\frac{f_T}{8\pi R_B C_{JC}}} \tag{3.3.4}$$

When designing for $f_{\max}$, it is not desirable to obtain the highest cutoff frequency f_T. To optimize $f_{\max}$ typically requires the use of relatively small values of base resistance and collector–base junction capacitance. They can be achieved by increasing base doping and lowering collector doping concentration. The resulting low base diffusion coefficient and large collector–base depletion width tends to increase the base and collector transit times, which reduces the cutoff frequency. The effect of a lower base resistance offsets a larger base transit time and larger collector junction capacitance. As long as the decreasing rate of $R_B C_{JC}$ is larger than that of f_T, $f_{\max}$ increases with decreasing base doping and collector dopings.

3.3.7. S-Parameters

Submicrometer emitter bipolar transistors produce small-signal responses that are difficult to characterize with existing s-parameter equipment. State-of-the-art probes and proper calibration techniques have proven essential for measuring the s-parameter of a single BJT test structure (van Wingen and Wolsheimer, 1987). However, s-parameter measurements cannot predict the test structure response of new BJT technologies in the preliminary development stage.

A new method of predicting the s-parameter test structure response from physical device simulation has been developed (Yuan and Eisenstadt, 1987b). This predicted s-parameter response is particularly useful when examining the performance of conceptual designs of submicron bipolar technologies. Submicrometer BJTs have significant dc, transient, and small-signal multidimensional effects that include lateral current in the collector, emitter crowding, and lateral injection between the base and emitter; these effects have been evaluated by a 2-D physical device simulator in the previous sections. The new method of predicting the s-parameter response provides a direct comparison between 2-D BJT simulations and measurement data from BJT test structures. Important usage of this simulated s-parameter response includes verifying BJT test structure s-parameter measurements and providing BJT characterization. The derived BJT test structure response can be used to confirm the accuracy of existing test structure measurements, potentially reducing the total number of test structures, measurements, and cost necessary to characterize a particular BJT process.

In order to get a complete characterization of a three-port BJT, three sets of two-port measurements must be taken, which generally requires three separate test structures. Since the three-port measurement is time consuming and the IC layout is intensive, often a single two-port BJT measurement is made. The s-parameter prediction technique can supplement an existing two-port test structure measurement so that a complete three-port BJT characterization is possible. The simulated s-parameter response also can be extended beyond s-parameter instrumentation frequency ranges.

This modeling technique is demonstrated using submicron BJT simulations from the PISCES 2-D physical device simulator (Strid, 1987). Other small-signal device simulations or characterizations can be substituted for the PISCES data. Simulated small-signal

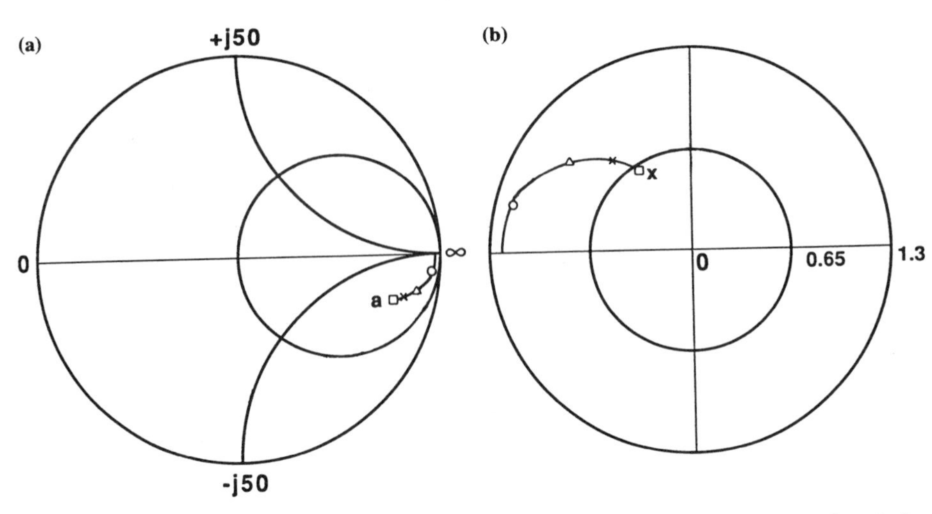

FIGURE 3.28. Polar graphs of (a) S_{11} interpreted from device simulation and (b) S_{21} interpreted from device simulation.

BJT y-parameters can be converted (via software) to s-parameters. s-Parameter measurements are preferred for high-frequency characterizations and have been demonstrated on chips at frequencies up to 50 GHz (Strid, 1987). In addition, s-parameters best represent a distributed circuit with high-frequency discontinuities, such as a BJT test structure model that includes the effects of the IC interconnects, discontinuities, and bond pads.

A 2-D simulation typical provides BJT y-parameter response up to the emitter contact, base contact, and collector contact. During y-parameter simulations the BJT is biased at $V_{BE} = 0.8$ V and $V_{CE} = 2.0$ V, the frequency is varied from 10 MHz to 7 GHz. The y-parameters are normalized by the distributed circuit admittance (frequency-dependent interconnect admittance) and then converted to s-parameters. Figure 3.28 shows a pair of polar graphs of s-parameters simulated for the BJT. The curves display the BJT S_{11} and S_{21} responses calculated from y-parameters. These s-parameter plots range from 10 MHz to 7 GHz and have markers at 1, 3, 5, and 7 GHz.

3.4. TRANSIENT OPERATION AND SIMULATION

Digital circuit designers are generally concerned with switching times such as rise time, fall time, propagation delay, and saturation time. The switching delay is a strong function of base resistance and junction capacitance (Asher, 1964). For submicron bipolar transistors, the delay time is on the order of 10 ps and is a difficult parameter to measure. In addition, the charge control model, such as the Gummel–Poon model, is derived based on the quasi-static approximation, in which minority carrier concentration is assumed to change instantaneously with the emitter–base junction voltage, so the model does not express actual transient phenomena. The use of 2-D device simulation provides a way to evaluate 2-D transient phenomena and non-quasi-static effects of the BJT.

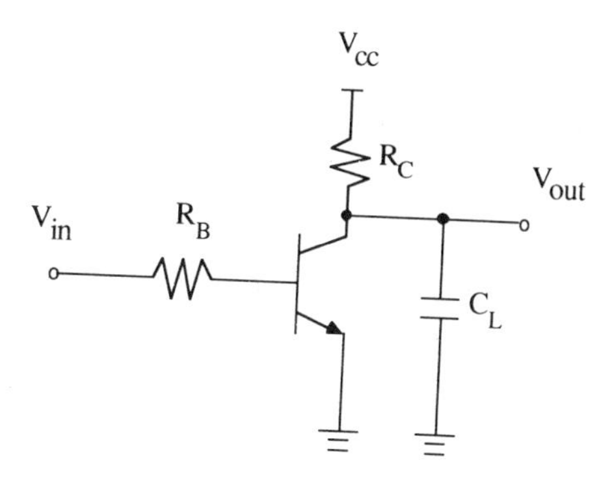

FIGURE 3.29. A schematic of bipolar circuit.

3.4.1. *Turn-on Transients*

The circuit used to operate a bipolar transistor in the switching mode is shown in Fig. 3.29. Consider the case when the external base resistance is zero. When a step voltage is applied, there will be a sudden increase in base current. The base current then decreases as charge diffuses away from the junction; simultaneously the collector current starts to rise.

A study of two-dimensional numerical simulation on the switch-on transient of advanced narrow-emitter bipolar transistors using the FIELDAY program is given by Chung (1988). The emphasis is placed on the effect of the "link-up" region between the intrinsic and extrinsic bases of the double-polysilicon bipolar transistor. The switch-on transient of the device is simulated by applying a ramped voltage source at the base. The voltage source ramps at a rate of 10 mV/ps from 0.5 to 0.85 V and then remains at 0.85 V. Figure 3.30 shows the distribution of the transient y components of electron current densities along the collector–base junction with spacing S between the emitter and base

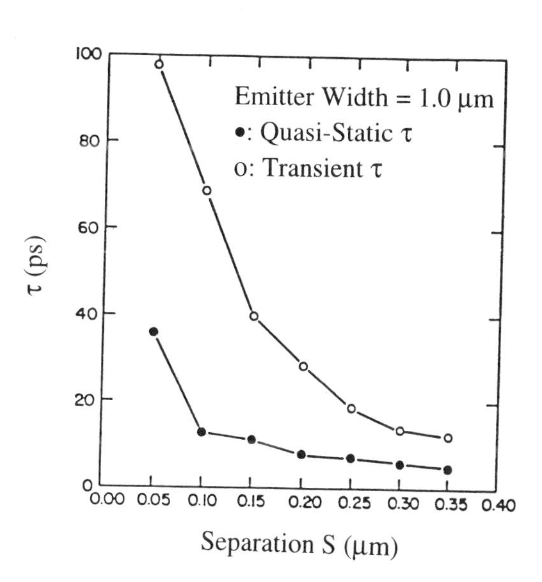

FIGURE 3.30. Quasi-static τ and transient τ as a function of spacing S (Chung, 1988).

contacts at $t = 45$ ps. More current is injected near the emitter edge, and the center of the emitter can be seen to turn on at a later time due to the finite RC time constant of the base resistance and capacitances associated with the junction. Although the center of the emitter for devices with smaller S turns on earlier, these devices are slower due to the reduced active area and increased charge storage in the thick base at the emitter edge.

The impact of the performance imposed by the extrinsic base encroachment combined with the nonuniform distribution during the transient can be revealed by plotting the quasi-static charge storage time constant (quasi-static τ) and the transient charge storage time constant (transient τ) as a function of S for a fixed W_E (= 1.0 μm) in Fig. 3.31. The quasi-static τ is obtained from extrapolation of the $\Delta Q_B/\Delta I_C$ versus $1/I_C$ under dc (quasi-static) conditions. The transient τ, which determines how fast the collector current rises under a constant base current drive when terminal voltages are varying slowly, is obtained from $I_B/(dI_c/dt)$ by switching the base drive to a constant current source at $t = 35$ ps after the initial voltage ramp. The transient τ increases much faster than the quasi-static τ as S is reduced, since the current flows primarily at the emitter edge during the initial turn-on transient. While reducing S from 0.30 to 0.20 μm causes an increase in the quasi-static τ of only 30% (from 6.0 to 7.8 ps), the transient τ degrades from 13.8 to 28.5 ps. Also, notice that the curve for the transient τ starts to tilt up as S is reduced to 0.25 μm and increases dramatically for S below 0.15 μm. Hence, for the profiles used, the optimum S should be placed around 0.25 μm with a lower bound at 0.15 μm (preferably at 0.20 μm) for an emitter width of 1.0 μm. Notice that for S below 0.1 μm, even the quasi-static τ increases rapidly.

Due to the nonuniform current distribution during the switch-on transient, the encroachment of the extrinsic base into the intrinsic base area causes much more severe

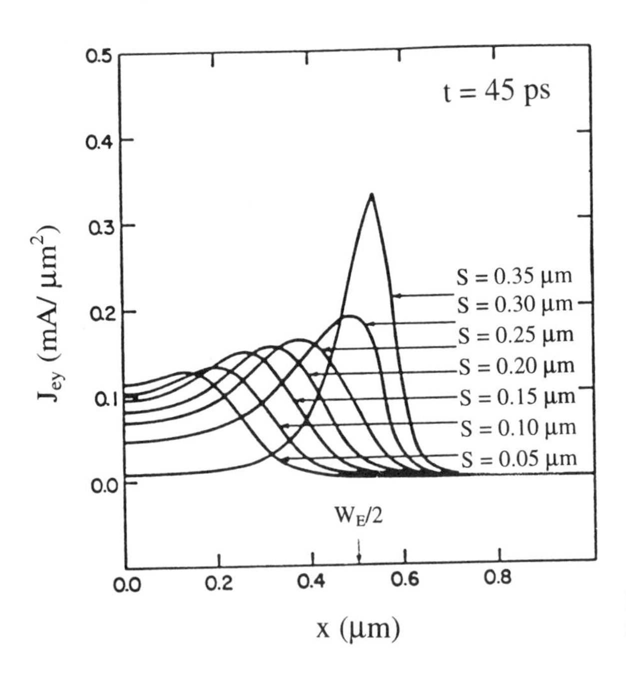

FIGURE 3.31. Electron concentration at $t = 45$ ps (after Chung).

degradation in the switching speed of the device than those predicted by either dc or quasi-static ac analyses. The design considerations and scaling implications imposed by this transient phenomenon are presented.

3.4.2. Turn-off Transients

When the bipolar transistor is switched from the on to the off state, minority carriers stored in the base cannot be removed instantaneously. The excess electron and hole charges in the base and collector regions must be removed from the junctions before the base–emitter and collector–base voltages fall from their saturated values down to zero voltage. For the bipolar transistor with low epilayer concentration, the charge stored in the epilayer is larger than that stored in the base. The saturation time depends on the complete removal of excess charge stored in the epicollector. The analysis during the turn-off transient is presented here. Similar analysis can be done for the turn-on transient.

The electron concentration in the base is determined by the continuity equation for electrons:

$$\frac{\partial n}{\partial t} = D_n \frac{\partial^2 n}{\partial x^2} + \mu_n E \frac{\partial n}{\partial x} + \mu_n n \frac{dE}{dx} - \frac{n}{\tau_n} \tag{3.4.1}$$

Assuming for simplicity that the distribution of the acceptor dopant in the base is an exponential gives

$$N_B(x) = N_B(0)e^{-\eta x/X_B}$$

where $N_B(0)$ is the base dopant concentration at the edge of the emitter–base space-charge region and η is the slope of base doping. The built-in field in the base is thus

$$E \approx \frac{kT}{q} \frac{1}{N_B(x)} \frac{dN_B(x)}{dx} = -\frac{kT}{q} \frac{\eta}{X_B}$$

The boundary conditions for the switch-off transient are assumed to be

$$\Delta n(0,t) = 0 \tag{3.4.2}$$

$$\Delta n(X_B, t) = 0 \tag{3.4.3}$$

and the initial condition is

$$n(x) = C_1 e^{\Phi_1 x} + C_2 e^{\Phi_2 x} \tag{3.4.4}$$

where

$$C_1 = \frac{-n(W_B) + n(0)e^{\Phi_2 X_B}}{e^{\Phi_2 X_B} - e^{\Phi_1 X_B}}$$

$$C_2 = \frac{n(W_B) - n(0)e^{\Phi_2 X_B}}{e^{\Phi_2 X_B} - e^{\Phi_1 X_B}}$$

$$\Phi_1 = \frac{\sqrt{\mu_n^2 E^2 + 4D_n/\tau_n} - \mu_n E}{2D_n}$$

$$\Phi_2 = \frac{-\sqrt{\mu_n^2 E^2 + 4D_n/\tau_n} - \mu_n E}{2D_n}$$

These boundary conditions imply that the junction voltages change instantaneously and neglect the series resistance and circuit impedance. Using (3.4.2)–(3.4.4), one solves Eq. (3.4.1) as

$$n(x,t) = \left[\sum_{s=1}^{\infty} H_s e^{-(s^2\pi^2 D_n/X_B^2)t} \sin\left(\frac{s\pi}{X_B}\right)x\right] e^{-(\mu_n Ex/2D_n - 1/\tau_n - \mu_n^2 E^2/4D_n)t} \tag{3.4.5}$$

where

$$H_s = -2\left\{\frac{C_1[s\pi(-1)^s e^{\Phi_3 X_B} - 1]}{(s\pi)^2 + (\Phi_3 X_B)^2} + \frac{C_2[s\pi(-1)^s e^{-\Phi_2 X_B} - 1]}{(s\pi)^2 + (\Phi_3 X_B)^2}\right\}$$

$$\Phi_3 = \frac{\sqrt{\mu_n^2 E^2 + 4D_n/\tau_n}}{2D_n}$$

This solution accounts for recombination and an aiding field in the base. From (3.4.5) the electron diffusion current plus drift current in the base as a function of time and position is given by

$$J_n(x,t) = qD_n \frac{\partial n(x,t)}{\partial x} + q\mu_n nE$$

$$= qD_n e^{-[1/\tau_n + \mu_n Ex/2D_n + \mu_n^2 E^2/4D_n]t} \sum_{s=1}^{\infty} H_s e^{-(s^2\pi^2 D_n/X_B^2)t} \frac{s\pi}{X_B} \cos\left(\frac{s\pi}{X_B}\right)x$$

$$-\frac{\mu_n E}{2D_n} \sum_{s=1}^{\infty} H_s e^{-(s^2\pi^2 D_n/X_B^2)t} \sin\left(\frac{s\pi}{X_B}\right)x$$

$$+ q\mu_n E \sum_{s=1}^{\infty} H_s e^{-(s^2\pi^2 D_n/X_B^2)t} \sin\left(\frac{s\pi}{X_B}\right) e^{-[\mu_n Ex/2D_n + 1/\tau_n + \mu_n^2 E/4D_n]t} \tag{3.4.6}$$

The electron current density at the emitter–base and collector–base junctions is

$$J_n(0,t) = qD_n e^{-[1/\tau_n + \mu_n E^2/4D_n]t} \sum_{s=1}^{\infty} H_s e^{-(s^2\pi^2 D_n/X_B^2)t} \frac{s\pi}{X_B} \tag{3.4.7}$$

$$J_n(X_B, t) = qD_n e^{-\mu_n E X_B/2D_n - (1/\tau_n + \mu_n^2 E/4D_n)t} \sum_{s=1}^{\infty} H_s e^{-(s^2\pi^2 D_n/X_B^2)t} \frac{s\pi}{X_B}(-1)^s \tag{3.4.8}$$

From (3.4.7) and (3.4.8) the electron charge flowing through the emitter terminal (Q_{nE}) and collector terminal (Q_{nC}) is

$$Q_{nE}(t) = -\int_0^t J_n(0,t)\, dx$$

$$= qD_n \sum_{s=1}^{\infty} \frac{H_s(s\pi/X_B)\{e^{-[1/\tau_n + \mu_n^2 E^2/4D_n + s^2\pi^2 D_n/X_B^2]t} - 1\}}{1/\tau_n + \mu_n^2 E^2/4D_n + s^2\pi^2 2D_n/X_B^2} \tag{3.4.9}$$

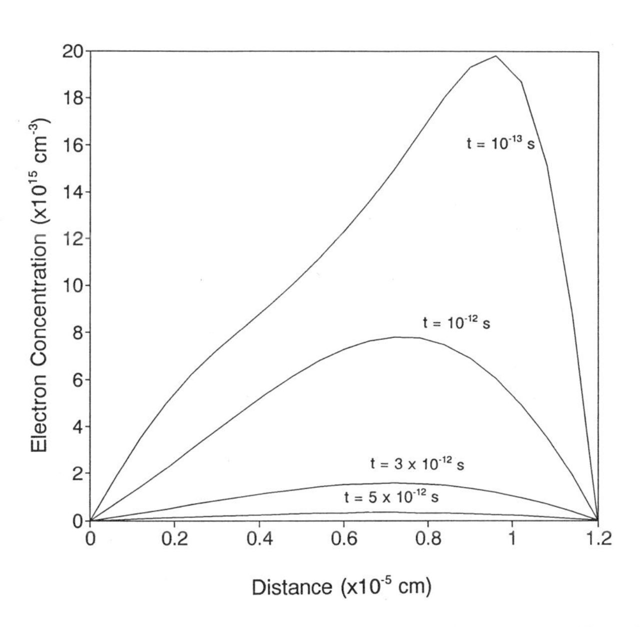

FIGURE 3.32. Electron concentration in the base during the switching-off transient.

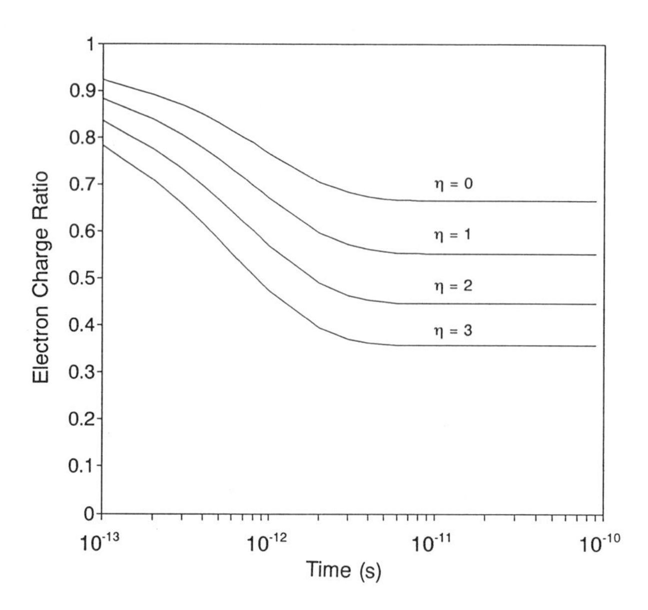

FIGURE 3.33. Electron charge ratio versus time from forward-active mode to cutoff.

$$Q_{nC}(t) = -\int_0^t J_n(X_B, t)\, dx$$

$$= qD_n e^{-\mu_n E X_B/2D_n} \sum_{s=1}^{\infty} \frac{H_s(s\pi/X_B)(-1)^s\{e^{-[1/\tau_n + \mu_n^2 E^2/4D_n + s^2\pi^2 D_n/X_B^2]t} - 1\}}{1/\tau_n + \mu_n^2 E^2/4D_n + s^2\pi^2 D_n/X_B^2} \tag{3.4.10}$$

The equations above are evaluated to examine time-dependent charge and minority carrier distribution in the quasi-neutral base. Figure 3.32 shows the electron concentration in the base during the switch-off transient. The bipolar transistor has peak base doping of 10^{18} cm^{-3}, base width of 0.12 μm, and built-in factor $\eta = 2$. The transistor is switching from 0.8 to 0 V. The stored electron charge flowing through emitter and collector terminals for forward-active ($V_{BE}(t{=}0) = 0.8$ V and $V_{BC}(t{=}0) = 0$ V) to cutoff and saturation ($V_{BE}(t{=}0) = 0.8$ V and $V_{BC}(t{=}0) = 0.8$ V) to cutoff is depicted in Figs. 3.33 and 3.34. For the forward-active to cutoff transient, the electron charge ratio ($Q_{nE}/(Q_{nE} + Q_{nC})$) close to steady state is consistent with the prediction of the partitioned charge model (ratio = 2/3) for the BJT with uniform base concentration ($\eta = 0$). As the base grading increases, more electrons are swept into the collector terminal. The electron charge ratio decreases with η. For the saturation-to-cutoff transient, $Q_{nC}(t)$ is always larger than $Q_{nE}(t)$ and the charge ratio is not a sensitive function of time. As $t \to \infty$, $Q_{nE}/(Q_{nE} + Q_{nC}) = 0.5$, 0.34, 0.2, and 0.11 for $\eta = 0, 1, 2, 3$, respectively. This lower charge ratio indicates that most electrons are removed from the collector terminal during the saturation-to-cutoff transient, especially for a BJT with a high built-in field.

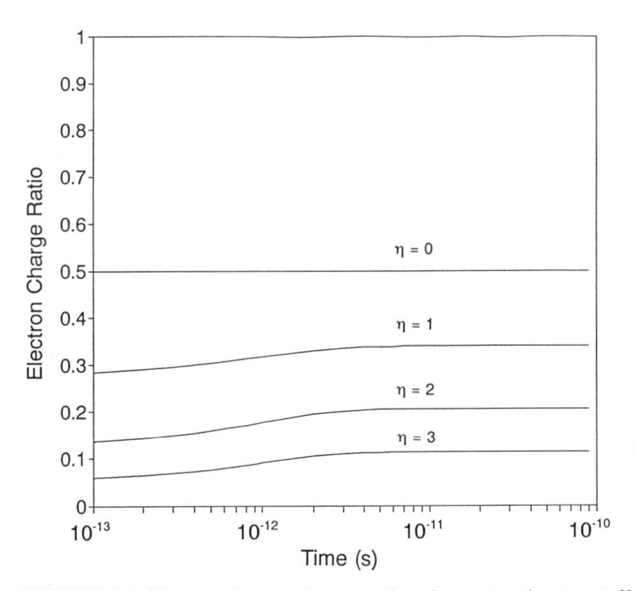

FIGURE 3.34. Electron charge ratio versus time from saturation to cutoff.

EXAMPLE 3.1

```
COMMENT  Bipolar transistor steady-state simulation
COMMENT  Define mesh points
 X.MESH  WIDTH=1.2  H1=0.1
 Y.MESH  DEPTH=0.4  H1=0.05
 Y.MESH  DEPTH=0.6  H1=0.05  H2=0.05
 REGION  NUM=1  SILICON
COMMENT  Electrodes: #1=Emitter #2=Base #3=Collector
 ELECTR  NUM=1  X.MAX=0.05  TOP
 ELECTR  NUM=2  X.MIN=0.9  TOP
 ELECTR  NUM=3  BOTTOM
COMMENT  Define doping profiles
PROFILE  N-TYPE  N.PEAK=2E16  UNIFORM
PROFILE  P-TYPE  N.PEAK=1E19  Y.PEAK=0  Y.JUNC=0.273
      +  X.PEAK=0  WIDTH=1.2
PROFILE  P-TYPE  N.PEAK=7E17  Y.PEAK=0.14  Y.CHAR=0.075  XY.RAT=1
      +  X.PEAK-0  WIDTH=1.2  XY.RAT=1
PROFILE  N-TYPE  N.PEAK=2E20  Y.PEAK=0  WIDTH=0  Y.JUNC=0.15
      +  X.CHAR=0.2  Y.MIN=0.0  Y.JUNC=0.5
PROFILE  N-TYPE  N.PEAK=5E19  Y.PEAK=1  Y.CHAR=0.14
COMMENT  Refine the mesh points
 REGRID  DOPING  LOG  RATIO=3  SMOOTH=1
 REGRID  DOPING  LOG  RATIO=3  SMOOTH=1
 REGRID  DOPING  LOG  RATIO=3  SMOOTH=1
COMMENT  Surface recombination velocity (3 × 10^4 cm/s) for polysilicon emitter
```

```
      CONTACT  NUM=1 SURF.REC vsurfn=3E4 vsurfp=3E4
      COMMENT  Physical models: concentration-dependent mobility, field-dependent mobility,
            +  concentration-dependent SRH recombination, Auger recombination, and
            +  band-gap narrowing
       MODELS  CONMOB FLDMOB CONSRH AUGER BGN
      COMMENT  Use Newton method to solve both minority and majority carriers
         SYMB  NEWTON CARRIERS=2
       METHOD  ITLIMIT=60
      COMMENT  Initialization
        SOLVE  INI
      COMMENT  Perform steady-state simulation
        SOLVE  ELEC=2 NSTEP=17 VSTEP=0.05
        SOLVE  ELEC=2 NSTEP=5 VSTEP=0.02
    CALCULATE  NAME=VBE A=V2
    CALCULATE  NAME=IC A=I3
    CALCULATE  NAME=IB A=I2
      PLOT.1D  X.AXIS=VBE Y.AXIS=IC Y.LOG POINTS TOP=1E-3 BOT=1E-12
+device=p/postscript x.length=11 y.length=11
      PLOT.1D  X.AXIS=VBE Y.AXIS=IB Y.LOG POINTS TOP=1E-3 BOT=1E-12
+device=p/postscript x.length=11 y.length=11
      PLOT.2D  FILL BOUND
+Title="Jp Vector"
+device=p/postscript x.length=11 y.length=11
       VECTOR  J.HOLE
        SOLVE  V1=0 V2=0.95 V3=0
        SOLVE  ELECT=3 NSTEP=10 VSTEP=0.1 ac freq=1E3
    CALCULATE  NAME=VCE A=V3
    CALCULATE  NAME=go A=V1 B=G33 DIFFEREN
    CALCULATE  NAME=gm A=V2 B=G32 DIFFEREN
      PLOT.1D  X.AXIS=VCE Y.AXIS=go X.LOG Y.LOG POINTS
+device=p/postscript x.length=11 y.length=11
      PLOT.1D  X.AXIS=VCE Y.AXIS=gm X.LOG Y.LOG POINTS
+device=p/postscript x.length=11 y.length=11
```

REFERENCES

Asher, K. G. (1964), *IEEE Trans. Electron Devices* **ED-13**, 497.

Bardeen, J. and W. H. Brattain (1948), *Phys. Rev.* **74**, 230.

Bouler, D. L. and F. A. Lindholm (1973), *IEEE Trans. Electron Devices* **ED-20**, 257.

Buhanan, D. (1969), *IEEE Trans. Electron Devices* **ED-16**, 117.

Castellan, G. M. and F. Seitz (1951), *Semiconductor Materials* (Butterworths, London).

Chuang, C. T., D. D. Tang, G. P. Li, and E. Hackbarth (1987), *IEEE Trans. Electron Devices* **ED-34**, 1519.

Chung, C. T. (1988), *IEEE Trans. Electron Devices* **ED-35**, 309.

Cressler, J. D., T. -C Chen, J. D. Warnock, D. D. Tang, and E. S. Yang (1990), *IEEE Trans. Electron Devices* **ED-37**, 680.

Crowell, C. R. and S. M. Sze (1966), *Appl. Phys. Lett.* **9**, 242.

de Graaff, H. C. (1973), *Solid-State Electron.* **16**, 587.

Di Carlo, A. and P. Lugli (1993), *IEEE Electron Device Lett.* **EDL-14**, 103.

Early, J. M. (1952), *Proc. IRE* **40**, 1401.

Ebers, J. J. and J. L. Moll (1954), *Proc. IRE* **42**, 1761.

Getreu, I. (1976), *Modeling the Bipolar Transistor* (Tektronix, Inc., Beaverton, OR).
Ghosh, H. N. (1965), *IEEE Trans. Electron Devices* **ED-11**, 513.
Gummel, H. K. and H. C. Poon (1970), *Bell Syst. Tech. J.* **49**, 827.
Jeong, H. and J. G. Fossum (1985), *IEEE Trans. Electron Devices* **ED-42**, 1103.
Kirk, C. T. (1962), *IRE Trans. Electron Devices* **ED-9**, 164.
Konaka, S., Y. Yamamoto, and T. Sakai (1986), *IEEE Trans. Electron Devices* **ED-33**, 626.
Kurkx, G. A. M. (1987), *IEEE Trans. Electron Devices* **ED-34**, 1939.
Li, G. P., E. Hackbarth, and T.-C Chen (1988), *IEEE Trans. Electron Devices* **ED-35**, 89.
Liou, J. J. (1987), *IEEE Trans. Electron Devices* **ED-34**, 2304.
Liou, J. J., F. A. Lindholm, and J. S. Park (1987), *IEEE Trans. Electron Devices* **ED-34**, 1752.
Liou, J. J. and J. S. Yuan (1990), *IEEE Trans. Electron Devices* **ED-37**, 2274.
Liu, T. M., T.-Z. Chiu, V. D. Archer, and H. H. Kim (1991), *IEEE Trans. Electron Devices* **ED-38**, 1845.
Lu, P.-F. (1990), *IEEE Trans. Electron Devices* **ED-37**, 762.
Lu, P.-F. and T.-C. Chen (1989), *IEEE Trans. Electron Devices* **ED-36**, 1182.
Neugroschel, A. (1987a), *IEEE Trans. Electron Devices* **ED-34**, 817.
Neugroschel, A. (1987b), *IEEE Trans. Electron Devices* **ED-34**, 817.
Ning, T. H., R. D. Issac, P. M. Solomon, D. D. Tang, H. N. Yu, G. C. Feth, and S. K. Weidmann (1981a), *IEEE Trans. Electron Devices* **ED-28**, 1010.
Ning, T. H., R. D. Issac, P. M. Solomon, D. D. Tang, H. N. Yu, G. C. Feth, and S. K. Weidmann (1981b), *IEEE Trans. Electron Devices* **ED-28**, 1010.
Ning, T. H. and D. D. Tang (1984), *IEEE Trans. Electron Devices* **ED-31**, 409.
Pawlikiewicz, A. H. and A. van der Ziel (1987), *IEEE Trans. Electron Devices* **ED-34**, 2009.
Rosenberg, H. M. (1963), *Low Temperature Solid-State Physics* (Oxford, London).
Roulston, D. J. (1990), *Bipolar Semiconductor Devices* (McGraw Hill, New York).
Sah, C. T., R. N. Joyce, and W. Shockley (1957), *Proc. IRE* **45**, 1288.
Shockley, W. (1949), *Bell Syst. Tech. J.* **28**, 435.
Slotboom, J. W. and H. C. de Graaff (1976), *Solid-State Electron.* **19**, 857.
Stork, J. M. C., D. L. Harame, B. S. Meyerson, and T. N. Nguyen (1989), *IEEE Trans. Electron Devices* **ED-36**, 1501.
Streetman, Ben G. (1990), *Solid State Electronic Devices*, 3rd ed. (Prentice Hall, New York).
Strid, E. (1987), *Proc. SPIE*, 795, Bay Pt., FL.
Sze, S. M. (1981), *Physics of Semiconductor Devices*, 2nd ed. (Wiley, New York).
Sze, S. M. (1988), *VLSI Technology*, 2nd ed. (McGraw Hill, New York).
Sze, S. M. and G. Gibbons (1966), *Appl. Phys. Lett.* **8**, 111.
Tang, D. D. (1985), *IEEE Trans. Electron Devices* **ED-32**, 2226.
Tang, D. D., T.-C. Chen, C.-T. Chuang, G. P. Li, M. C. Stork, M. B. Ketchen, E. Hackbarth, and T. H. Ning (1987), *IEEE Electron Device Lett.* **EDL-8**, 174.
Thurmond, C. D. (1975), *J. Electrochem. Soc.* **122**, 1133.
van der Ziel, A. and D. Agouridis (1966), *Proc. IEEE (Lett.)*, **54**, 411.
Van Overstaeten, R. and D. De Man (1970), *Solid-State Electron.* **13**, 583.
van Wingen, P. and E. Wolsheimer (1987), *Proc. BCTM*, Minneapolis, MN.
Verret, D. P. and J. E. Brighton, (1987), *IEEE Trans. Electron Devices* **ED-34**, 2297.
Webster, W. M. (1954), *Proc. IRE* **42**, 914.
Whitter, R. J. and D. A. Tremere (1969), *IEEE Trans. Electron Devices* **ED-16**, 39.
Woo, J. C. S. and J. Plummer (1987), *IEDM*, 401.
Yuan, J. S. (1992), *Phys. Stat. Sol. (a)* **134**, 575.
Yuan, J. S. and W. R. Eisenstadt (1988a), *Solid-State Electron.* **31**, 1725.
Yuan, J. S. and W. R. Eisenstadt (1988b), *IEEE Trans. Electron Devices* **ED-35**, 1633–1639.
Yuan, J. S. and J. J. Liou (1989), *Solid-State Electron.* **32**, 623.
Yuan, J. S., J. J. Liou, and W. R. Eisenstadt (1988), *IEEE Trans. Electron Devices* **ED-37**, 1055.
Zanoni, E., E. F. Crabbé, J. M. C. Stork, P. Pavan, G. Verzellesi, L. Vendrame, and C. Canali (1992), *Int. Electron Device Meeting*, 927.
Zanoni, E., E. F. Crabbé, J. M. C. Stork, P. Pavan, G. Verzellesi, L. Vendrame, and C. Canali (1993), *IEEE Electron Device Lett.* **EDL-14**, 69.

4

Junction Field-Effect Transistors

The junction field-effect transistor (JFET) was first analyzed by Shockley in 1952 (Shockley, 1952). Because its conduction process involves predominately one kind of carrier, the JFET is called a unipolar transistor to distinguish it from the bipolar transistor. Based on Shockley's theoretical treatment, the first working JFET was reported by Dacey and Ross (1953). The junction field-effect transistor has several key features. First, it has no surface effects, such as interface traps occurring at the oxide–semiconductor interface. As a result, the noise level associated with the current fluctuation due to capture and release of free carriers at the surface effects is very low. Second, the isolated two gate terminals of a JFET allow two different input signals to be applied simultaneously for signal-mixing purposes. Third, the carrier transport in a JFET is made up predominantly of the majority carriers, and thus the switching speed of a JFET is not limited by the minority carrier charge storage such as that in a bipolar junction transistor. The main drawbacks of JFETs are that the structure is not compact and the JFET is a low-current-handling device and is thus not suitable for power amplification.

4.1. CONCEPT AND THEORY

An ion-implanted, bipolar-process-compatible *p*-channel JFET structure is shown in Fig. 4.1a. A simplified structure of the JFET under the following normal bias conditions is given in Fig. 4.1b: $V_{GS} > 0$ and $V_{SD} > 0$. In the figure, the shaded regions are the space-charge regions (or the depletion regions) associated with the top and bottom gate regions, V_{GS} is the gate-to-source applied voltage, V_{SD} is the source-to-drain applied voltage, and I_{SD} is the steady-state source-to-drain current. The current I_{SD} is controlled by the undepleted, current-carrying region (referred to as the channel) sandwiched between the two depletion regions associated with the top and bottom gate *n–p* junctions (Fig. 4.1b). The shape of the conducting channel is a function of V_{GS} and V_{SD}. If V_{GS} is maintained constant, the channel acts like a resistor for small V_{SD}, and the resulting variation of I_{SD} with V_{SD} is linear. As V_{SD} increases, the top and bottom depletion regions progressively widen going down the channel from the source to drain, resulting in a loss of conductive volume in the channel and an increased

a

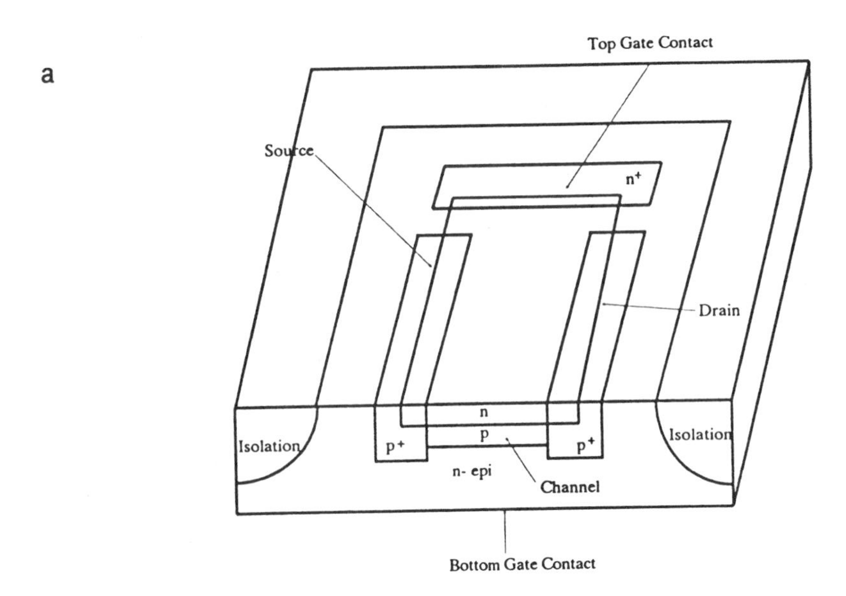

b

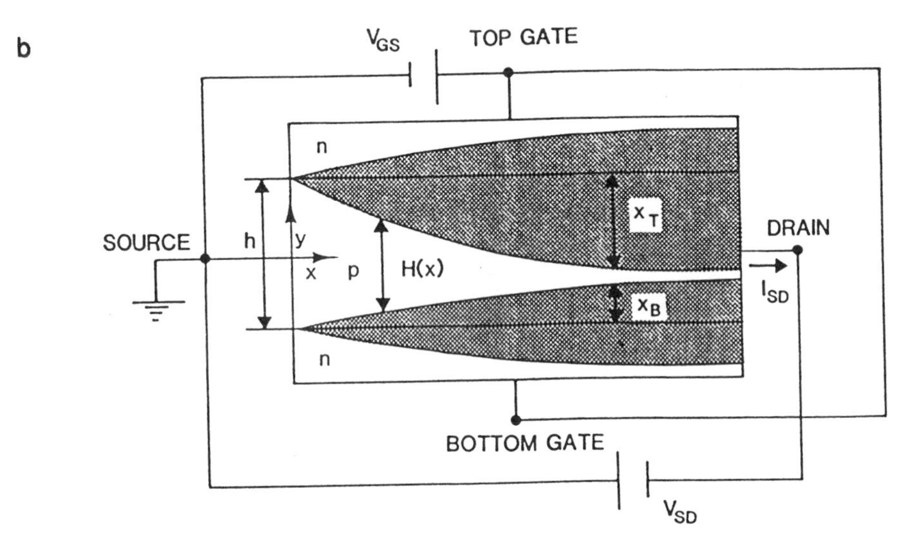

FIGURE 4.1. (a) Schematic of a *p*-channel JFET structure. The *p*-type channel and the *n*-type top gate are normally formed by ion implantation (b) Simplified JFET structure illustrating the bias conditions and the depletion regions associated with the top and bottom gate junctions.

source-to-drain resistance. Increasing the drain voltage further causes the channel to narrow more and more until eventually the top and bottom depletion regions touch near the vicinity of the drain region (referred to as channel pinch-off). When the channel is pinched off at the drain end, where $V_{SD} \equiv V_{SDS}$ (the saturation source-to-drain voltage) and $I_{SD} \equiv I_{SDS}$ (the saturation source-to-drain current), the slope of I_{SD}–V_{SD} characteristics becomes approximately zero. For V_{SD} in excess of V_{SDS}, I_{SD} saturates; that is, it remains approximately constant at I_{SDS}.

When V_{SD} remains constant, increasing V_{GS} widens the depletion regions of the top and bottom gate and narrows the channel; therefore, I_{SD} decreases. The conducting channel near the drain will be pinched off if V_{GS} is continuously increased. The channel can still conduct at or beyond pinch-off, but I_{SD} becomes nearly constant. The I_{SD} will approach zero if V_{GS} is sufficiently large such that the entire channel is pinched off (referred to as channel cutoff). In this case, $V_{GS} = V_P$ with $V_{SD} = 0$, where V_P is called the pinch-off voltage. The pinch-off voltage is related to the saturation source-to-drain voltage by $V_{SDS} = V_P - V_{GS}$.

We first discuss the basic theory by considering a JFET having symmetrical top and bottom gates. An important approximation, called the gradual channel approximation (GCA), has frequently been used in conventional JFET modeling. It states that the carrier density related to the variation of the electric field E_x in the x direction is much smaller than that related to the variation of the electric field E_y in the y direction (i.e., $dE_x/dx << dE_y/dy$). Therefore, the electric field in the depletion region may be considered in the y direction only, and the electric field in the channel may be considered in the x direction only. This separation allows one to solve the two-dimensional problem in a one-dimensional manner. Such an approximation suffices for long-channel JFETs, but may fail if the channel length is less than about 2 μm. Other approximations used conventionally include the two space-charge regions associated with the top and bottom gate junctions are assumed depleted from free carriers, and that the transition between the space-charge regions and the channel region is abrupt.

Since the hole distribution in the p-type channel is assumed uniform, the gradient of holes is zero. Thus, the source-to-drain current density J_{SD} consists of drift current only:

$$J_{SD} = qp\mu_p E \approx qpN_A dV_i/dx \tag{4.1.1}$$

where E is the electric field in the x direction, N_A is the average doping concentration in the channel, and V_i is the electrostatic potential in the channel. For a channel width W, the drain current I_{SD} becomes

$$I_{SD} = WH(x)q\mu_p N_A dV_i/dx \tag{4.1.2}$$

where $H(x)$ is the undepleted channel height (Fig. 4.1b), which is a function of V_i $(0 \le V_i \le V_{SD})$ and V_{GS}. Integrating both sides of (4.1.2) from $x = 0$ to $x = L$ (the channel length) and noting that the sum of the two depletion region thicknesses at the pinch-off point equals the channel height h (Fig. 4.1b), the JFET static current–voltage model can be derived as (Shockley, 1952)

$$I_{SD} = \frac{W}{L} qN_A\mu_p \int_0^{V_{SD}} H(x)\, dV_i$$

$$= G_0 V_{SD} - G_0 \frac{2}{3\sqrt{V_p + \psi}} \left[(V_{SD} + \psi + V_{GS})^{1.5} - (\psi + V_{GS})^{1.5} \right] \tag{4.1.3}$$

and $G_0 = qhW\mu_p N_A/L$. Here ψ is the built-in potential of the symmetrical top and bottom gate junctions, and $V_P = q(h/2)^2 N_A/2\varepsilon - \psi$. Equation (4.1.3) describes the drain current as a function of drain and gate voltages before the channel at the drain end is pinched off ($V_{SD} < V_{SDS}$). Note that this is different from pinching off the entire channel (channel cutoff). When V_{SD} is increased beyond V_{SDS}, the JFET is operated in the saturation region in which the drain current is nearly constant with respect to V_{SD}. When this occurs, $I_{SD} \approx I_{SDS}$ and can be derived from (4.1.3) by replacing V_{SD} with $V_{SDS} = V_P - V_{GS}$ (Cobbold, 1970):

$$I_{SDS} = G_0(V_P - V_{GS}) - G_0 \frac{2}{3\sqrt{V_p + \psi}} \left[(V_P + \psi)^{1.5} - (\psi + V_{GS})^{1.5} \right] \tag{4.1.4}$$

The current–voltage behavior described in (4.1.3) and (4.1.4) can also be characterized by the following two empirical equations, which have been used in the SPICE circuit simulator (Massobrio and Antognetti, 1993):

$$I_{SD} = (W/L)\beta[2(V_P - V_{GS})V_{SD} - V_{SD}](1 + \lambda V_{SD}) \quad \text{for } V_{SD} < V_{SDS} \quad \text{(linear)} \tag{4.1.5}$$

$$I_{SD} = I_{SDS}(1 + \lambda V_{SD}) \quad \text{for } V_{SD} < V_{SDS} \quad \text{(saturation)} \tag{4.1.6}$$

where $I_{SDS} = (W/L)\beta(V_P - V_{GS})^2$, β is the quadratic transfer parameter (the slope of the $\sqrt{I_{SDS}}$ versus V_{GS} curve), and λ is the channel modulation coefficient. Note that V_{SDS}, which is the onset source–drain voltage of the saturation region, is equal to $V_{GD} = V_P$ ($V_{GD} = V_{GS} + V_{SD}$ is the gate-to-drain voltage) because $V_{SDS} = V_P - V_{GS}$. Conventionally, ψ and V_P in (4.1.5) and (4.1.6) are extracted from measurement. Since I_{SDS} given in (4.1.6) follows the relation $(V_P - V_{GS})^2$, the drain current in the saturation region decreases quadratically as V_{GS} is increased and becomes zero when $V_{GS} = V_P$. For $V_{GS} > V_P$ (channel cutoff), which is called the subthreshold region, the model gives $I_{SDS} = 0$. As discussed later, however, a small drain current can exist in the subthreshold region (Brewer, 1975) and thus needs to be included in the model for better accuracy.

The above model usually gives satisfactory predictions to JFETs having a relatively long channel length, but fails when the channel length is about 2 μm or less (Wong and Liou, 1992). Analyzing the short-channel JFET is more complex simply because the *x*-component (*x* direction is defined from the edge of the drain region horizontally into the source region) electric field in the channel is not relatively small compared to the *y* component (*y* direction is defined from the surface of the top gate vertically into the bottom gate), and the quasi-two-dimensional gradual-channel approximation used in modeling the long-channel device no longer suffices. Not only can this large *x*-direction field alter the direction of the overall field in the channel, it can also cause the *x*-direction free-carrier drift velocity to saturate (Kennedy and O'Brien, 1970). Because a large *y*-direction electric field is required to sustain the top and bottom gate depletion regions, the shift of the electric field in the channel from the y direction toward the x direction removes the necessary condition for forming the depletion regions. This leads to a unique phenomenon in the short-channel JFET in which the channel remains open (no channel pinch-off) under a bias condition that will cause channel pinch-off in the long-channel JFET. This, however, is not saying that channel pinch-off can never occur in the short-channel JFET. It all depends on the relative magnitudes of the y-direction electric field, caused largely by the gate voltage and, to a lesser extent, the drain voltage, and the x-direction electric field, which is caused solely by the drain voltage. In fact, as shown later, it is possible to have channel pinch-off in a short-channel JFET if both the gate and drain voltages are relatively large.

4.2. MEDICI SIMULATION

4.2.1. DC Characteristics of Long- and Short-Channel JFETs

In this section, we analyze and compare in detail the long- and short-channel JFETs' steady-state characteristics using MEDICI. Two p-channel JFETs with channel lengths of 6.3 and 1.3 μm are considered. The MEDICI file of a short-channel JFET is shown in Example 4.1. Figure 4.2 shows the doping profile of the JFETs used in simulation. It consists of a top gate layer with 2×10^{18} cm^{-3} peak concentration and 0.5-μm thickness, a channel layer with 8×10^{16} cm^{-3} peak concentration and 0.4-μm thickness, and a bottom gate layer with 10^{15} cm^{-3} uniform concentration and 1-μm thickness. Note that while submicron-gate MESFETs have been realized from recent processes, it is relatively difficult to fabricate the bipolar-process-compatible JFET with a submicron channel length. This is because the drain and source regions of the JFET are formed during the base diffusion of the bipolar processing, and the lateral diffusion in the drain and source is somewhat difficult to control. Channel doping is usually carried out by ion implantation (Hartgring, 1982).

4.2.1.1. Linear Region (before channel pinch-off) The bias conditions considered here are such that the two depletion regions in the conducting channel do not touch each other. Figures 4.3a,b show the simulated electric field vectors in the long- and short-channel JFETs, respectively, biased at $V_{GS} = 1$ V and $V_{SD} = 1$ V. In these figures, the dashed lines indicate the top and bottom gate metallurgical junctions, and the dotted lines represent the edges of the depletion regions in the channel. Figures 4.4a,b illustrate the hole concentration contour in the channel of long- and short-channel JFETs, respectively, biased at $V_{GS} = 1$ V and $V_{SD} = 1$ V.

4.2.1.2. Saturation Region (beyond channel pinch-off) When the drain voltage is sufficiently large, edges of two depletion regions in the channel touch each other and the channel is pinched off. Figures 4.5a,b show the electric field vectors and the edges of the depletion regions in the channel of the long- and short-channel JFETs at $V_{GS} = 1$ V and $V_{SD} = 4$ V, respectively. Note that the channel of the short-channel JFET remains open (Fig. 4.5b) at such a V_{SD}, a bias condition large enough to cause channel pinch-off in the long-channel JFET (Fig. 4.5a). As discussed earlier, this results from the fact that the y-direction field necessary to form the depletion regions is now reduced because the

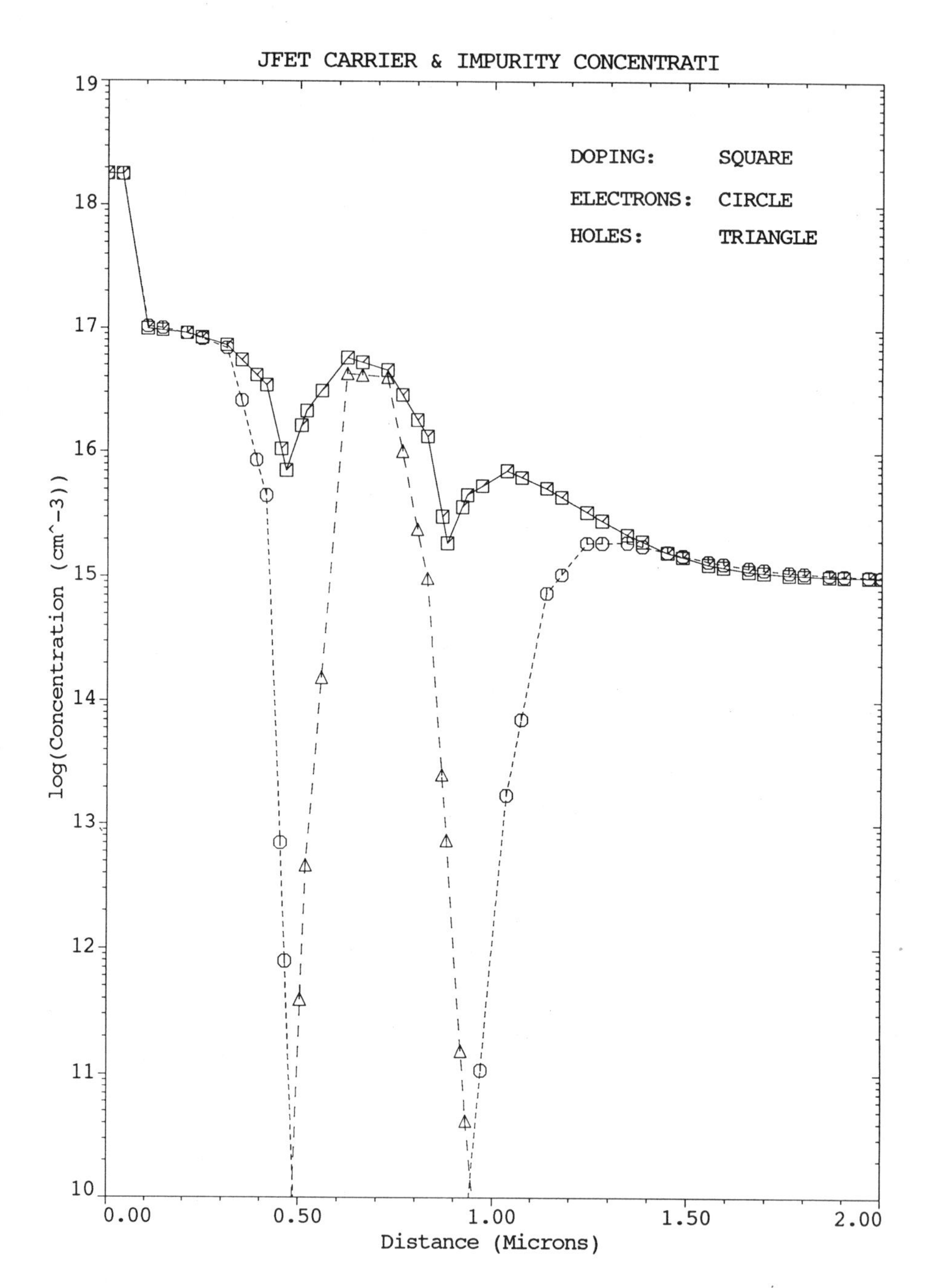

FIGURE 4.2. JFET's impurity doping profile used in MEDICI simulation. The profile is assumed varied only in the y direction (shown) and is constant in the x and z directions.

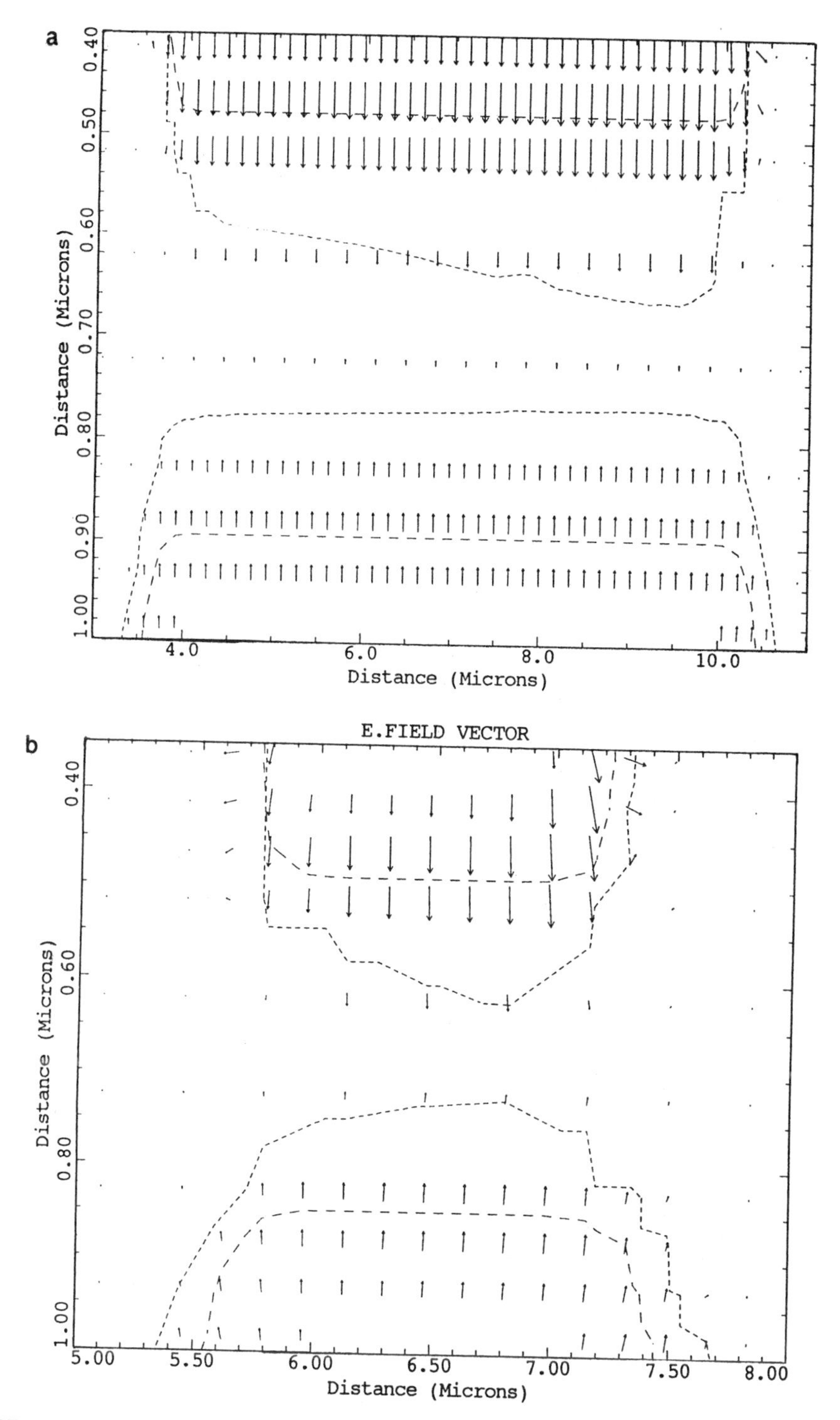

FIGURE 4.3. Electric field vectors (arrows) and the edges of the depletion regions (dotted lines) of the (a) long-channel JFET and (b) short-channel JFET biased in the linear region ($V_{GS} = 1$ V and $V_{SD} = 1$ V).

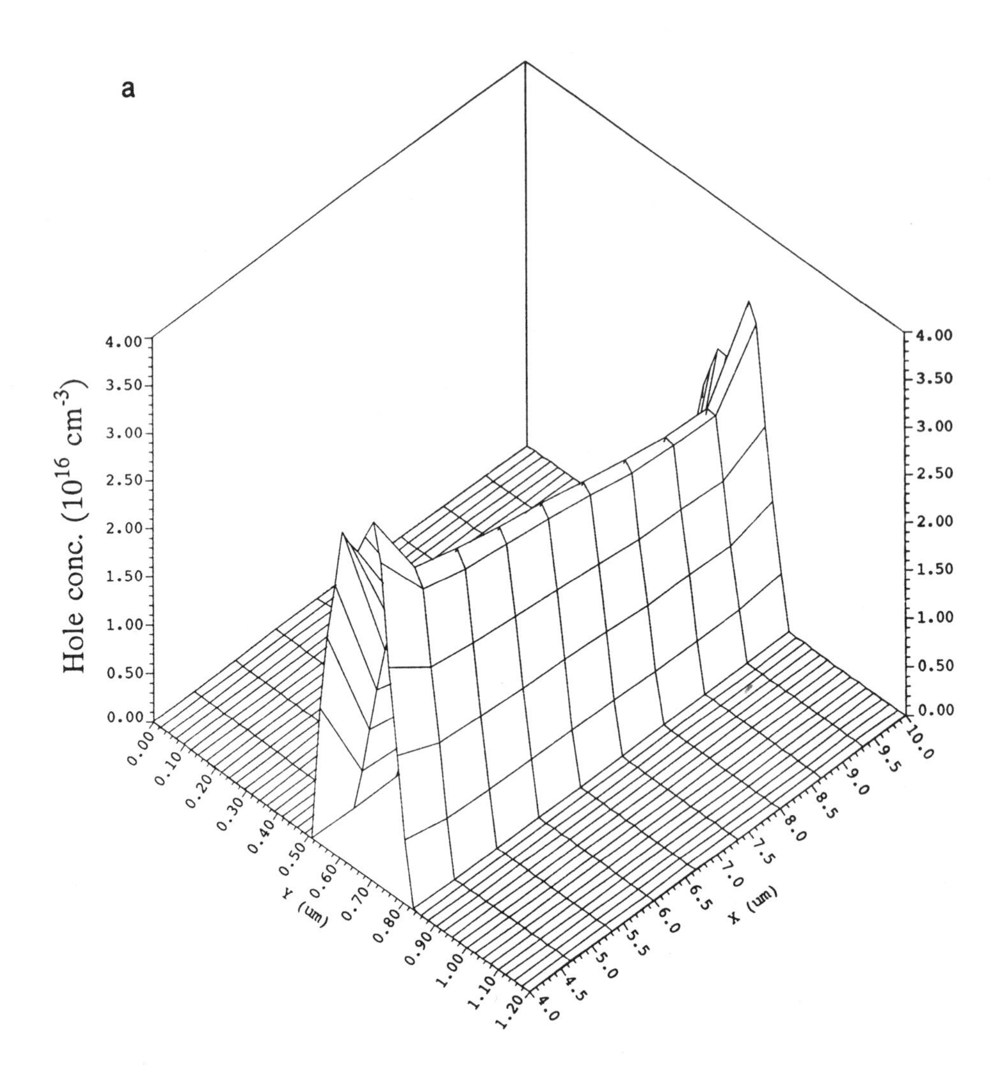

FIGURE 4.4. Hole concentration contours in the channel of (a) long-channel JFET and (b) short-channel.

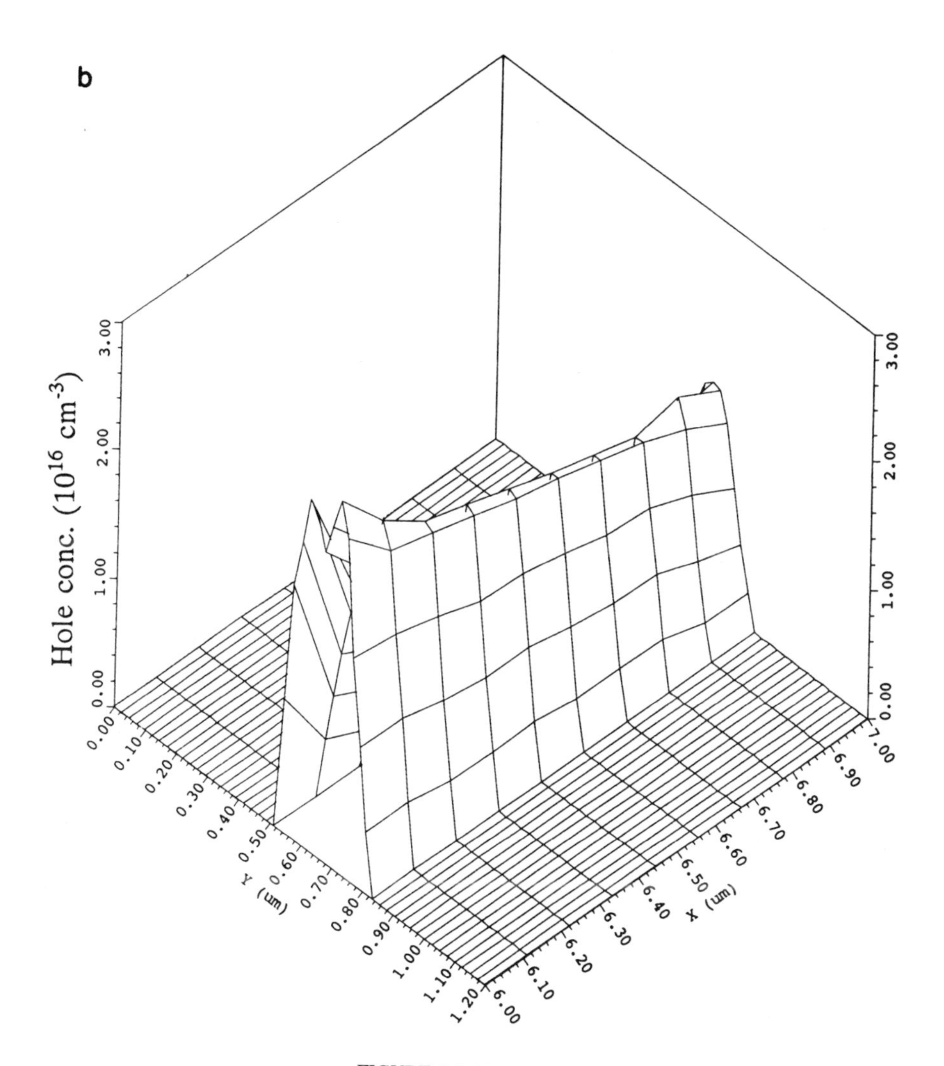

FIGURE 4.4. (Continued)

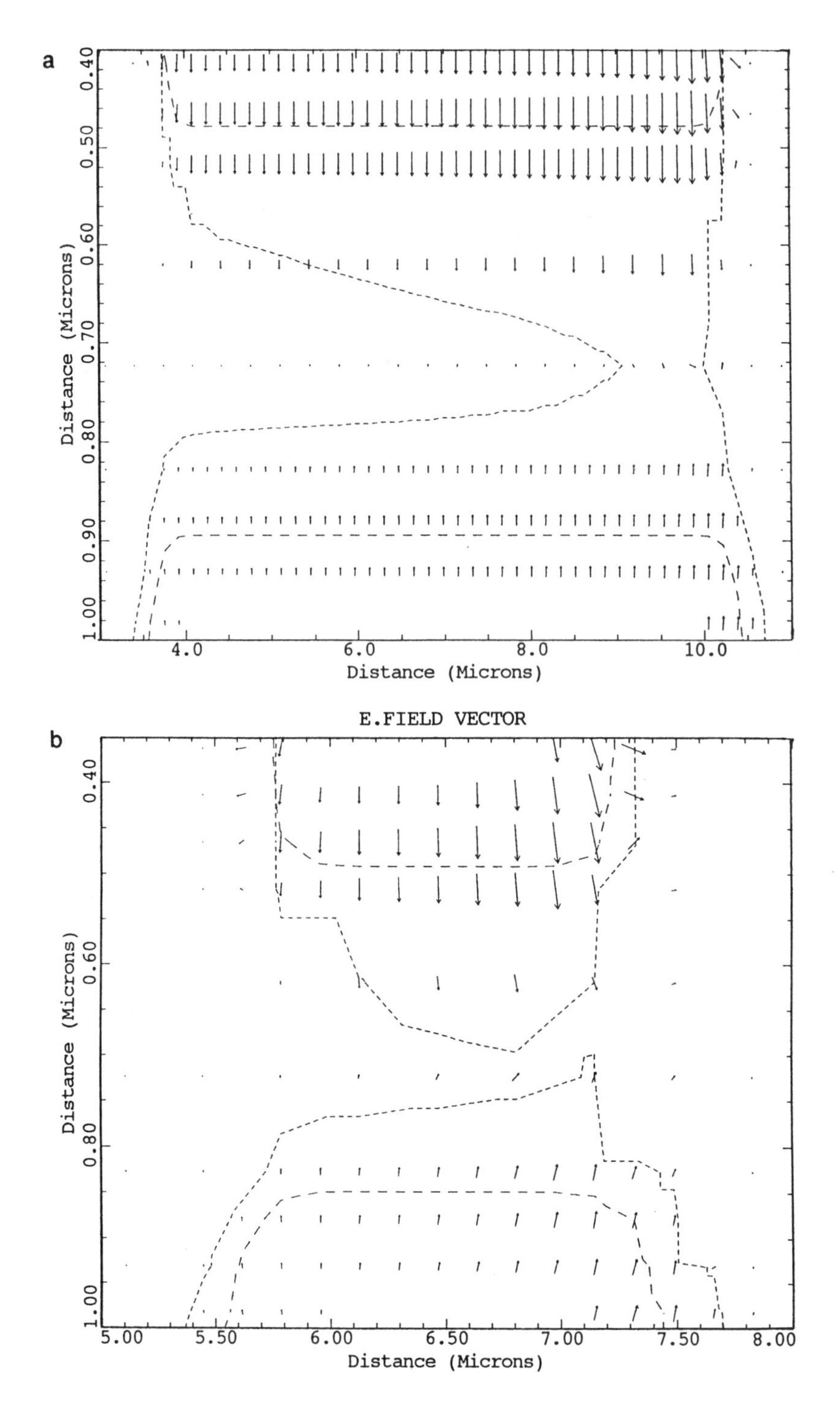

FIGURE 4.5. Electric field vectors and the edges of the depletion regions of the (a) long-channel JFET and (b) short-channel JFET in the saturation region ($V_{GS} = 1$ V and $V_{SD} = 4$ V).

overall field is altered toward the x direction, as evidenced in Fig. 4.5b. Also, the x-direction electric field in the channel near the drain end has exceeded the critical field, suggesting that the x-direction free-carrier velocity has saturated in the region that is supposed to be pinched-off if the channel were long. As the drain voltage increases further, the length in which velocity saturation occurs widens and the channel remains open. Figures 4.6a,b give the hole contours in the channel of long- and short-channel JFETs, respectively. Note that the hole concentration in the short-channel JFET is not decreased significantly toward the drain region due to the fact that channel pinch-off does not occur in such a device.

4.2.1.3. Cutoff Region Another sequence of bias conditions is now considered. The drain voltage is kept constant and the reversed-bias gate voltage is increased. For zero drain voltage, the increased gate voltage will increase the depletion regions uniformly throughout the channel. Thus, the entire channel region can be pinched off (channel cutoff) if the gate voltage is sufficiently large. In such a case, the hole current in the channel vanishes and the device is turned off. Figures 4.7a,b show the electric field vectors and edges of the depletion region in the channel of long- and short-channel JFETs, respectively, operated under the cutoff region (e.g., $V_{GS} = 3$ V and $V_{SD} = 1$ V). The entire channel region becomes a depletion region in the long- and short-channel JFETs. Also, the hole concentration in the channel approaches zero, as evidenced by the hole concentration of the short-channel JFET in Fig. 4.8. This will lead to nearly zero drain current, as would be the case for a long-channel JFET. For the short-channel JFET, however, a small drain current can still flow in the channel due to the space-charge-limited flow. Figures 4.9a,b show the square-root drain saturation current versus gate voltage characteristics of the long- and short-channel JFETs, respectively. For the long-channel JFET, the drain saturation currents for two different V_{SD} are very similar. This is because the channel is long and the channel length modulation effect is not important. The difference in the drain saturation currents for different V_{SD} becomes clearer in the short-channel JFET (Fig. 4.10b). Note that the current exhibits exponential-tail behavior near the cutoff region.

The current–voltage characteristics of the long- and short-channel JFETs are plotted in Figs. 4.10a,b, respectively. Note that the drain current of the short-channel JFET does not show the saturation behavior seen in the long-channel counterpart, as a result of the no-pinch-off phenomenon discussed previously. Also, while the long-channel JFET has a zero drain voltage in the cutoff region, a small drain current still exists in the short-channel JFET. Furthermore, this current shows current–voltage characteristics (i.e., current is proportional to V_{SD}^2) different from those currents in other regions. To explain this, let us consider a metal–insulator–metal system, which is similar to the JFET in cutoff region. Since the current is drift only,

$$J = q\mu nE$$

where J is the electron current density. The space-charge density in the insulator is qn, so

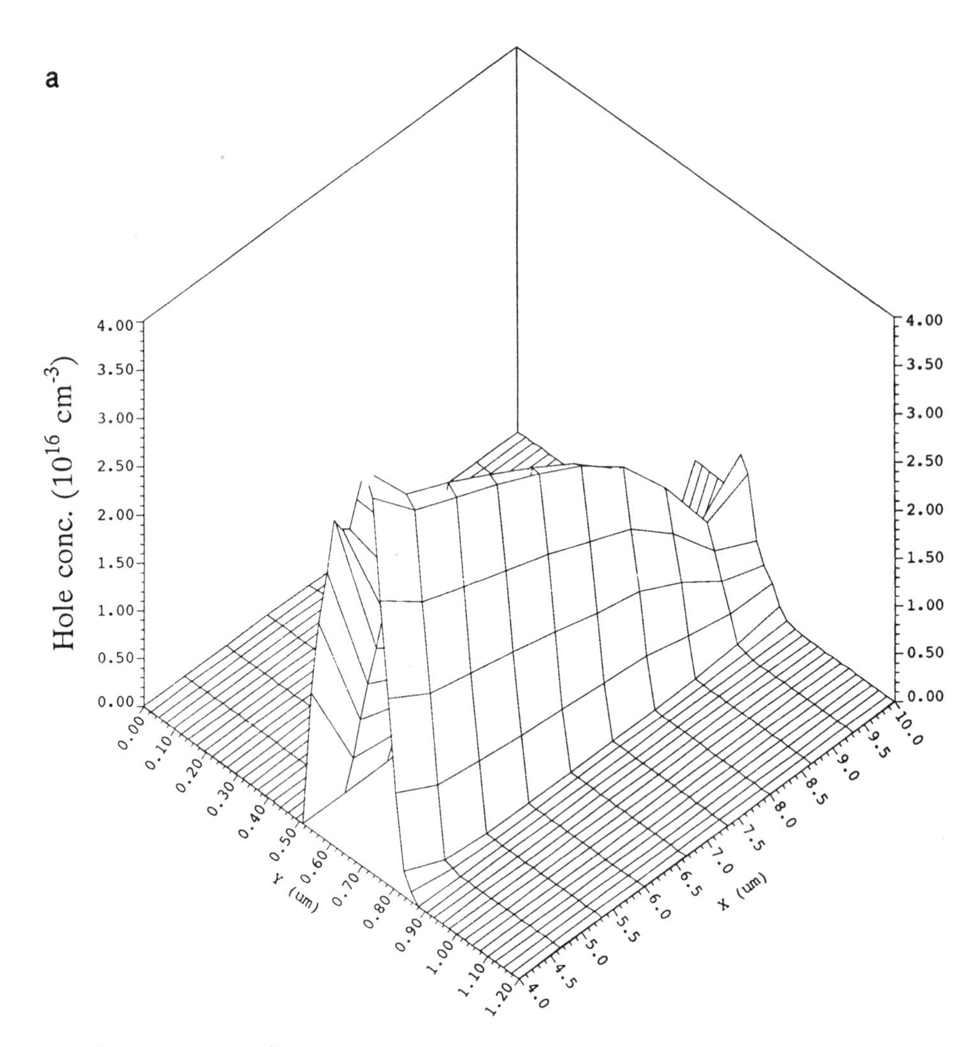

FIGURE 4.6. Hole concentration contours in the channel of (a) long-channel JFET and (b) short-channel JFET biased in the saturation region ($V_{GS} = 1$ V and $V_{SD} = 4$ V).

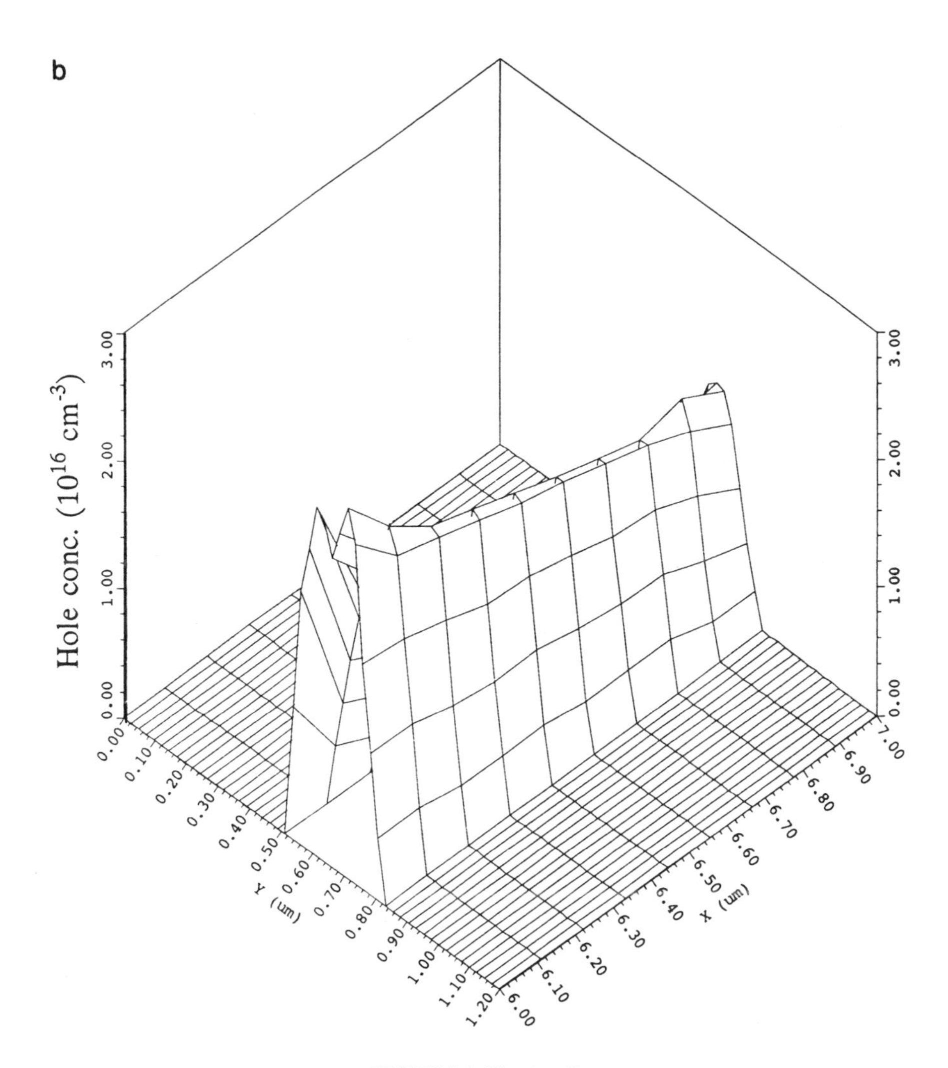

FIGURE 4.6. (Continued)

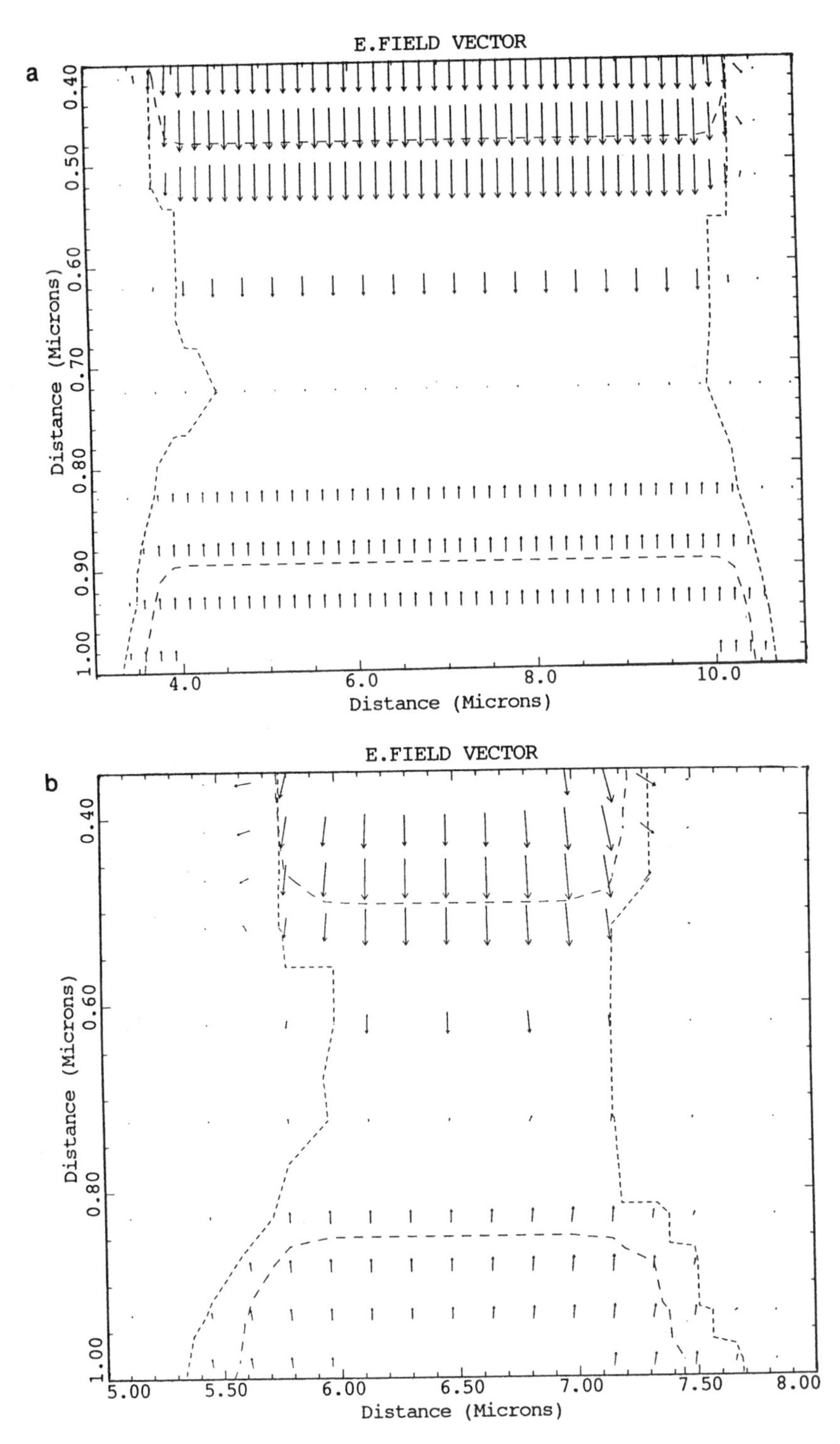

FIGURE 4.7. Electric field vectors (arrows) and the edges of the depletion regions (dotted lines) of the (a) long-channel JFET and (b) short-channel JFET biased in the cutoff region ($V_{GS} = 3$ V and $V_{SD} = 1$ V).

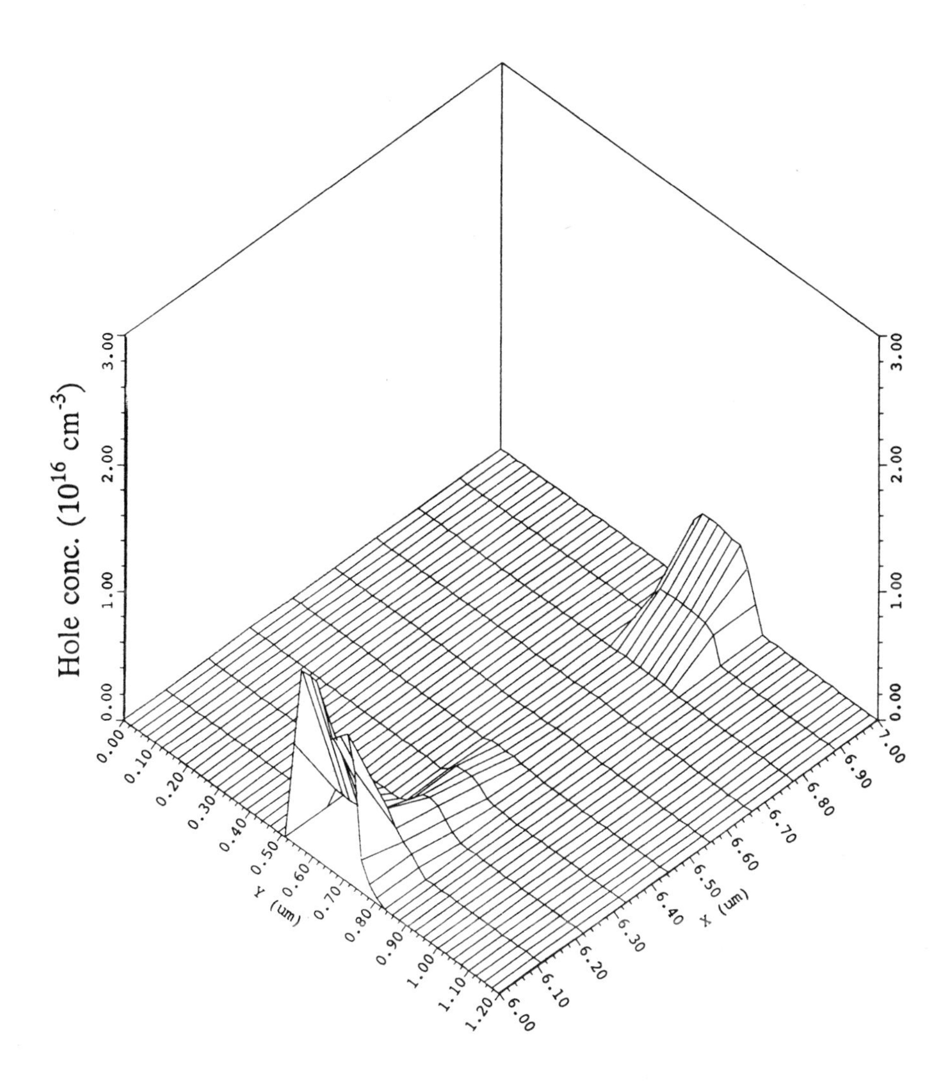

FIGURE 4.8. Hole concentration contour in the channel of the short-channel JFET operated in the cutoff region.

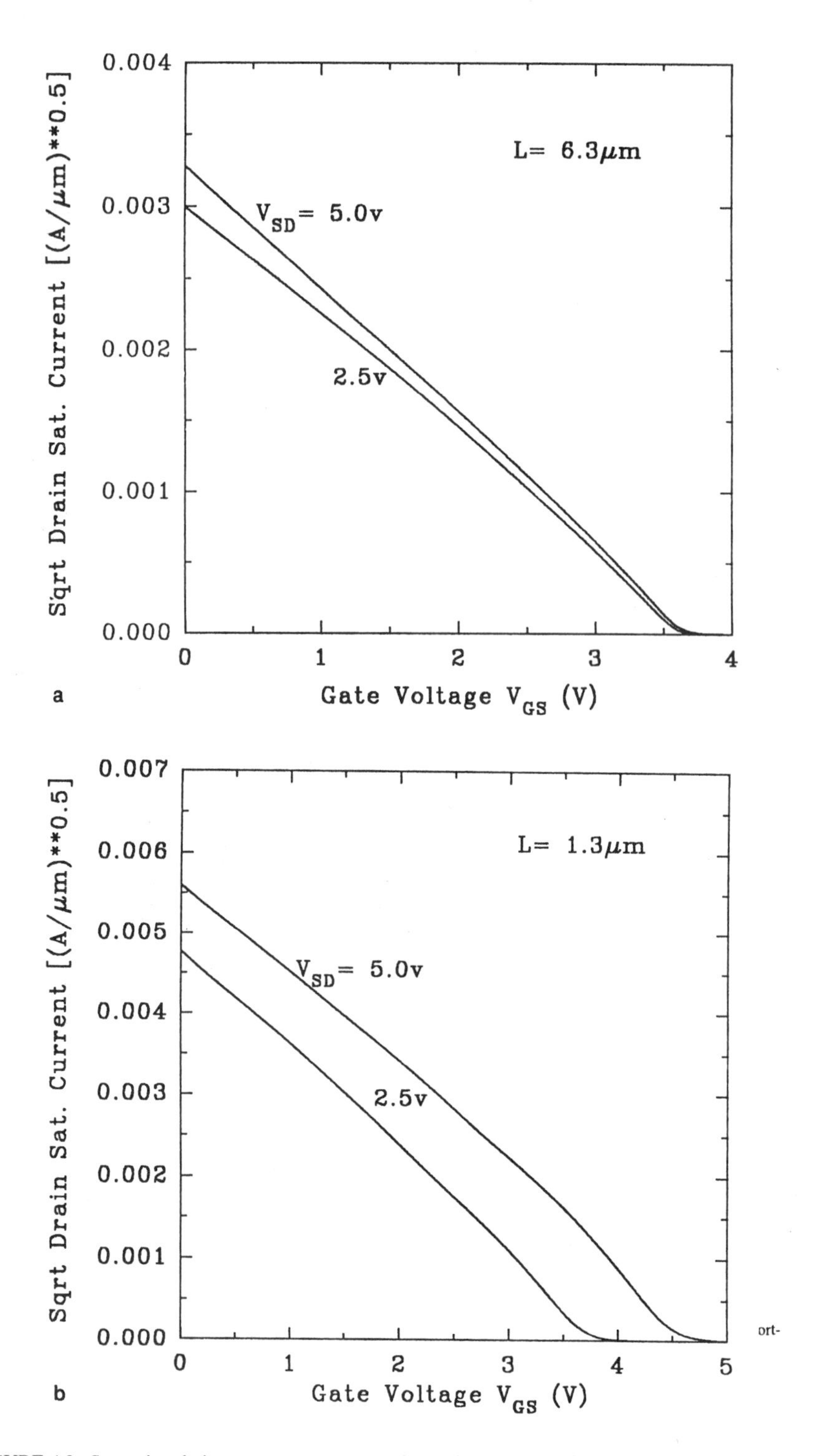

ort-

FIGURE 4.9. Saturation drain current versus gate voltage characteristics for (a) long-channel and (b) short-channel JFETs.

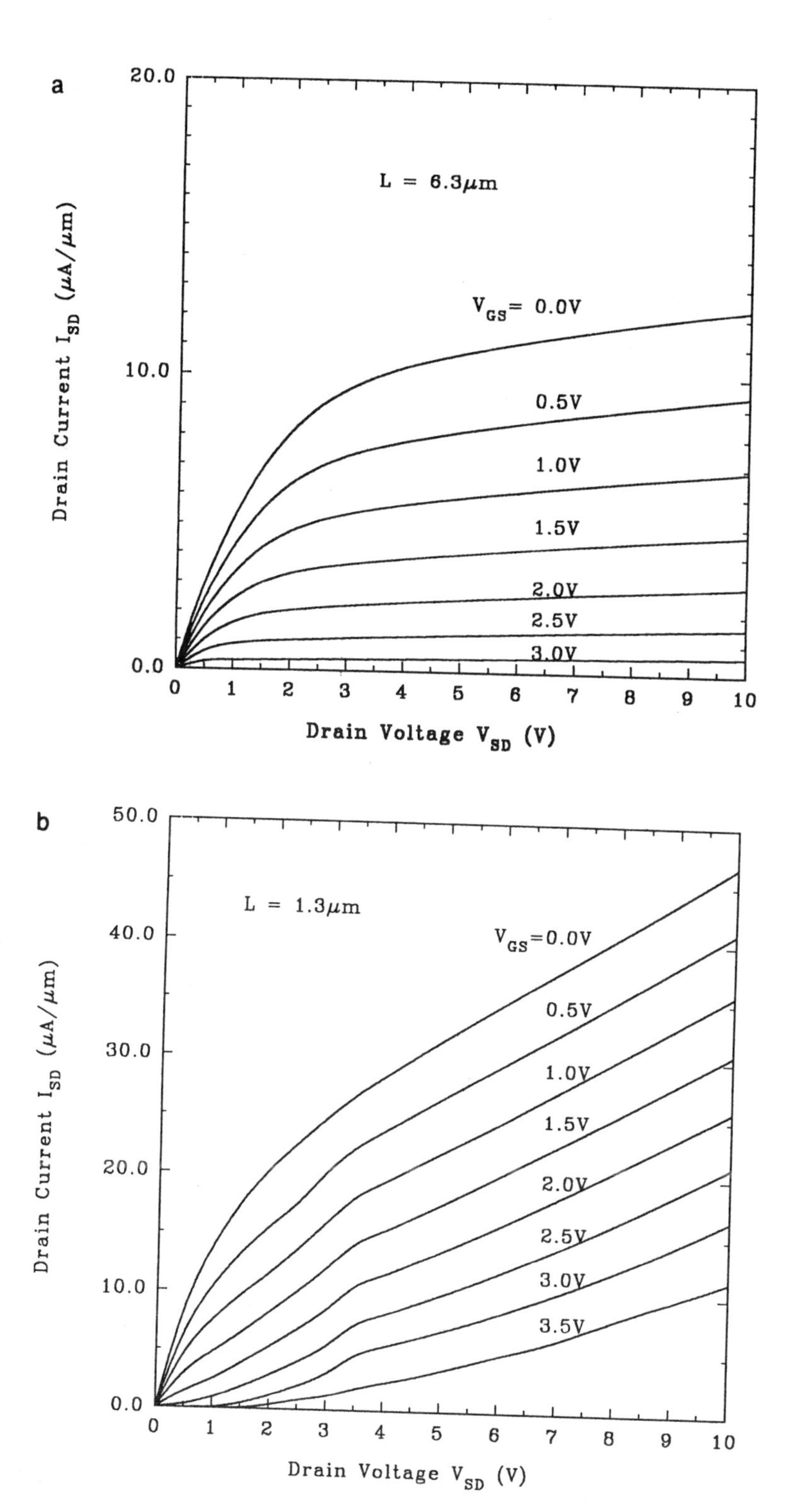

FIGURE 4.10. Drain current versus drain voltage characteristics for the (a) long-channel JFET and (b) short-channel JFET.

$$\frac{dE}{dx} = -\frac{qn}{\varepsilon}$$

This equation indicates that the electric field E depends on n, and thus injection of n into the insulator is limited by n itself. Combining the equations and integrating the result yield

$$E = \sqrt{\frac{2Jx}{\mu\varepsilon}}$$

The voltage V across the insulator is determined by the integral of E from $x = 0$ to $x = L$ (insulator length). This gives

$$J = \frac{9\mu\varepsilon V^2}{8L^3}$$

This current, which is called the space-charge-limited current, is proportional to V^2 and becomes very small if L is relatively large.

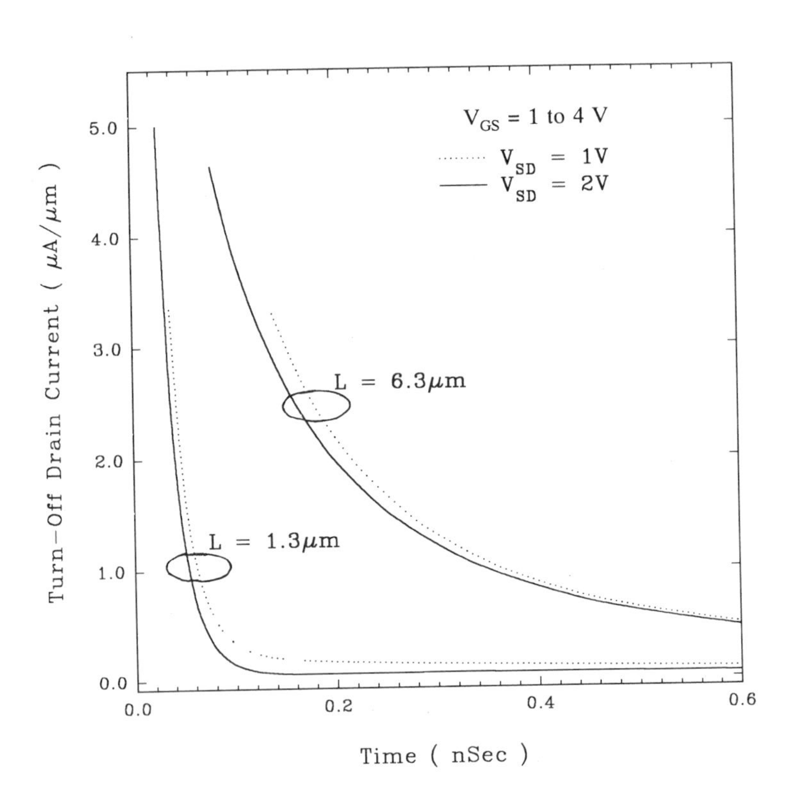

FIGURE 4.11. Drain current turn-off transient behavior simulated for long- and short-channel JFETs at two drain voltages.

4.2.2. Transient Characteristics of JFETs

The JFET turn-off transient is investigated using MEDICI. Figure 4.11 shows the turn-off drain current characteristics of the 1.3- and 6.3-μm JFETs at $V_{SD} = 1$ V and 2 V and a gate voltage changing from 1 to 4 V instantaneously at $t = 0$. The results indicate that the drain current will decay faster if V_{SD} is larger or the channel length is decreased. This is because a larger V_{SD} reduces the free-carrier charge in the channel and, therefore, gives rise to a faster turn-off speed. The same concept is applied to a shorter-channel JFET, which has less charge storage in the channel, and hence a smaller turn-off time.

Figures 4.12a,b show the electric field contour in the 1.3-μm JFET taken at $t = 4 \times 10^{-13}$ s and $t = 2 \times 10^{-10}$ s, respectively. Clearly, the electric field at the top gate–channel junction increases substantially from $t = 4 \times 10^{-13}$ s to $t = 2 \times 10^{-10}$ s, which then turns off the JFET. Note that the electric field at the bottom gate–channel junction is small and insensitive to the change of gate voltage.

Figure 4.13a plots the drain current versus time for the long-channel JFET subject to a step transient at the drain contact (0 to 2 V at $t = 0$) for two gate voltages (1 and 2 V). There is an overshoot of the current occurring around $t = 1.5$ ns. A similar trend is also found in the short-channel JFET (Fig. 4.13b). This appears to be caused by a surge of free carriers passing through the channel of the JFET when the drain voltage is suddenly increased. Such a surge subsides after a short time, and the current reaches its steady-state value afterward.

4.2.3. Small-Signal Characteristics of JFETs

The JFET behavior subject to a small-signal excitation (i.e., a dc bias plus a small sinusoidal component) will now be presented. Figures 4.14a,b show the transconductance, g_m $(= dI_{SD}/dV_{GS})$, characteristics of the long- and short-channel JFETs, respectively. For a fixed V_{SD}, g_m is calculated at different V_{GS}, which give rise to different I_{SD}. The results show that g_m is increased with increasing I_{SD} or V_{SD}. Also, transconductance increases as channel length decreases.

Figures 4.15a,b show the output conductance, g_o $(= dI_{SD}/dV_{SD})$, characteristics of the long- and short-channel JFETs, respectively. Here g_o decreases with increasing I_{SD} and V_{GS} because the geometry of the conducting channel decreases as V_{GS} increases, which results in a small conductance of the JFET. Note that g_o does not approach zero at large I_{SD} in the short-channel JFET due to the no-pinch-off phenomenon in such a device.

Figures 4.16a,b plot the three intrinsic capacitances between the drain and source (C_{DS}), the gate and source (C_{GS}), and the gate and drain (C_{GD}). Note that C_{DS} decreases to zero as V_{GS} is increased beyond about 5 V. This results because the channel is cut off at such a high gate voltage and the charge storage in the channel between the drain and source regions is zero.

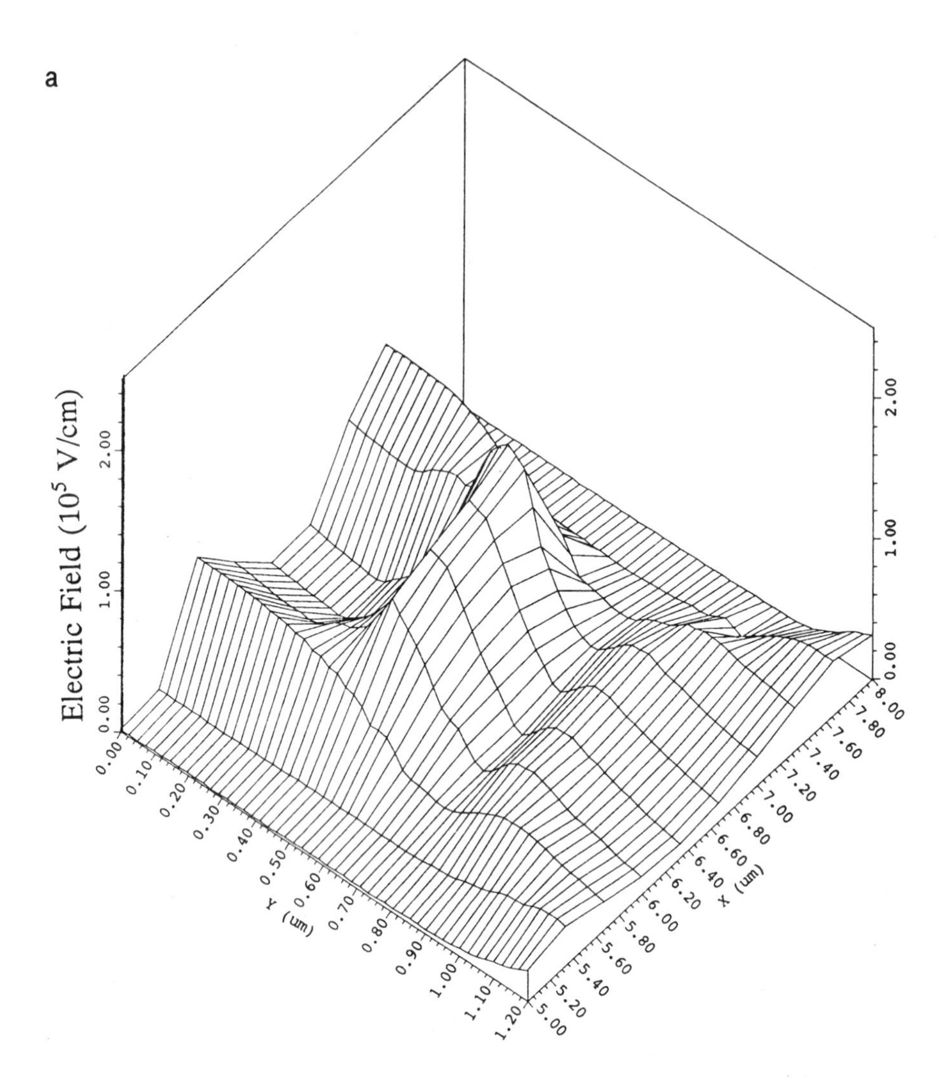

FIGURE 4.12. Electric field contours in the short-channel JFET simulated at $V_{GS} = 1$ V, $V_{SD} = 1$ V, and (a) $t = 0.4$ ps and (b) $t = 200$ ps.

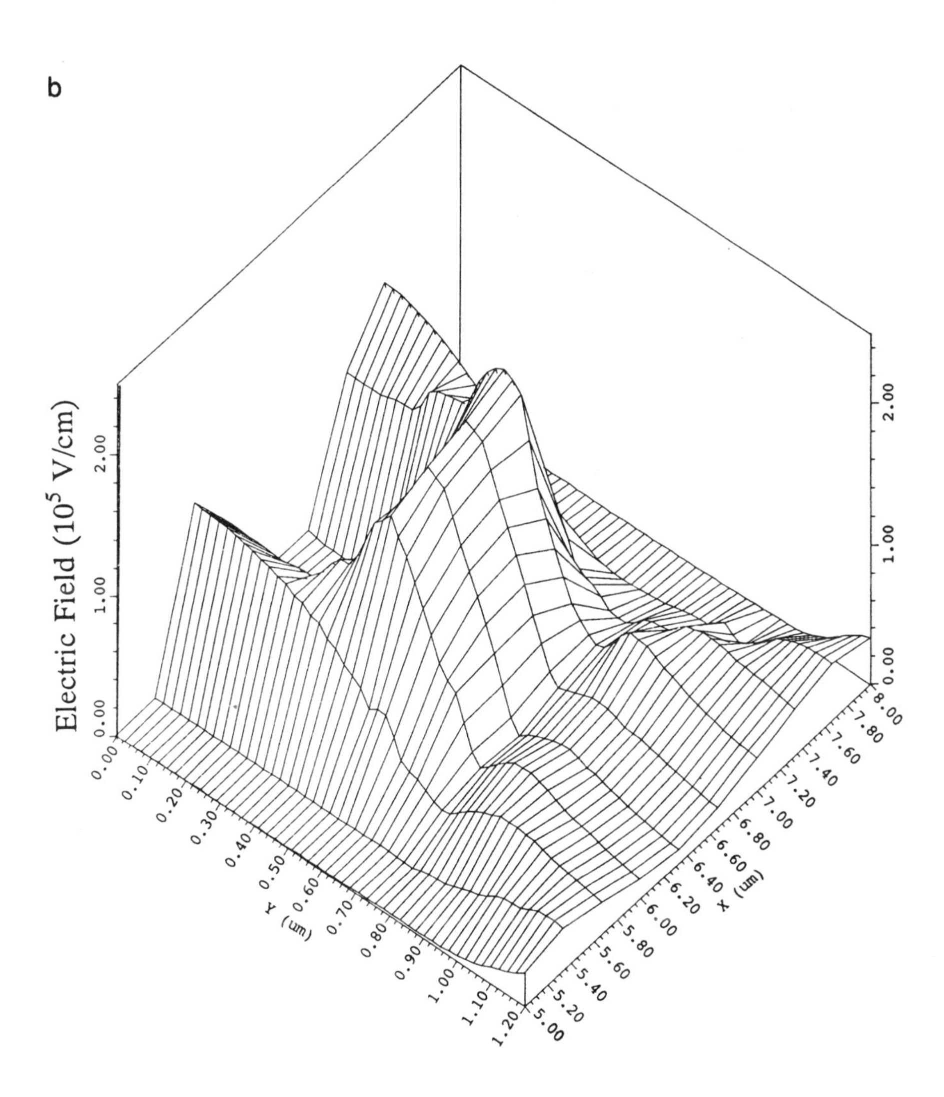

FIGURE 4.12. (Continued)

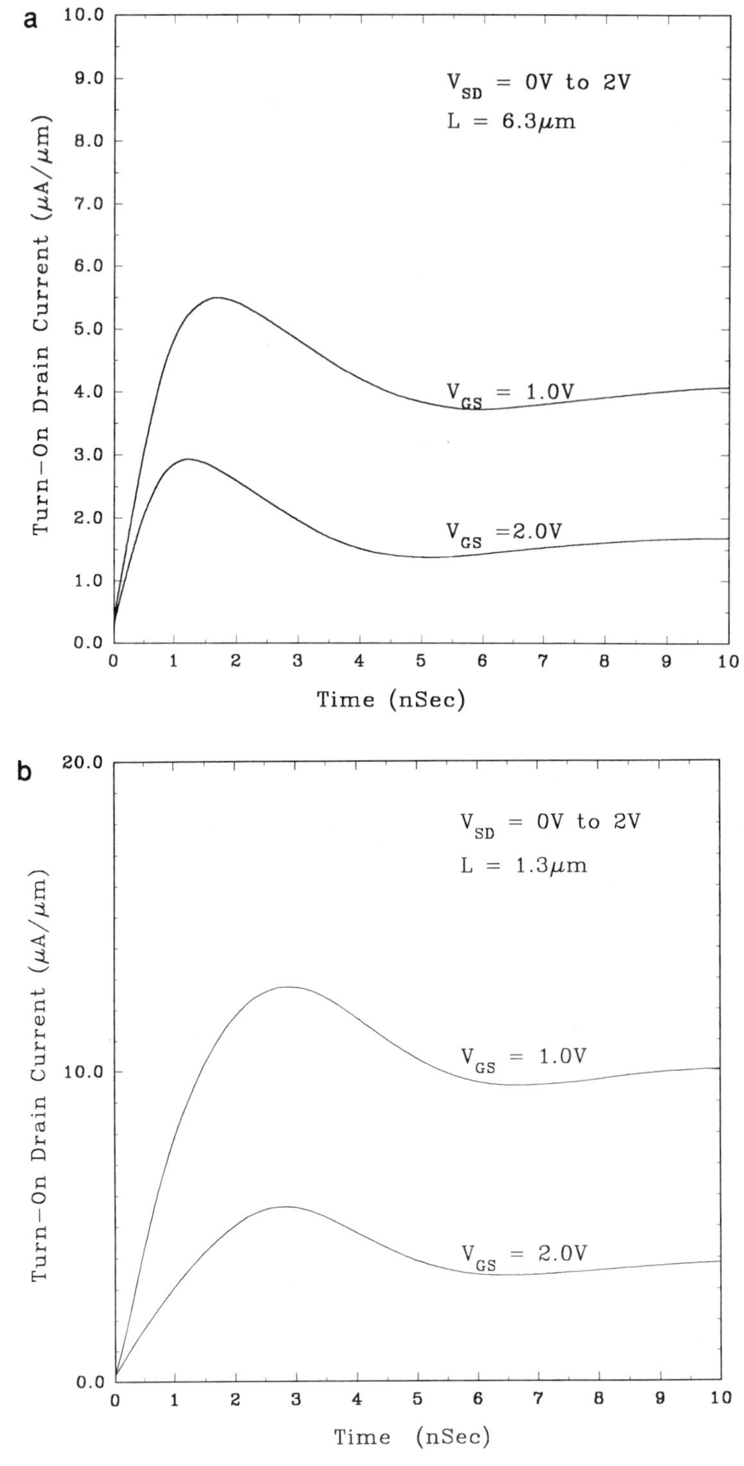

FIGURE 4.13. Drain current transient behavior for (a) long-channel JFET and (b) short-channel JFET subject to a step-up (0 to 2 V) drain voltage.

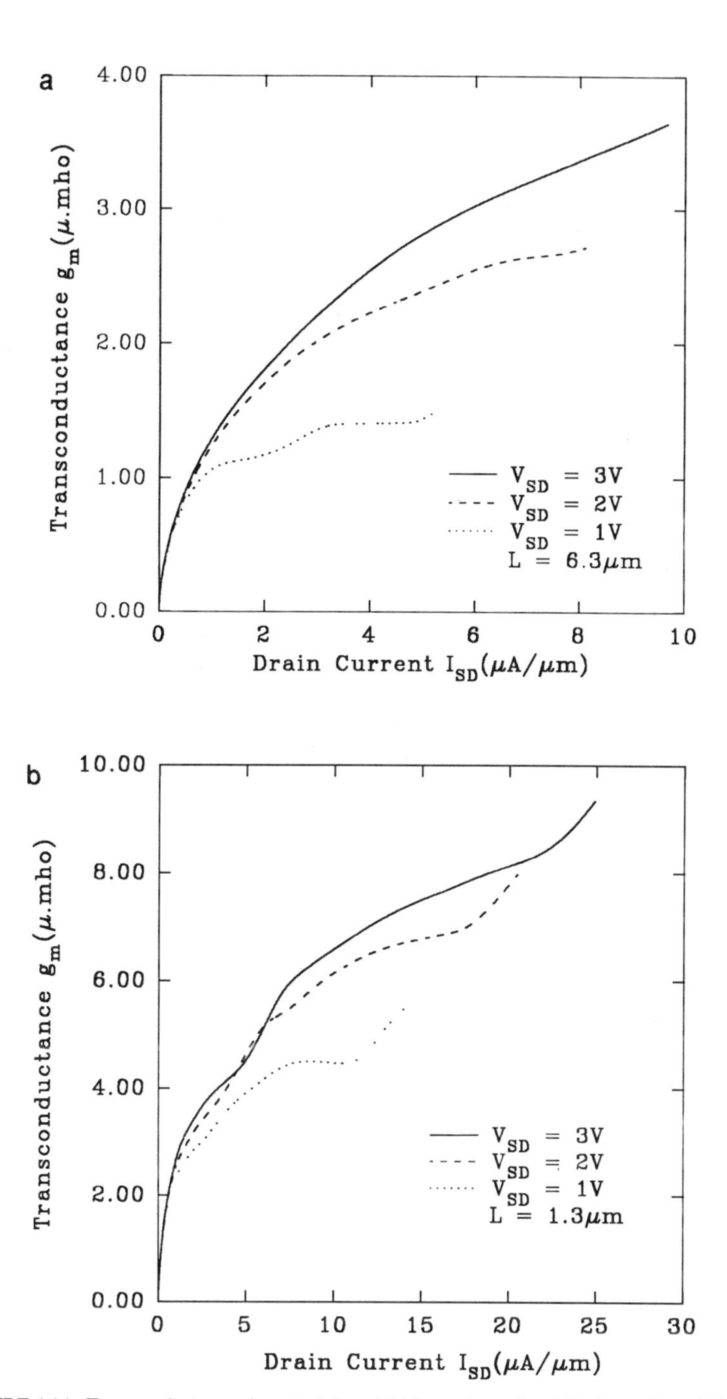

FIGURE 4.14. Transconductance characteristics of (a) long-channel and (b) short-channel JFETs.

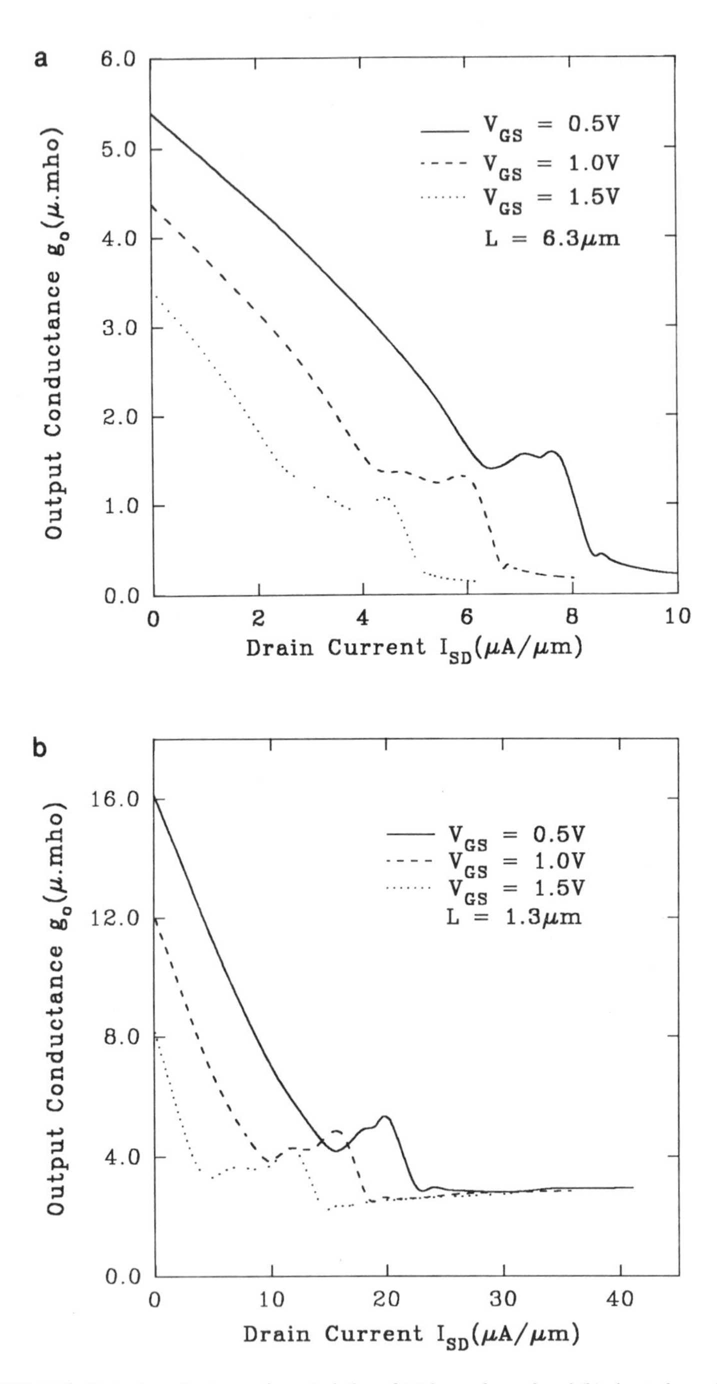

FIGURE 4.15. Output conductance characteristics of (a) long-channel and (b) short-channel JFETs.

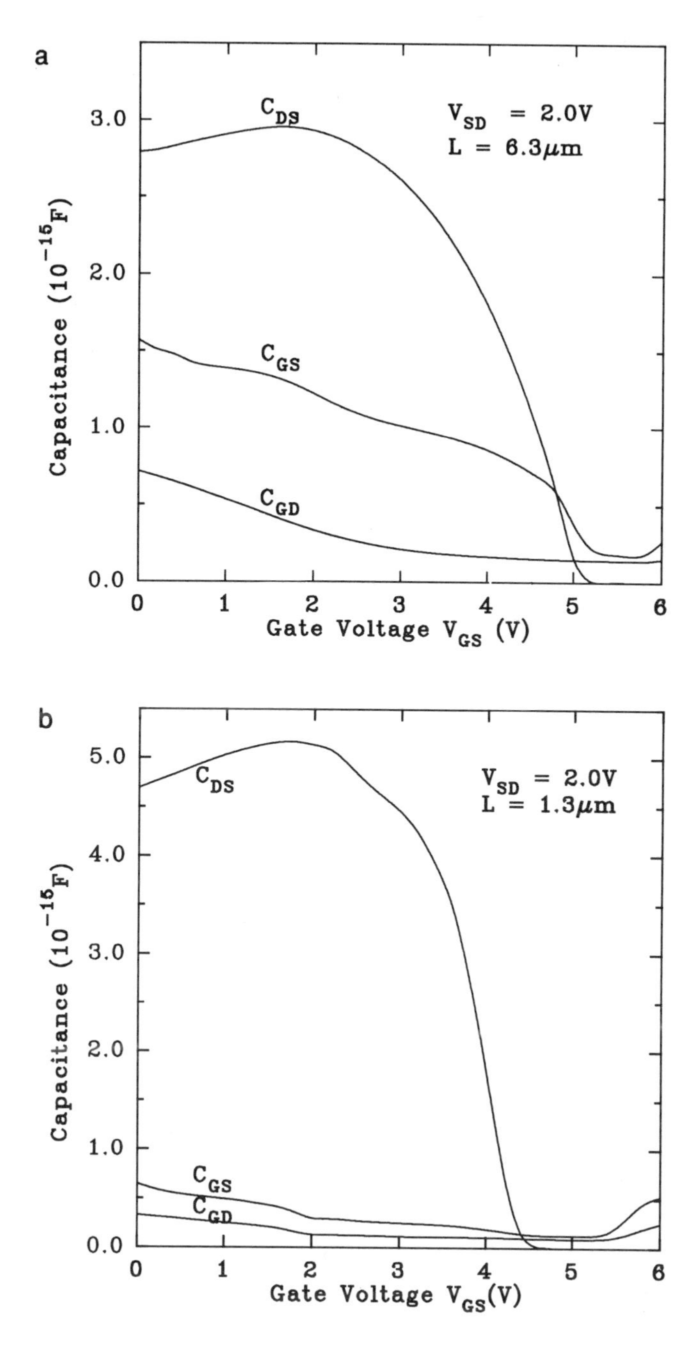

FIGURE 4.16. Intrinsic capacitances versus gate voltage characteristics for the (a) long-channel and (b) short-channel JFETs.

EXAMPLE 4.1

```
COMMENT   p-Channel JFET
   MESH   RECTANG NX=48 NY=30
 X.MESH   N=1 L=0
 X.MESH   N=48 L=16
 Y.MESH   N=1 L=0
 Y.MESH   N=30 L=3
 REGION   NUM=1 IX.MIN=1 IX.MAX=48 IY.MIN=1 IY.MAX=30 silicon
COMMENT   Electrode #1=Top gate, #2=Bottom gate, #3=Source #4=Drain
 ELECTR   NUM=1 X.MIN=6 X.MAX=7 Y.MIN=0 Y.MAX=0.05
 ELECTR   NUM=2 X.MIN=14 X.MAX=15 Y.MIN=0.0 Y.MAX=0.05
 ELECTR   NUM=3 X.MIN=3.5 X.MAX=4.5 Y.MIN=0.0 Y.MAX=0.1
 ELECTR   NUM=4 X.MIN=8.5 X.MAX=9.5 Y.MIN=0.0 Y.MAX=0.1
COMMENT   Impurity profile definition
PROFILE   N-TYPE N.PEAK=1E15 UNIFORM
PROFILE   P-TYPE N.PEAK=9.9E16 Y.MIN=0.62 Y.CHAR=0.175 X.MIN=5
      +   WIDTH=3 XY.RATIO=0.75
PROFILE   N-TYPE N.PEAK=1E17 Y.MIN=0.02 Y.CHAR=0.63 X.MIN=5
      +   WIDTH=3 XY.RATIO=0.6
PROFILE   P-TYPE N.PEAK=1E19 Y.PEAK=0.0 Y.JUNC=1.4 X.PEAK=3
      +   WIDTH=2 XY.RATIO=0.75
PROFILE   P-TYPE N.PEAK=1E19 Y.PEAK=0.0 Y.JUNC=1.4 X.PEAK=8
      +   WIDTH=2 XY.RATIO=0.75
PROFILE   N-TYPE N.PEAK=1E19 Y.PEAK=0.0 Y.CHAR=0.2 X.PEAK=14
      +   WIDTH=2 XY.RATIO=0.75
COMMENT   MESH regrid
 REGRID   DOPING LOG RATIO=3 SMOOTH=1 DOPF=JFETDS
 REGRID   DOPING LOG RATIO=3 SMOOTH=1 DOPF=JFETDS
COMMENT   MODEL definition
 MODELS   CONMOB FLDMOB SRFMOB CONSRH
   SYMB   CARRIERS=0
 METHOD   ICCG DAMPED
MOBILITY  SILICON vsatp=8.34E6 vsatn=1.07E17
COMMENT   Initial condition
  SOLVE   INI
COMMENT   POTENTIAL regrid
 REGRID   POTEN IGN=1 RATIO=.05 SMOOTH=1 DOPF=JFETDS
 REGRID   POTEN IGN=1 RATIO=.05 SMOOTH=1 DOPF=JFETDS
COMMENT   Use Newton's method and solve both electrons and holes
   SYMB   NEWTON CARRIERS=2 MIN.DEGR
COMMENT   Calculate the drain current characteristics
  SOLVE   V3=0.0 ELEC=3 VSTEP=-0.5 NSTEP=20 OUTF=J1
PLOT.1D   Y.AXIS=I3 X.AXIS=V3 POINTS BOT=1E-9 LEFT=1E-9 COLOR=2
```

REFERENCES

Brewer, R. J. (1975), *Solid-State Electron.* **18**, 1013.

Cobbold, R. S. C. (1970), *Theory and Application of Field-Effect Transistors* (Wiley, New York).
Dacey, G. C. and I. M. Ross (1953), *Proc. IRE* **41**, 970.
Hartgring, C. D. (1982), *Solid-State Electron.* **25**, 233.
Kennedy, D. P. and R. R. O'Brien (1970), *IBM J. Res. Dev.* **14**, 95.
Massobrio, G. and P. Antognetti (1993), *Semiconductor Device Modeling with SPICE*, 2nd ed. (McGraw-Hill, New York).
Shockley, W. (1952), *Proc. IRE* **40**, 1365.
Wong, W. W. and J. J. Liou (1992), *IEEE Trans. Electron Devices* **ED-39**, 2576.

5

Metal–Oxide Semiconductor Field-Effect Transistors

The metal–oxide semiconductor field-effect transistor (MOSFET) is the most widely used semiconductor device in very-large-scale-integrated (VLSI) circuits because of its compactness and low power consumption (Tsividis, 1987; Nicollian and Brews, 1982; Pierret, 1983). The principle of the surface field-effect transistor was first proposed in the early 1930s by Lilienfeld (1930) and Heil (1935). It was subsequently studied by Shockley and Pearson (1948) in the late 1940s. In 1960, Kahng and Atalla (1960) proposed and fabricated the first MOSFET, using a thermally oxidized silicon structure. The basic device characteristics have been subsequently studied by Ihantola and Moll (Ihantola, 1961; Ihantola and Moll, 1964), Sah (1964), and Hofstein and Heiman (1963). Unlike the bipolar junction transistor, the current in the MOSFET is transported predominantly by the majority carriers, and the MOSFET is a four-terminal device with source, drain, gate, and substrate terminals.

Figure 5.1 is an n-channel MOSFET, with the x direction pointing vertically from the surface into the bulk and the y direction pointing horizontally from the source region to the drain region. Two n^+ regions are formed in the p-type silicon by thermal diffusion or ion implantation. A thin silicon dioxide layer separates the gate contact from the conducting channel, which is underneath the gate and sandwiched between the drain and source regions. The gate contact is made of metal or, in modern MOSFETs, heavily doped polysilicon. The mask gate length L_m is larger than the channel length L defined by metallurgical junctions due to the lateral diffusion of the source and drain dopants (Fig. 5.1). Furthermore, a recent study (Narayaron *et al.*, 1994) has suggested that the effective channel length L_{eff} is larger than L. This results because, when the channel is inverted, the free-carrier inversion channel actually extends into the source and drain regions; it is not just limited to the region defined by the source and drain metallurgical junctions. To simplify the analysis, however, $L \approx L_{\text{eff}}$ will be used for MOSFET modeling discussed in this chapter.

The source-to-drain current of the MOSFET is controlled by the gate voltage V_G and drain voltage V_D. When a positive voltage is applied to the gate terminal, a vertical electric

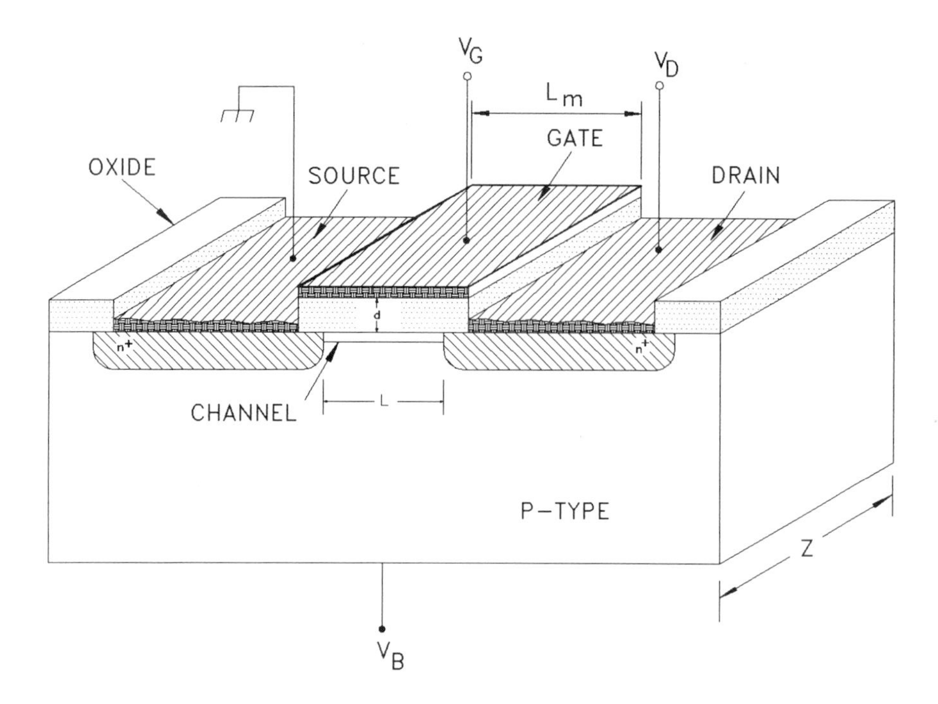

FIGURE 5.1. A two-dimensional n-channel MOSFET structure.

field is induced across the oxide layer into the semiconductor. This field then repels the holes from the $Si–SiO_2$ interface. If the gate voltage exceeds a critical voltage, called the threshold voltage V_{TH}, a thin layer of electrons (inversion layer) is created near the interface, which then connects the n-type source and drain regions. If a positive voltage is applied to the drain terminal, the drain contact will drain the free carriers out of the device, and to maintain charge neutrality the source region will supply the free carriers to the device. Thus, under such a bias condition, the MOSFET is similar to a resistor, and its current–voltage characteristics exhibit a linear relation (linear region). As the drain voltage is increased to a critical voltage, called the saturation drain voltage, however, the inversion layer near the drain junction vanishes (channel pinch-off) due to the presence of a large field. Under this condition, the drain current does not increase considerably with increasing V_D, and the MOSFET is said to operate in the saturation region.

5.1. CURRENT–VOLTAGE CHARACTERISTICS

The threshold voltage needed for n-channel inversion in the MOSFET is given by (Liou, 1994)

$$V_{TH} = V_{FB} + 2\phi_B + V_y(y) + \frac{\sqrt{2q\varepsilon N_A\,[2\phi_B + V_y(y) - V_B]}}{C_{ox}} \tag{5.1.1}$$

where V_{FB} is the flat-band voltage, ϕ_B is the bulk potential, $V_y(y)$ is the voltage drop along the channel (i.e., $V_y(0) = 0$ and $V_y(L) = V_D$), V_B is the voltage applied to the body terminal, and C_{ox} is the oxide capacitance per unit area.

When the channel is inverted, the drain current density J_D can, in general, be described by drift–diffusion theory as

$$J_D(x, y) = q\mu_n n(x, y)E + qD_n \nabla n(x, y) \tag{5.1.2}$$

If only the y-direction J_D is considered, then

$$J_D(y) \approx q\mu_n n(x, y)E_y + qD_n \frac{\partial n}{\partial y} \tag{5.1.3}$$

or

$$J_D(y) \approx \mu_n n \frac{\partial E_{Fn}}{\partial y} \tag{5.1.4}$$

Here E_y is the y-direction electric field and E_{Fn} is the quasi-Fermi energy for electrons. Integrating both sides of (5.1.4) with respect to dx and dz (z is the third dimension) gives the drain current I_D:

$$I_D = Z\mu_n \frac{dE_{Fn}}{dy} \int_0^{X_d} n\, dx = Z\mu_n \frac{dE_{Fn}}{dy} \frac{Q_n(y)}{q} \tag{5.1.5}$$

where Z is the channel width, X_d is the depletion region thickness, and Q_n is the inversion charge density. Note that E_{Fn} has both diffusion and drift components ($E_{Fn} = E_{ch} + E_i$, where E_{ch} and E_i are the chemical and electrostatic energies related to the diffusion and drift tendencies, respectively).

We now proceed to find $Q_n(y)$. The energy-band diagram at an arbitrary y is shown in Fig. 5.2. Note that there are two quasi-Fermi energies, E_{Fp} and E_{Fn}, which are separated by $qV_y(y)$. Also, $V_y(y) = 0$ in the bulk region ($x \geq X_d$) and the position of E_{Fp} is independent of V_y. Thus, in the channel,

$$p(x, y) = N_A e^{-q\psi(x,y)/kT} \quad \text{and} \quad n(x, y) = \frac{n_i^2}{N_A} e^{q[\psi(x,y) - V_y(y)]/kT} \tag{5.1.6}$$

where ψ is the electrostatic potential with respect to the conduction-band edge E_C. Putting this into Poisson's equation yields the electric field $E_S(y)$ at the interface ($x = 0$):

$$E_s(y) = \frac{\sqrt{2}kT}{qL_D} F_y[\psi_S(y), V_y, \nu] \tag{5.1.7}$$

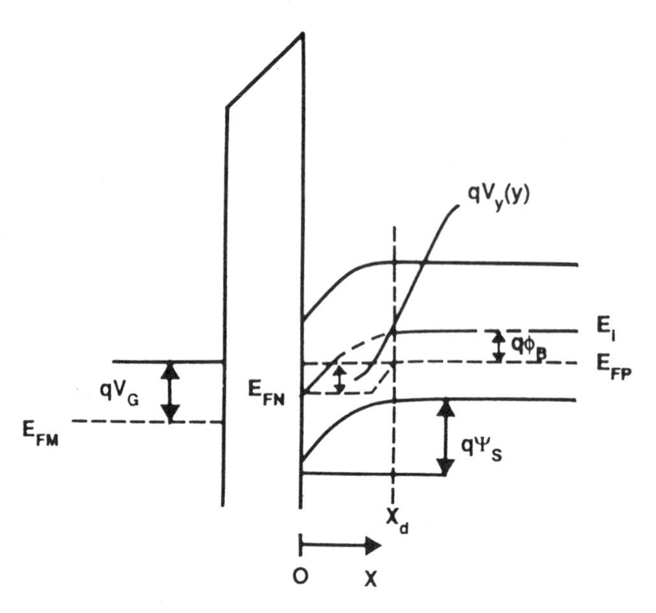

FIGURE 5.2. Energy-band diagram of the MOSFET at a particular y position illustrating the two quasi-Fermi levels E_{Fn} and E_{Fp} in the surface region.

$$F_y[\psi_S(y), V_y(y), \nu] = \sqrt{e^{-q\psi_S/kT} + \frac{q\psi_S}{kT} - 1 + \nu^2 e^{-qV_y/kT}[e^{q\psi_s/kT} - \frac{q\psi_S}{kT}e^{qV_y/kT} - 1]} \tag{5.1.8}$$

where L_D is the extrinsic Debye length, and $\nu = n_i/N_A$. $Q_n(y)$ can be expressed as

$$Q_n(y) = -q\int_0^{X_d(y)} n(x, y)\, dx = -q\int_{\psi_C(y)}^{0} \frac{n(\psi)\, d\psi}{d\psi/dx} \tag{5.1.9}$$

Since $d\psi/dx = -E$ and E has the same form as E_S given in (5.1.7), provided ψ_S is replaced by ψ, (5.1.9) is rewritten as

$$Q_n(y) = q\int_{\psi_S(y)}^{0} \frac{n_i^2/N_A e^{q(\psi - V_y)/kT}\, d\psi}{(\sqrt{2}kT/qL_D)F_y(\psi, V_y, \nu)} \tag{5.1.10}$$

Equations (5.1.5) and (5.1.10) can be used to calculate numerically the inversion-layer charge Q_n and drain current I_D for $V_D > 0$ and for all inversion cases, including weak inversion, moderate inversion, and strong inversion.

5.1.1. Strong Inversion

For the case of strong inversion ($V_G > V_{TH}$), a well-defined inversion layer exists in the channel, and the device is practically an n^+–n–n^+ resistor (for now, let us assume channel pinch-off does not occur). Applying a positive voltage to the drain terminal thus gives rise to a large y-direction voltage drop and a large y-direction electric field throughout the inverted channel. As a result, the drain current I_D is predominantly the drift current (Tsividis, 1987; Nicollian and Brews, 1982; Pierret, 1983; Sah, 1964). This, together with the assumption that the electron quasi-Fermi energy E_{Fn} is flat across the surface region, gives

$$\frac{\partial E_{Fn}(x,y)}{\partial y} = \frac{dE_{Fn}(0,y)}{dy} = \frac{d[E_{\text{ch}}(0,y) + E_i(0,y)]}{dy}$$

$$= \frac{dE_i(0,y)}{dy} = q\frac{d\psi_S(y)}{dy} \tag{5.1.11}$$

Note that $\psi_S(y)$ is defined as $\psi(x=0,y)$ and

$$\psi_S(y) = \psi_S(V_D = 0) + V_y(y) \tag{5.1.12}$$

Combining (5.1.5), (5.1.11), and (5.1.12) yields

$$I_D = Z\mu_n Q_n(y)\frac{dV_y(y)}{dy} \tag{5.1.13}$$

Integrating both sides of (5.1.13) from $y = 0$ to $y = L$ yields

$$I_D = \frac{\mu_n Z}{L}\int_0^{V_D} Q_n(y)\,dV_y \tag{5.1.14}$$

where μ_n is the average electron mobility in the inversion channel, which is roughly half the value of bulk mobility, and $Q_n(y) = Q_s - Q_d$:

$$Q_s = -C_{\text{ox}}[V_G - V_B - V_{\text{FB}} - \psi_S(y)] \quad \text{and} \quad Q_d = -qX_dN_A \tag{5.1.15}$$

Here Q_s is the total semiconductor charge and Q_d is the semiconductor depletion charge. In strong inversion,

$$\psi_S = 2\phi_B + V_y(y) - V_B \quad \text{and}$$

$$Q_d = Q_{d\max} = -\sqrt{2q\varepsilon_s N_A[2\phi_B + V_y(y) - V_B]} \tag{5.1.16}$$

Thus,

$$Q_n = -C_{\text{ox}}[V_G - V_{\text{FB}} - 2\phi_B - V_y(y)] + \sqrt{2q\varepsilon_s N_A[2\phi_B + V_y(y) - V_B]} \tag{5.1.17}$$

Inserting (5.1.7) into (5.1.5), we obtain

$$I_D = \frac{\mu_n Z}{L}\left\{C_{\text{ox}}\left(V_G - V_{\text{FB}} - 2\phi_B - \frac{V_D}{2}\right)V_D\right.$$

$$\left. - \frac{2}{3}\sqrt{2q\varepsilon N_A}\left[(2\phi_B + V_D - V_B)^{3/2} - (2\phi_B - V_B)^{3/2}\right]\right\} \tag{5.1.18}$$

If $V_y(y)$ in the square-root term of (5.1.17) is neglected and V_B is assumed to be zero, then (5.1.18) reduces to the well-known level-1 MOS model:

$$I_D = \frac{\mu_n Z}{L}\left\{C_{\text{ox}}\left(V_G - V_{\text{FB}} - 2\phi_B - \frac{V_D}{2}\right)V_D - 2V_D\, \varepsilon q N_A\, \phi_B\right\}$$

$$= \frac{\mu_n C_{\text{ox}} Z}{L}\left[(V_{GS} - V_T)V_D - \frac{1}{2}V_D^2\right] \tag{5.1.19}$$

Equations (5.1.18) and (5.1.19) can be used to describe the I_D–V_D relation when V_D is relatively small such that $Q_n(y = L)$ is not zero (the channel is not pinched off). This region is called the linear region.

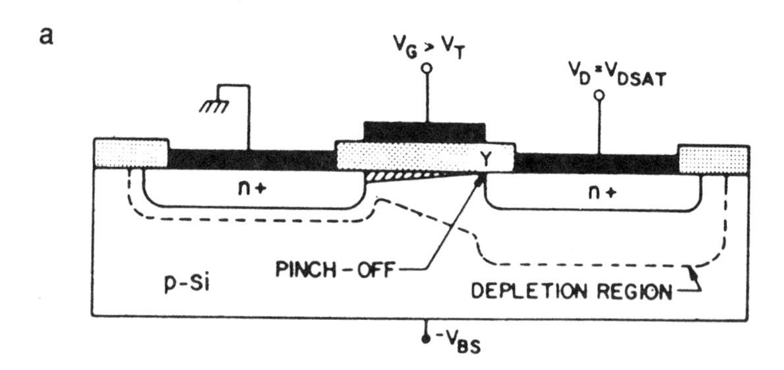

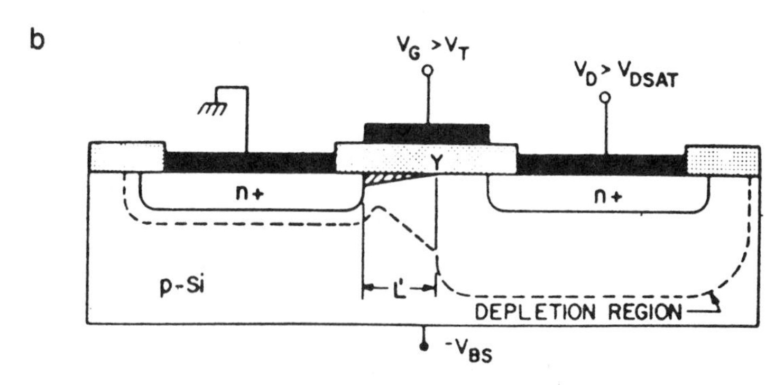

FIGURE 5.3. MOSFET structure operated at (a) onset of saturation (Y indicates the pinch-off point) and (b) beyond saturation with an effective length L'.

As V_D is increased, however, the inversion-layer charge near the drain region can become zero due to the increased electric field at that point (Fig. 5.3a). This is called channel pinch-off. For an even larger V_D, the effective channel length L' becomes smaller than the physical channel length L (Fig. 5.3b), a phenomenon referred to as channel-length modulation. The drain voltage that causes $Q_n(y = L) = 0$ (channel pinch-off) is defined as the saturation drain voltage $V_{D\text{sat}}$. As V_D is increased beyond $V_{D\text{sat}}$, the drain current will remain nearly constant (saturation region), and (5.1.19) needs to be modified to describe the drain current in such a region. Letting $Q_n(L) = 0$ and $V_y(L) = V_{D\text{sat}}$ in (5.1.17), we can derive $V_{D\text{sat}}$ as

$$V_{D\text{sat}} = V_{GS} - V_{\text{FB}} - 2\phi_B - \frac{q\varepsilon N_A[\sqrt{1 + 2C_{\text{ox}}^2(V_G - V_{\text{FB}} - V_B)/q\varepsilon N_A} - 1]}{C_{\text{ox}}^2} \quad (5.1.20)$$

Replacing V_D in (5.1.18) by this $V_{D\text{sat}}$ yields the model for the saturation drain current $I_{D\text{sat}}$. Putting this into (5.1.19), we obtain the conventional $I_{D\text{sat}}$ model:

$$I_{D\text{sat}} = \frac{\mu_n C_{\text{ox}} Z}{2L}(V_{GS} - V_T)^2 \quad (5.1.21)$$

for $V_D \geq V_{D\text{sat}}$ and $V_G \geq V_{\text{TH}}$.

The effect of channel-length modulation, which is important for MOSFETs with channel length smaller than 2 μm, can be included in $I_{D\text{sat}}$ as

$$I_{D\text{sat}} = \frac{\mu_n C_{\text{ox}} Z}{2L} \frac{(V_{GS} - V_T)^2}{1 - \Delta L/L} \quad \text{for } V_D > V_{D\text{sat}} \quad (5.1.22)$$

where $\Delta L = L - L'$ and

$$\Delta L = \sqrt{\frac{2\varepsilon}{qN_A}}\left(\sqrt{\phi_D + V_D - V_{D\text{sat}}} - \phi_D\right) \quad (5.1.23)$$

$$\phi_D = \frac{\varepsilon}{2qN_A}\left(\frac{V_D}{L}\right)^2 \quad (5.1.24)$$

It is important to point out that in small-geometry MOSFETs, the value of V_{TH} is also influenced by the channel length and width (Tsividis, 1987).

Figure 5.4 compares the $I_D - V_D$ characteristics calculated from the more accurate model (Eqs. (5.1.18) and (5.1.20)) and from the conventional model (Eqs. (5.1.19) and (5.1.21)) for a MOSFET with $N_A = 2 \times 10^{16}$ cm^{-3}, d = 87 nm, V_{FB} = –0.5 V, and $\mu_n Z/L = 1.2 \times 10^4$ cm^{2}V^{-1} s^{-1} (Muller and Kamins, 1986). As shown in the figure, both models predict the correct trend of the MOSFET, but the conventional model overestimates considerably the drain current if V_G is relatively large. The I_D versus V_G characteristics in strong inversion are illustrated in Fig. 5.5.

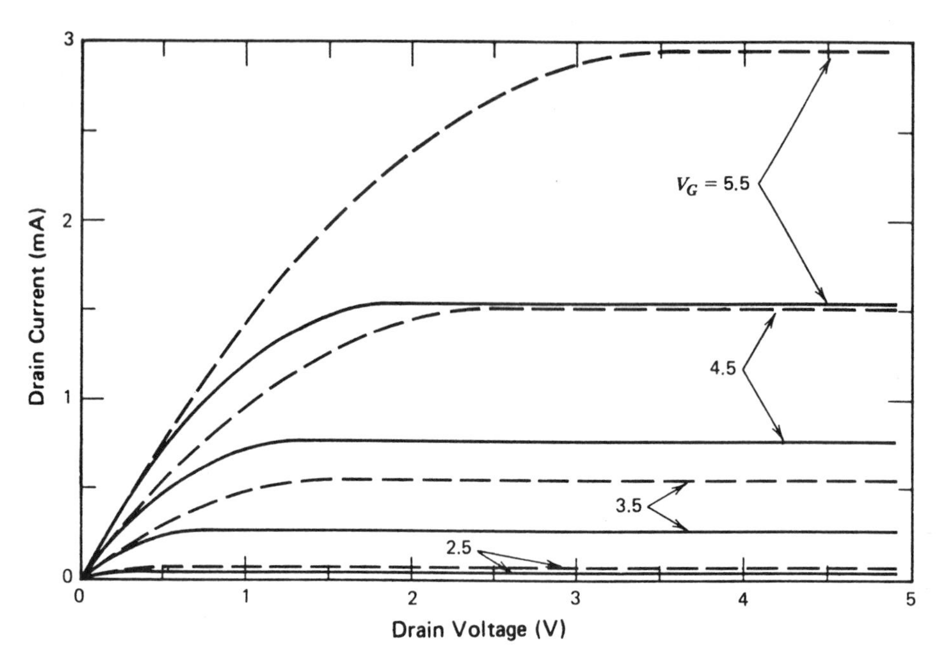

FIGURE 5.4. I_D–V_D characteristics calculated from the conventional model (dashed lines) and the more accurate model (solid lines) (Muller and Kamins, 11).

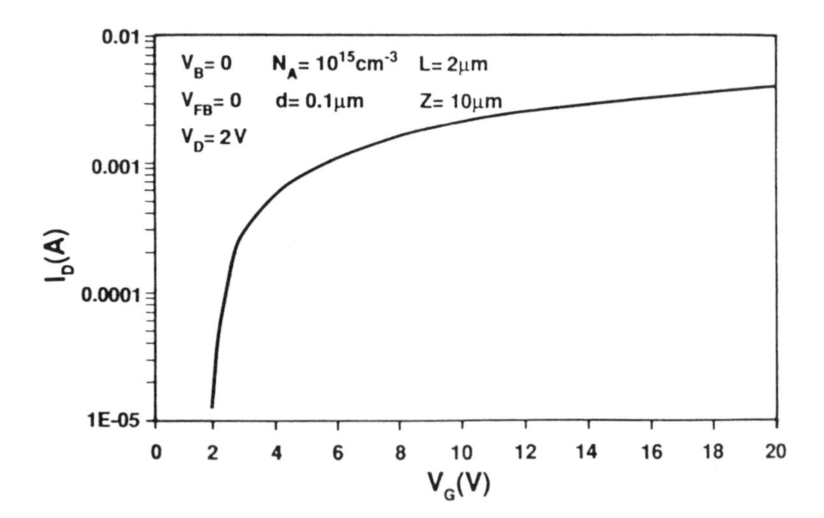

FIGURE 5.5. Drain current versus gate voltage calculated from the strong-inversion approximation.

5.1.2. Weak Inversion

Strong inversion assumes $I_D = 0$ if $V_G < V_{TH}$. As shown in Fig. 5.6, however, the MOSFET does not turn off abruptly at $V_G = V_{TH}$, and a small drain current (called subthreshold current) exists when V_G is a few kT/q below V_{TH} (subthreshold region) (Tsividis, 1987). In general, the drain current is

$$I_D = Z\mu_n \frac{dE_{Fn}}{dy} \frac{Q_n(y)}{q} \tag{5.1.25}$$

and the electron concentration in the channel is

$$n = n_i e^{(E_{Fn} - E_i)/kT} \tag{5.1.26}$$

The drift-only assumption ($dE_{Fn}/dy \approx qdV_y/dy$) used in deriving the strong-inversion I_D model is no longer valid here (Pao and Sah, 1966). In weak inversion, which occurs if $2\phi_B > \phi_S > \phi_B$, it can be assumed that the potential difference between any two y points at the surface is zero, and thus the y-direction electric field in the channel is zero (Tsividis, 1987). Physically, this is no inversion layer charge, and the device resembles an n^+–p–n^+ bipolar transistor, where the source plays the role of an emitter, the drain plays the role of a collector, and the region of the p-type substrate in between behaves like a base. Unlike the strong inversion case in which the electric field exists throughout the "n^+–n–n^+ resistor," the electric field is nearly absent in the weakly inverted channel, as is the case in the base region of an n^+–p–n^+ bipolar transistor. This is due to the fact that the voltage drops almost entirely across the n–p junctions; outside the junctions the voltage drop is

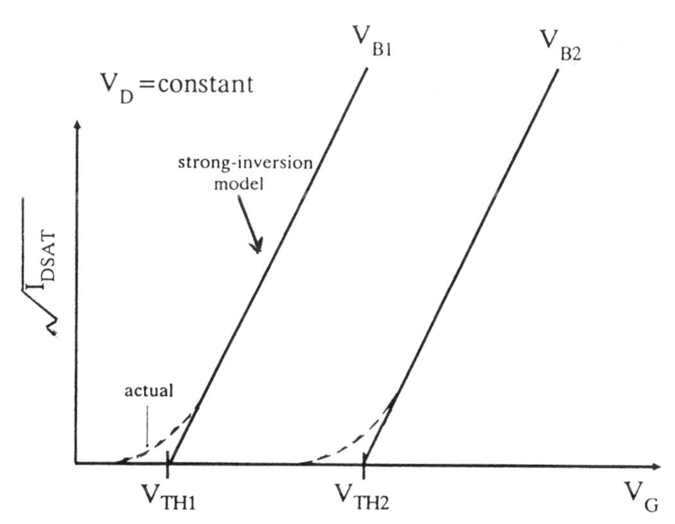

FIGURE 5.6. Saturation drain current versus gate voltage, showing discrepancies of the strong-inversion model and the actual characteristics including subthreshold behavior. Also shown is the effect of different body voltages (V_{B1} and V_{B2}) on the current, where $|V_{B1}| < |V_{B2}|$.

zero and the electric field vanishes. Hence, if there is current flow in the channel, it must be caused by the diffusion tendency.

Assuming that the drift tendency is negligibly small along the y direction $[dE_{Fn}(y)/dy >> dE_i(y)/dy]$, (5.1.26) can be rewritten as

$$\frac{dn(x, y)}{dy} = \frac{n(x, y)}{kT} \frac{dE_{Fn}}{dy} \tag{5.1.27}$$

Integrating both sides of (5.1.27) from $x = 0$ to $x = X_d$, we have

$$\frac{dE_{Fn}}{dy} = \frac{kT}{Q_n(y)} \frac{dQ_n(y)}{dy} \tag{5.1.28}$$

Combining (5.1.25) and (5.1.28) yields

$$I_D = \frac{kTZ\mu_n}{q} \frac{dQ_n(V_{Dy})}{dy} \tag{5.1.29}$$

We next assume that $Q_n(y)$ varies linearly with y and that $Q_n(L) << Q_n(0)$. This leads to

$$I_D \approx \frac{kTZ\mu_n}{q} \frac{Q_n(0)}{L} \tag{5.1.30}$$

To find $Q_n(0)$, we use the general equation derived in (5.1.10):

$$Q_n(0) = q \int_{\psi_s(y=0)}^{0} \frac{(n_i^2/N_A)e^{q(\psi - V_{y=0})/kT}\, d\psi}{(\sqrt{2}kT/qL_D)F_y(\psi, V_y, \nu)|_{y=0}} \tag{5.1.31}$$

Putting $V_y(y = 0) = 0$ into $F_y(\psi, V_y, \nu)$ and noting that ψ is a positive but small number in weak inversion and ψ is very small, we have

$$F_y(y = 0) \approx \sqrt{\frac{q\psi}{kT}} \tag{5.1.32}$$

Thus, (5.1.31) becomes

$$Q_n(0) = \frac{q^2 n_i^2 L_D}{\sqrt{2}kTN_A} \int_{\psi_s(0)}^{0} \frac{e^{q\psi/kT}\, d\psi}{\sqrt{q\psi/kT}} \tag{5.1.33}$$

Since the square root of $q\psi/kT$ depends much less on ψ than the term $\exp(q\psi/kT)$, it can be assumed that $\sqrt{q\psi/kT}$ is constant with respect to ψ and that $\psi \approx \psi_s(0)/2$. This results in

$$Q_n(0) = \frac{-qn_i^2 L_D}{\sqrt{2}N_A}\sqrt{\frac{2kT}{q\psi_S(0)}}\left(1 - e^{q\psi_s(0)/kT}\right) \tag{5.1.34}$$

and $\psi_S(0)$ can be related to V_G by

$$V_G = V_{\text{FB}} + \psi_S(0) + V_y - \frac{Q_d + Q_n}{C_{\text{ox}}}$$

$$\approx V_{\text{FB}} + \psi_S(0) - \frac{Q_n(0)}{C_{\text{ox}}} + \frac{\sqrt{2q\varepsilon_s N_A[\psi_S(0) - V_B]}}{C_{\text{ox}}} \tag{5.1.35}$$

Combining (5.1.30), (5.1.34), and (5.1.35), we can calculate the subthreshold I_D–V_G characteristics. Note that I_D calculated from the model is independent of V_D. This is expected because, similar to the collector current in a bipolar transistor being nearly independent of the base–collector voltage, the gradient of electrons in the weakly inverted channel is not affected by V_D (Troutman, 1974). Of equal importance is that the MOSFET will operate predominantly in the saturation region (beyond channel pinch-off). This is because there is very little inversion charge in the channel, and even a very small drain voltage can cause the channel to be pinched off at the drain junction.

Figure 5.7 shows the I_D–V_G characteristics calculated from the weak-inversion model. The preceding model describes quite accurately the dc behavior of the MOSFET, provided the device has a relatively long channel length (e.g., $L > 2$ μm) or a relatively wide channel width (e.g., $Z > 10$ μm). For short- or narrow-channel MOSFETs, additional device physics need to be included in the model (Tsividis, 1987; Liou, 1994). For

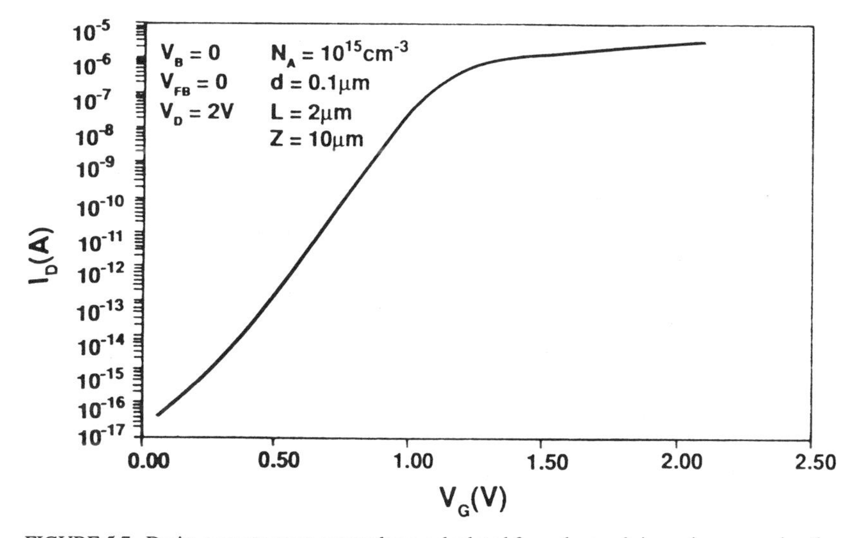

FIGURE 5.7. Drain current versus gate voltage calculated from the weak-inversion approximation.

instance, the threshold voltage of the MOSFET decreases and increases with decreasing L and Z, respectively, which then affect the current–voltage characteristics. Also, in a short-channel MOSFET, the linear and saturation regions are less distinguishable due to the significance of the channel-length modulation in such a device.

5.2. MEDICI SIMULATION

5.2.1. DC Characteristics

The effect of the drain voltage V_D on the inverted free-carrier concentration in the channel is first investigated. Figures 5.8a–c show the two-dimensional electron concentration contours in an n-channel MOSFET with a mask channel length $L_m = 2$ μm. The device is biased at a V_G larger than V_{TH} (e.g., $V_G = 3$ V) and three V_D, namely 0, 2, and 4 V. At $V_D = 0$, the electron density in the channel is uniform. As V_D is increased, the electron concentration contour in the channel near the drain region is more crowded, indicating a reduction in the electron density in the region.

The effect of V_G on the electron concentration is shown in Fig. 5.9, which plots the electron density at the SiO_2–Si interface for three V_G. The electron density underneath the gate increases significantly from $V_G = 0$ to 1 V, which is about equal to V_{TH}. A smaller increase in electron density is found when V_G is increased beyond 1 V.

Figure 5.10a shows the I_D versus V_G characteristics of the same MOSFET previously used. The device is biased in the linear region (i.e., V_D is small) and with a zero body voltage. From the curve, the threshold voltage of the MOSFET is estimated to be 1 V. Figure 5.10b illustrates the I_D versus V_D characteristics of the MOSFET in which the linear and saturation regions are clearly distinguishable. Note that the current is in units of μA/μm, where μm is the length of the third dimension.

The MOSFET will have a smaller threshold voltage V_{TH} if the channel length is reduced. This is evidenced by the I_D versus V_G curve for a MOSFET with $L_m = 0.75$ μm (Fig. 5.11a). Reducing the channel length also increases the drain current level and the transconductance of the MOSFET, (Fig. 5.11b). For example, at $V_G = 4$ V, the 0.75-μm MOSFET has a saturation drain current of about 500 μA/μm and a transconductance of about 400 μA/μm-V, compared to about 220 μA/μm and 100 μA/μm-V, respectively, in the 2-μm MOSFET. This increase results primarily from the fact that the electric field, and thus the drift tendency, is increased when the channel length is decreased. Saturation and linear regions also become less distinguishable in the submicron MOSFET due to the significant channel-length modulation effect in such a device.

Increasing the background doping concentration increases the threshold voltage but decreases the drain current level and transconductance of the MOSFET (Figs. 5.12a,b). Note that the drain current exhibits a kink behavior in the bias region, which separates the linear and saturation regions.

FIGURE 5.8. Electron concentration contours in a 2-μm MOSFET biased at $V_G = 3$ V and (a) $V_D = 0$, (b) $V_D = 2$ V, and (c) $V_D = 4$ V.

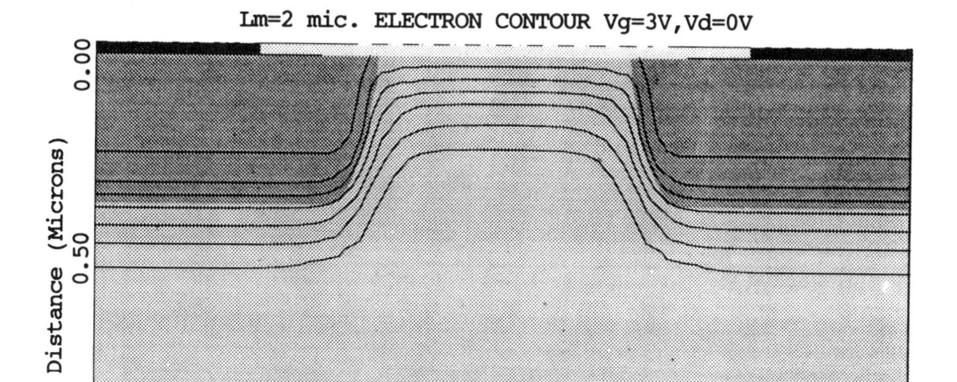

Lm=2 mic. ELECTRON CONTOUR Vg=3V,Vd=0V
Distance (Microns)
0.00
0.50
1.00
0.00
1.00
2.00
3.00
4.00
5.00
Distance (Microns)
a

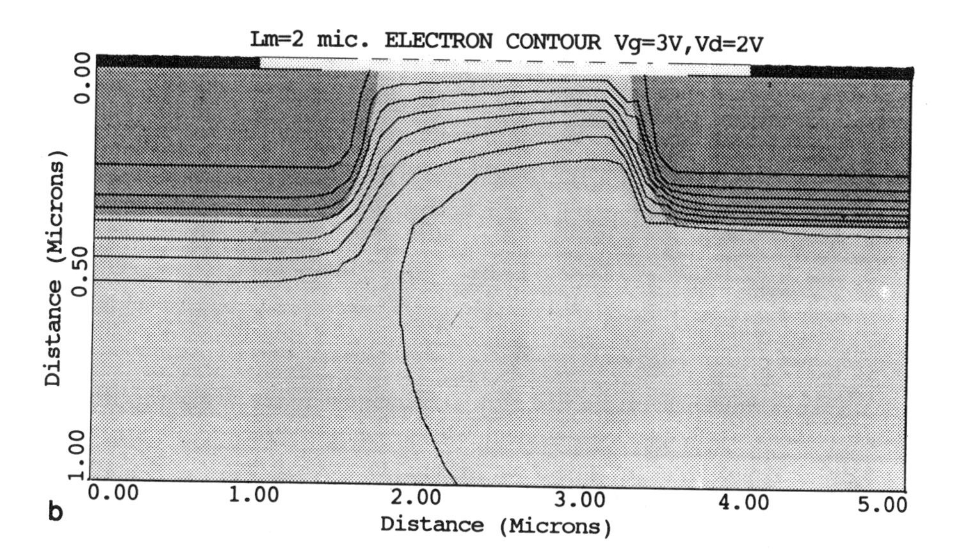

Lm=2 mic. ELECTRON CONTOUR Vg=3V,Vd=2V
Distance (Microns)
0.00
0.50
1.00
0.00
1.00
2.00
3.00
4.00
5.00
Distance (Microns)
b

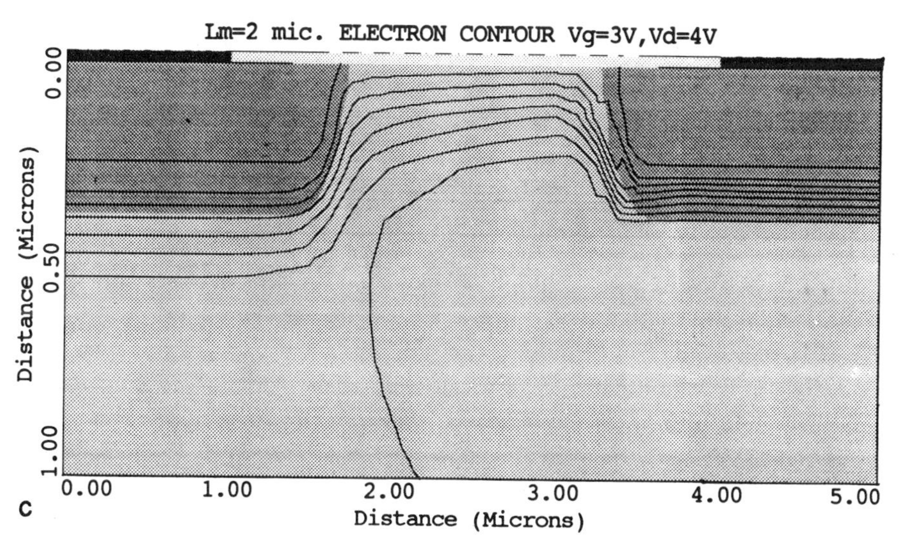

Lm=2 mic. ELECTRON CONTOUR Vg=3V,Vd=4V
Distance (Microns)
0.00
0.50
1.00
0.00
1.00
2.00
3.00
4.00
5.00
Distance (Microns)
c

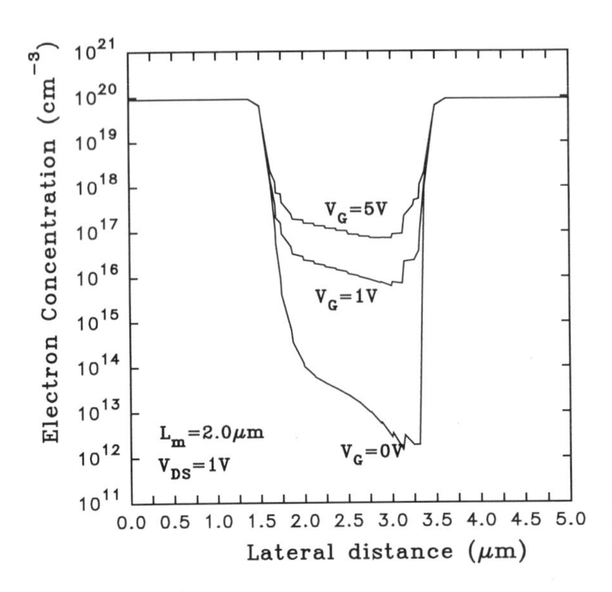

FIGURE 5.9. Electron concentration at the Si–SiO_2 interface of the MOSFET biased at different gate voltages.

The electric field in the 0.75-μm channel is also increased significantly as V_G is increased, as evidenced in the three-dimensional electric field contours in Figs. 5.13a,b simulated at two V_G. At $V_G = 0$, the peak field results from the largely reverse-biased drain junction, and the field in the rest of the channel is small due to a small V_G. As V_G is increased, the field in the channel is increased and becomes fairly uniform. For $V_G = 0$, because of the relatively small field, the current transport in the channel is carried out primarily by the diffusion tendency, and the MOSFET operates in the weak-inversion region. This observation is consistent with the I_D–V_G characteristics shown in Fig. 5.11a.

The two-dimensional electron current contours and vectors in the 0.75-μm MOSFET are given in Figs. 5.14a,b. The results indicate that the current transport in the channel occurs primarily in the region near the Si–SiO_2 interface.

We next study the effect of body voltage V_B on the MOSFET dc performance. Figure 5.15 plots the I_D–V_D curves of the 0.75-μm MOSFET biased with three V_B. When V_B becomes more negative, the drain current at a particular V_G is reduced. This is due to the fact that a more negative V_B causes the source–substrate and drain–substrate junctions to be more reverse biased, thus resulting in a lower inversion concentration in the channel and a lower drain current. The body effect can be better illustrated by the electric field distribution in the MOSFET. Figures 5.16a,b show the electric field contours in the MOSFET biased at $V_B = 0$ and $V_B = -3$ V. It can be seen that the electric field underneath the source and drain regions is increased as V_B becomes more negative. Also, the electron concentration contour in the MOSFET is more crowded in the channel and drain–source regions and less spread out into the substrate as V_B becomes more negative (Fig. 5.17a,b).

5.2.2. Transient and AC Characteristics

In this section, transient and ac characteristics of MOSFETs are simulated and discussed. Figure 5.18 shows the turn-off transients (V_G decreases abruptly from 3 V to

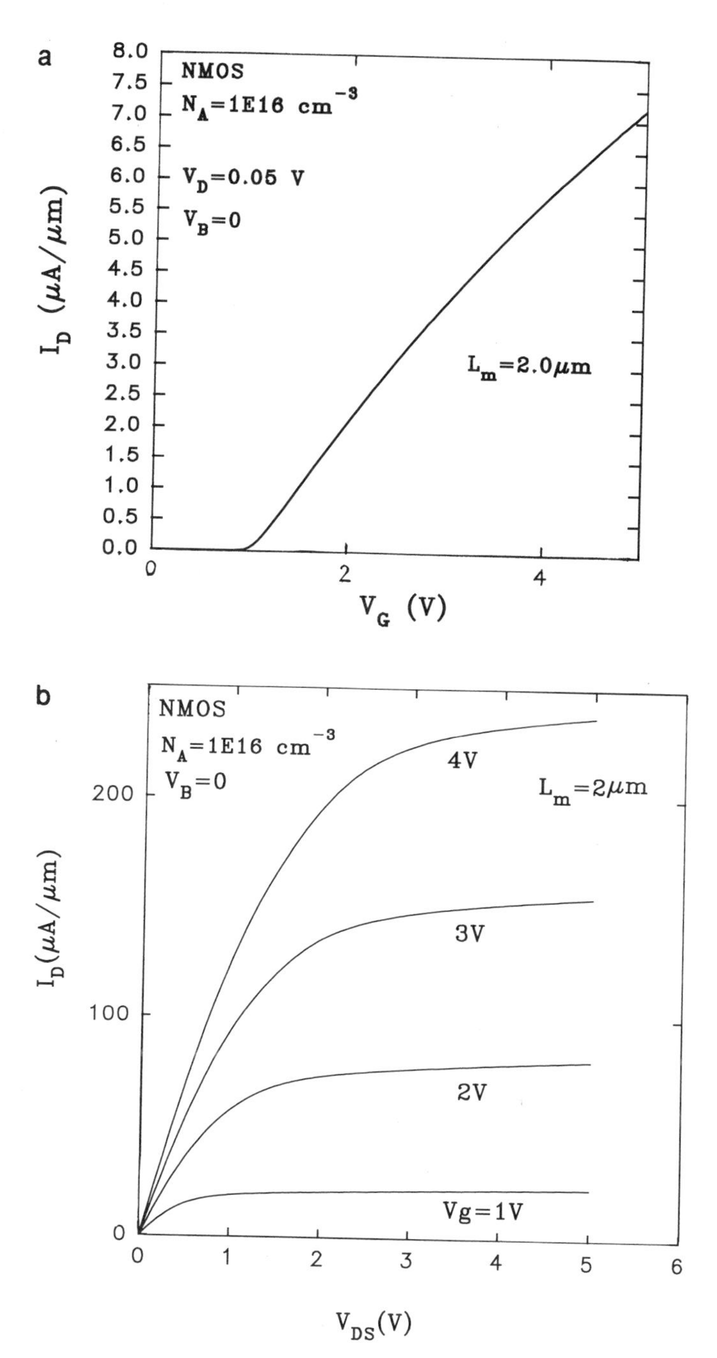

FIGURE 5.10. (a) I_D versus V_G characteristics, and (b) I_D versus V_D characteristics simulated for a MOSFET with a mask length of 2 μm and substrate doping concentration of 10^{16} cm^{-3}.

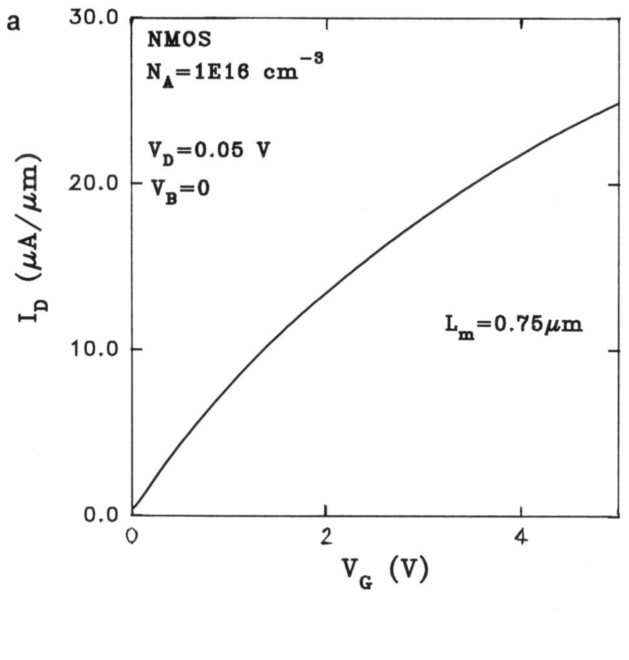

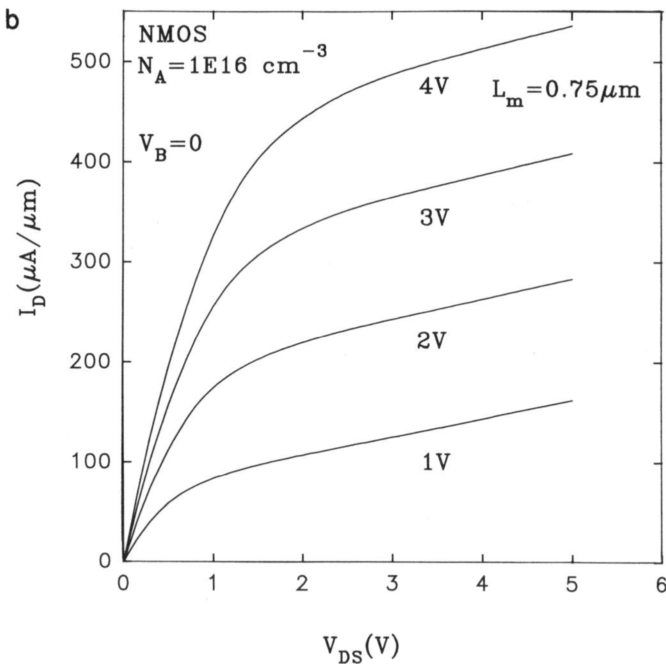

FIGURE 5.11. (a) I_D versus V_G characteristics, and (b) I_D versus V_D characteristics simulated for a MOSFET with a mask length of 0.75 μm and substrate doping concentration of 10^{16} cm^{-3}.

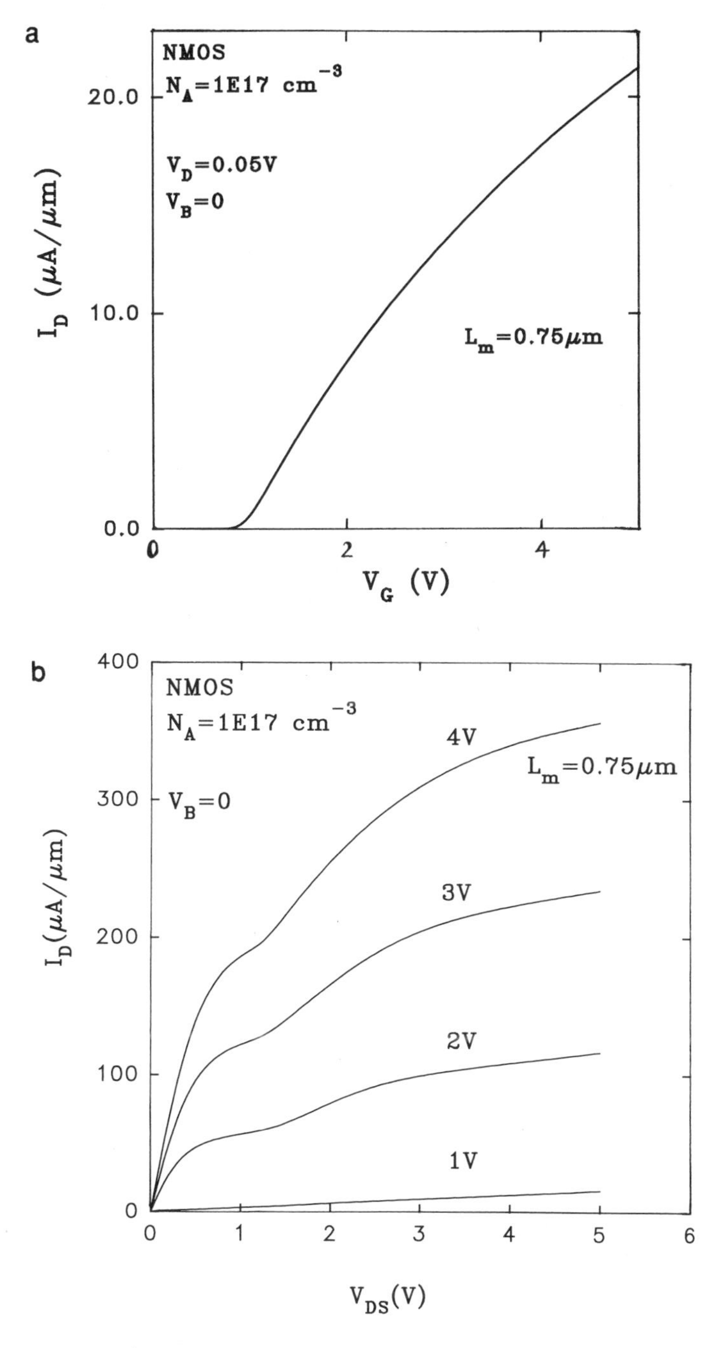

FIGURE 5.12. (a) I_D versus V_G characteristics, and (b) I_D versus V_D characteristics simulated for a MOSFET with a mask length of 0.75 μm and substrate doping concentration of 10^{17} cm^{-3}.

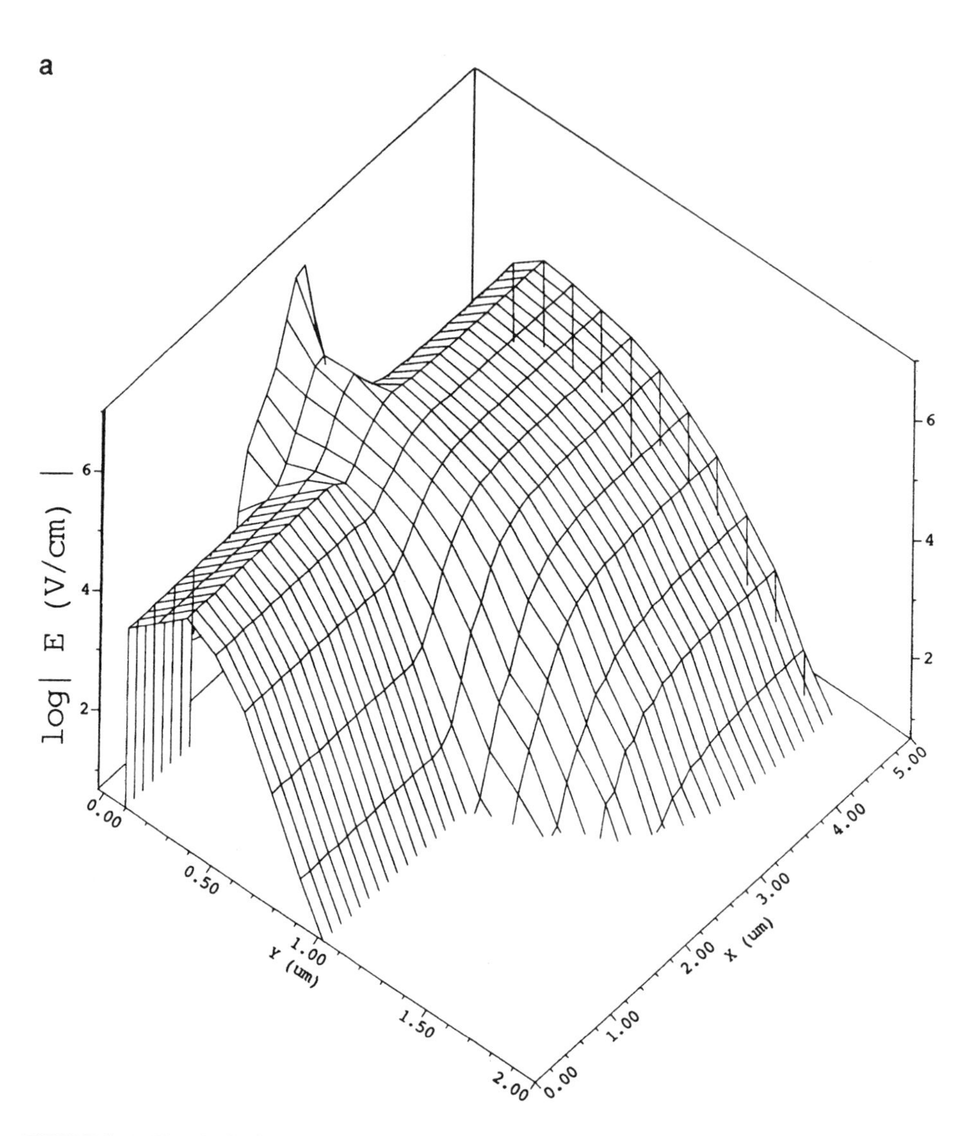

FIGURE 5.13. Electric field contours in the 0.75-μm MOSFET biased at $V_D = 5$ V and (a) $V_G = 0$ and (b) $V_G = 5$ V.

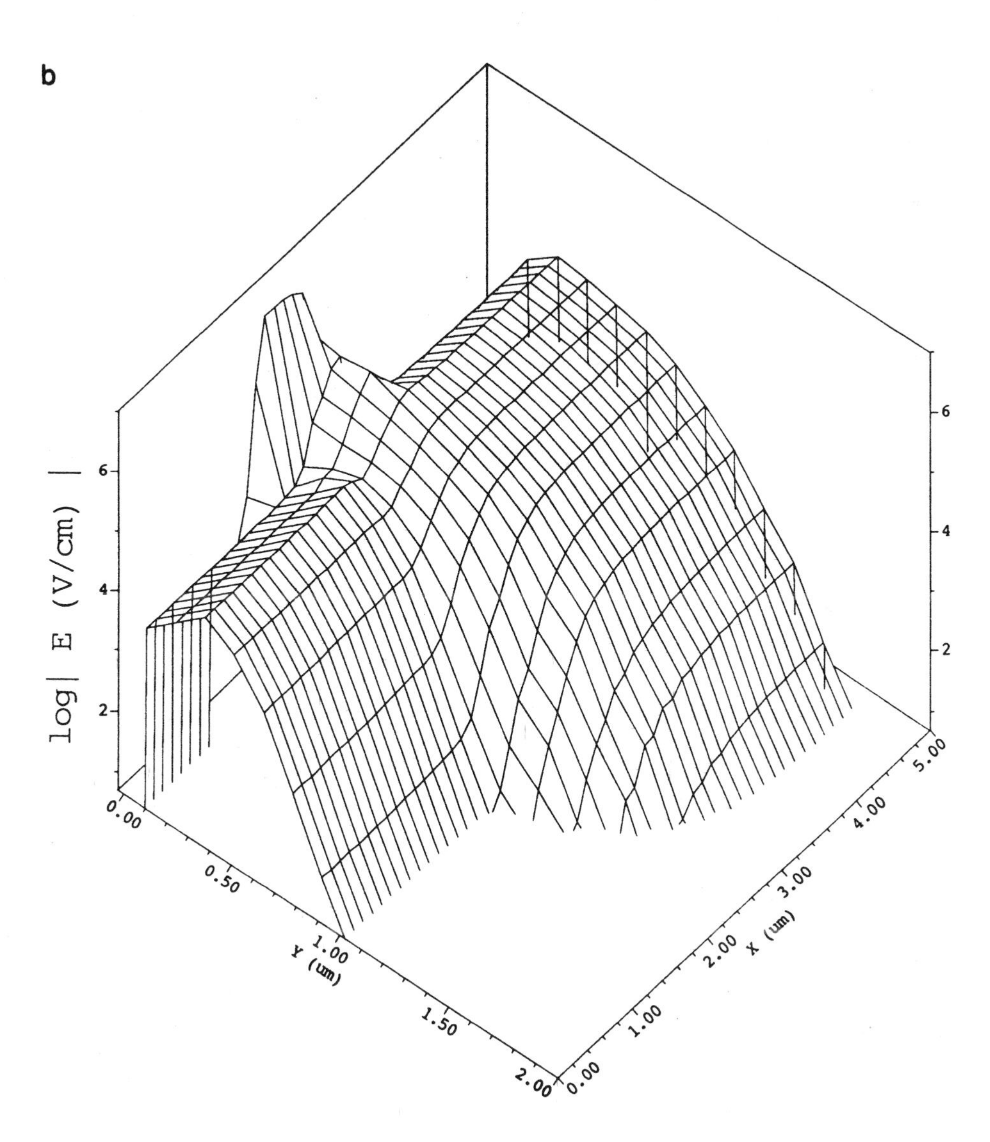

FIGURE 5.13. (Continued)

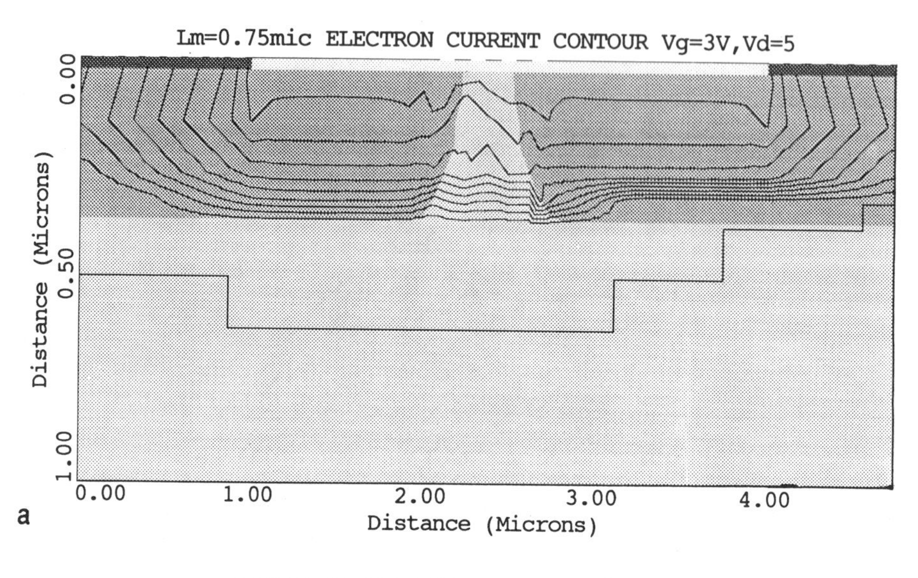

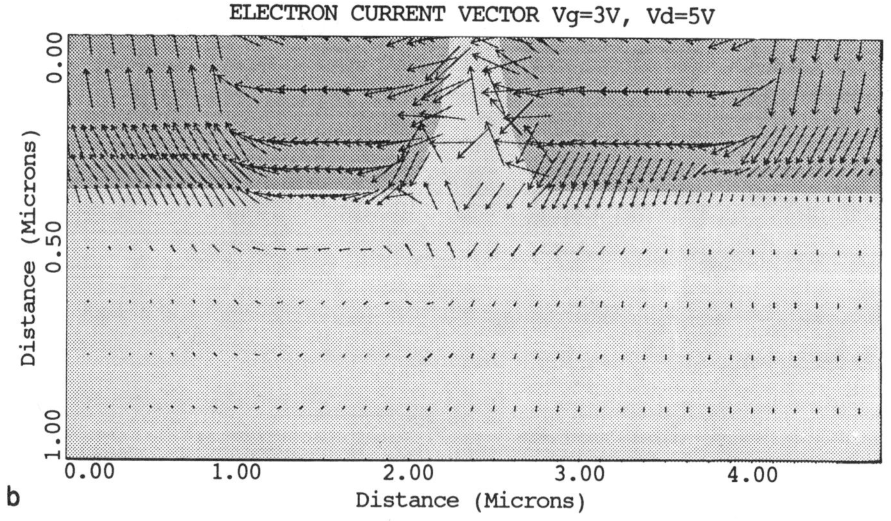

FIGURE 5.14. (a) Electron current contours and (b) electron current vectors simulated for a 0.75-μm MOSFET.

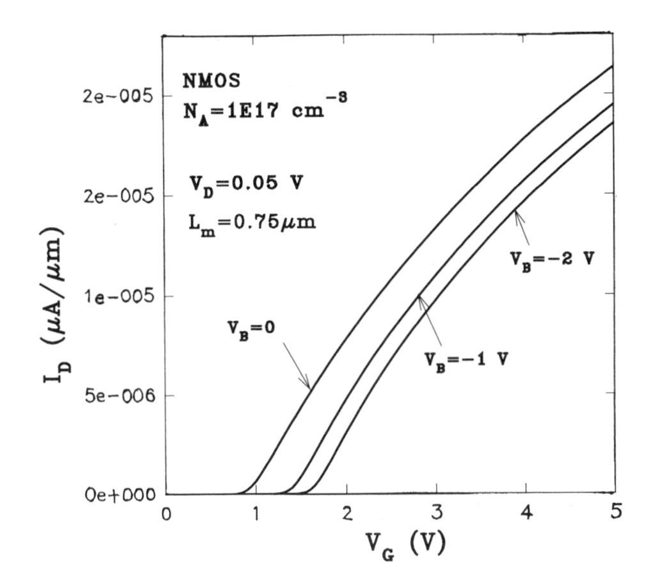

FIGURE 5.15. Effect of different body voltages V_B on I_D versus V_G characteristics.

0 at $t = 0$) of 1-μm and 0.75-μm MOSFETs. It can be seen that the drain current of the MOSFET with a longer channel decays faster initially but has a longer tail afterward compared to a MOSFET with a shorter channel length. The MEDICI input file for the transient simulation of a 1-μm *n*-MOSFET is given in Example 5.1 to demonstrate the mesh point definition, physical model selection, and switching setup.

Figure 5.19 shows the turn-off transients of a 1-μm MOSFET biased with two drain voltages (e.g., $V_D = 2$ and 5 V). The MOSFET turns off within 1.5 ps, and a longer transient response is found for 2 V. This results because a larger V_D gives rise to a larger electric field near the drain region. This, during turn-off transient, will remove the electrons from the channel more quickly and thus gives rise to a shorter turn-off time. The electron density contours and electric field contours in such a device simulated at two times (e.g., 0.1 and 2 ps) are shown in Figs. 5.20a,b and 5.21a,b, respectively. At $t = 0.1$ ps (initial transient, Fig. 5.20a), the electron density in the channel is still very high because the electric field beneath the gate cannot response to the change of V_G instantaneously and is still high at this time (Fig. 5.21a). After a couple of picoseconds, both the electron density and electric field in the channel decrease to their steady-state values (Figs. 5.20b and 5.21b).

Next, the MOSFET behavior subject to a small-signal excitation is evaluated. Figure 5.22 shows the transconductance g_m $(= dI_D/dV_G)$ of a 1-μm MOSFET simulated at three V_D's. The transconductance increases rapidly at small I_D and becomes nearly constant at a relatively large I_D. Also, g_m increases with increasing V_D.

The output conductance g_o $(= dI_D/dV_D)$ versus I_D characteristics of the MOSFET are given in Fig. 5.23. The output conductance decreases with increasing I_D, except for the transition region between the linear and saturation region in which g_o increases slightly

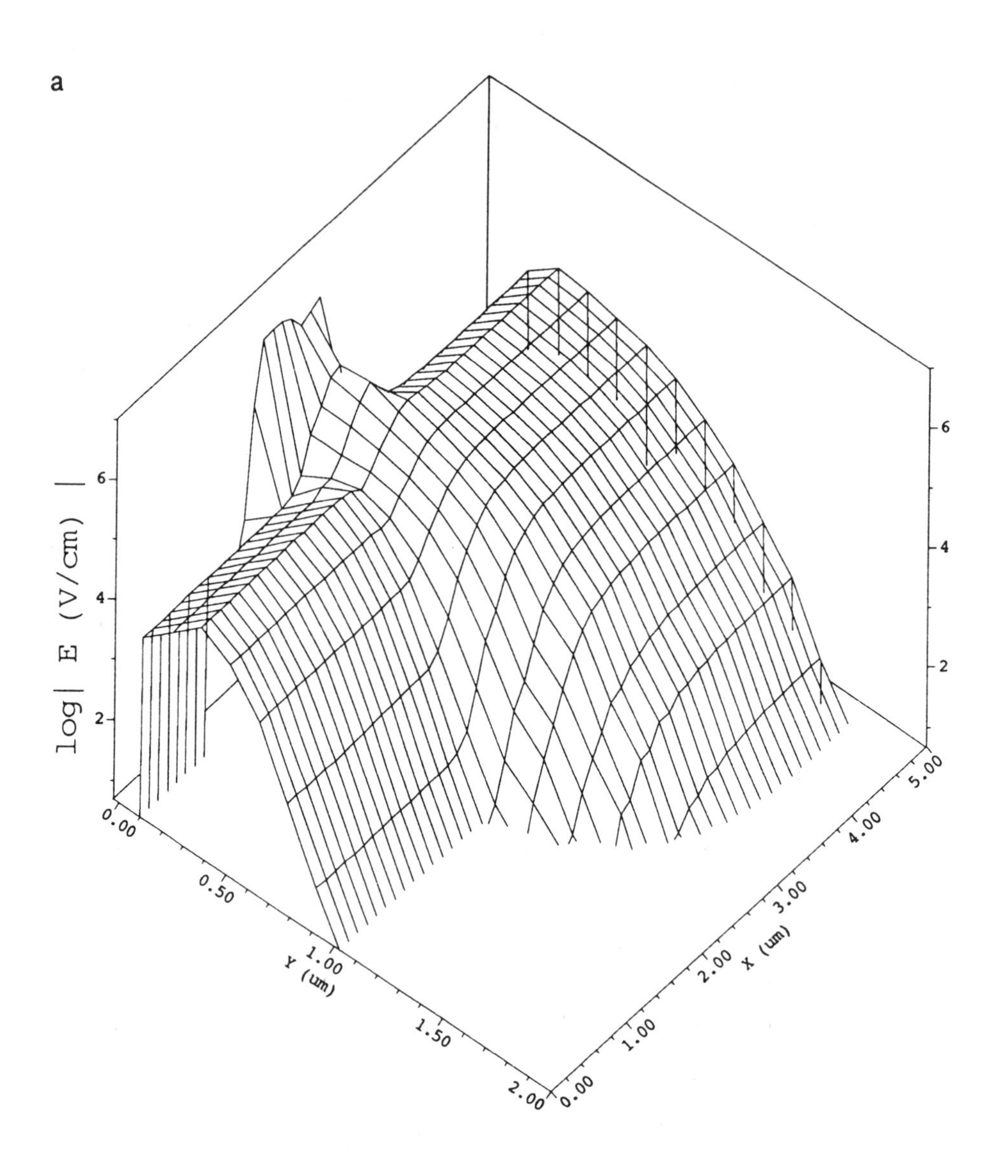

FIGURE 5.16. Electric field contours in the 0.75-μm MOSFET biased at $V_D = 5$ V, $V_G = 3$ V, (a) $V_B = 0$, and (b) $V_B = -3$ V.

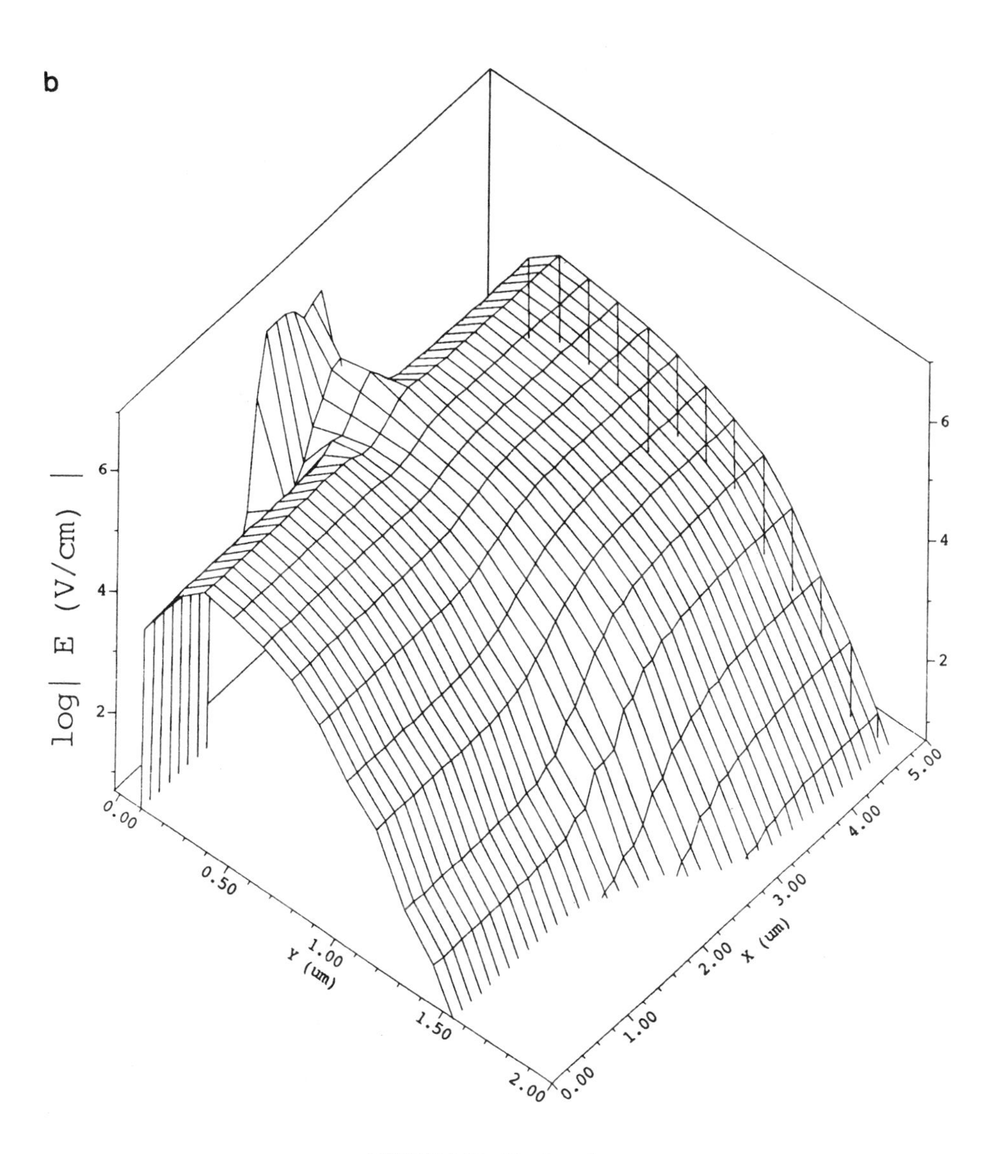

FIGURE 5.16. (Continued)

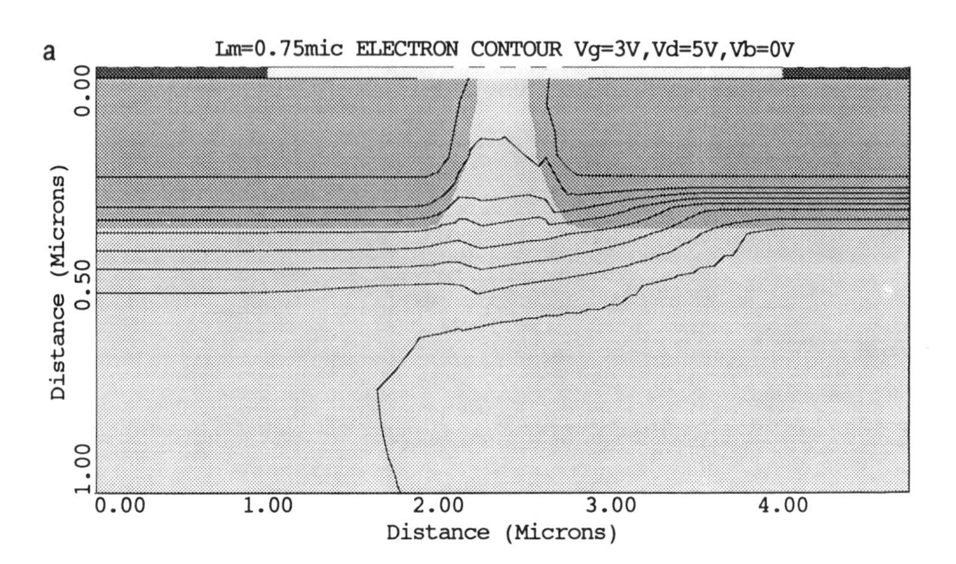

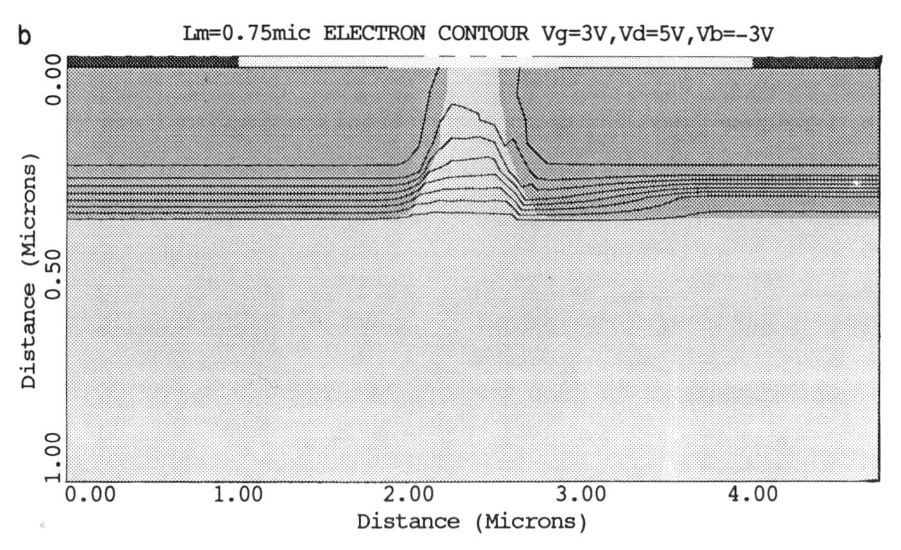

FIGURE 5.17. Electron density contours in the 0.75-μm MOSFET biased at $V_D = 5$ V, $V_G = 3$ V, (a) $V_B = 0$, and (b) $V_B = -3$ V.

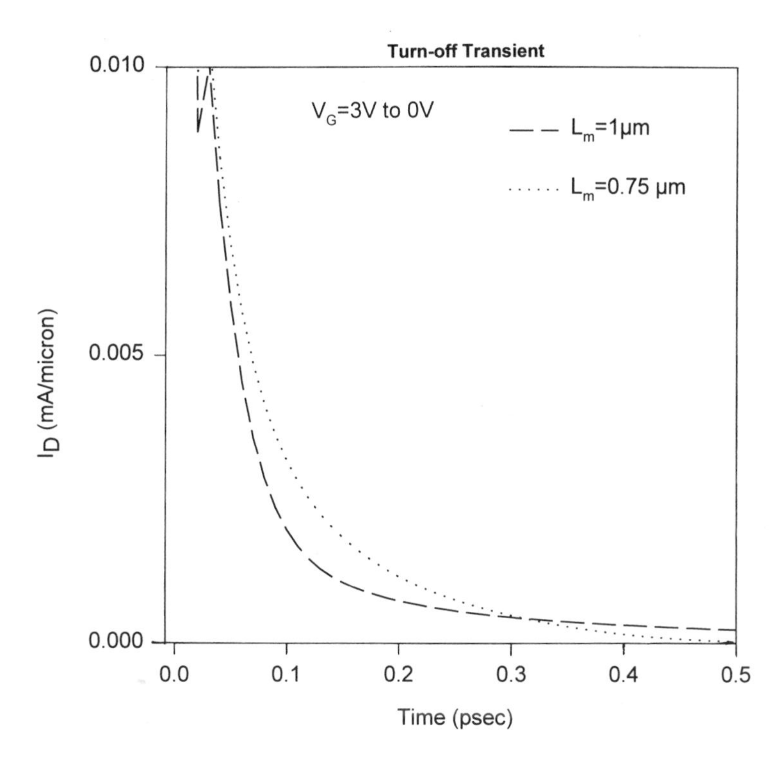

FIGURE 5.18. Drain current versus time characteristics of 0.75-μm and 1-μm MOSFETs subject to a step-down (from 3 V to 0 at $t = 0$) gate voltage.

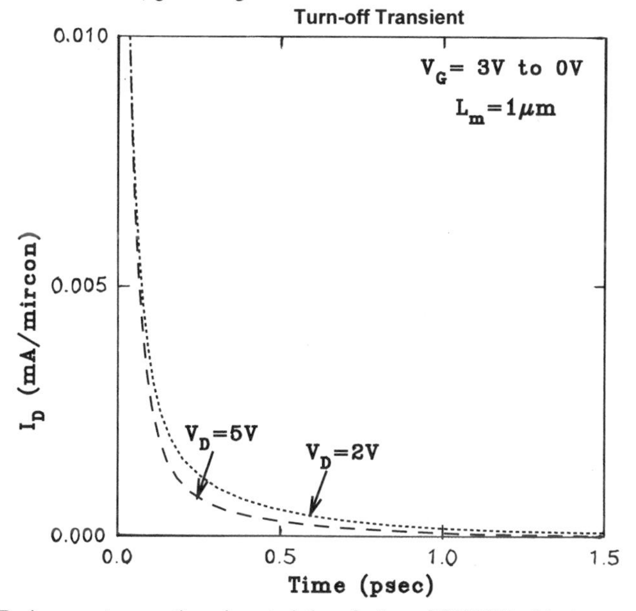

FIGURE 5.19. Drain current versus time characteristics of a 1-μm MOSFET subject to a step-down (from 3 V to 0 at $t = 0$) gate voltage.

a

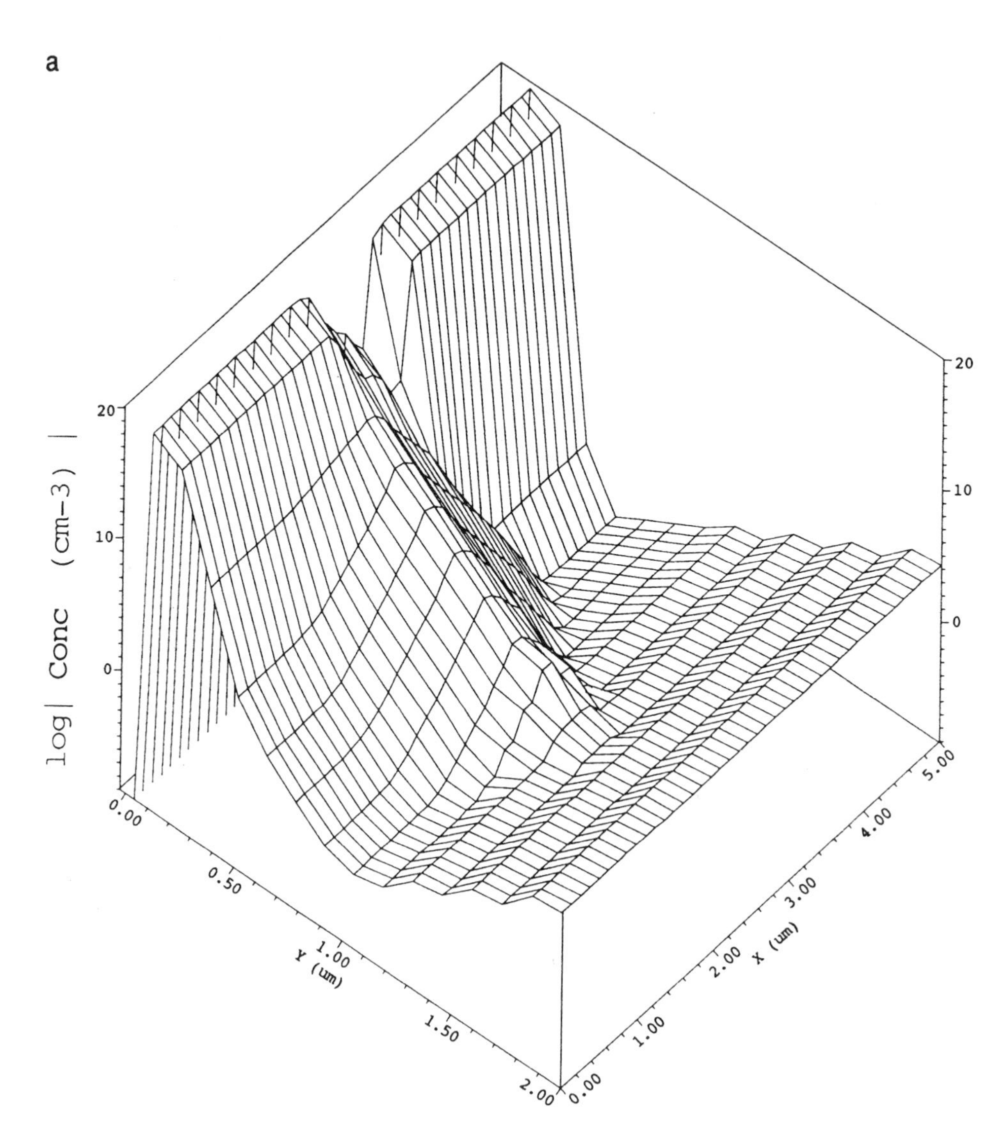

FIGURE 5.20. Electron density contours simulated at $V_D = 5$ V at (a) $t = 0.1$ ps, and (b) $t = 2$ ps.

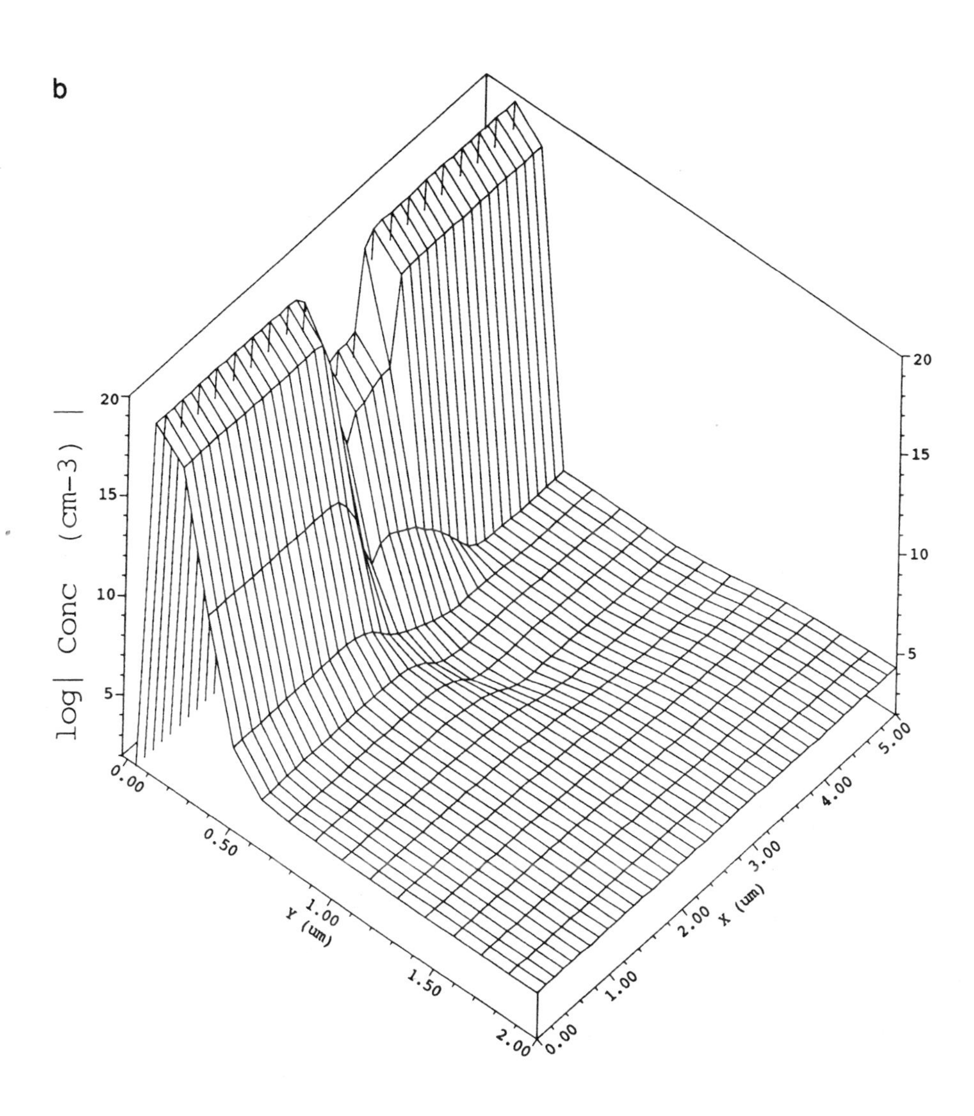

FIGURE 5.20. (Continued)

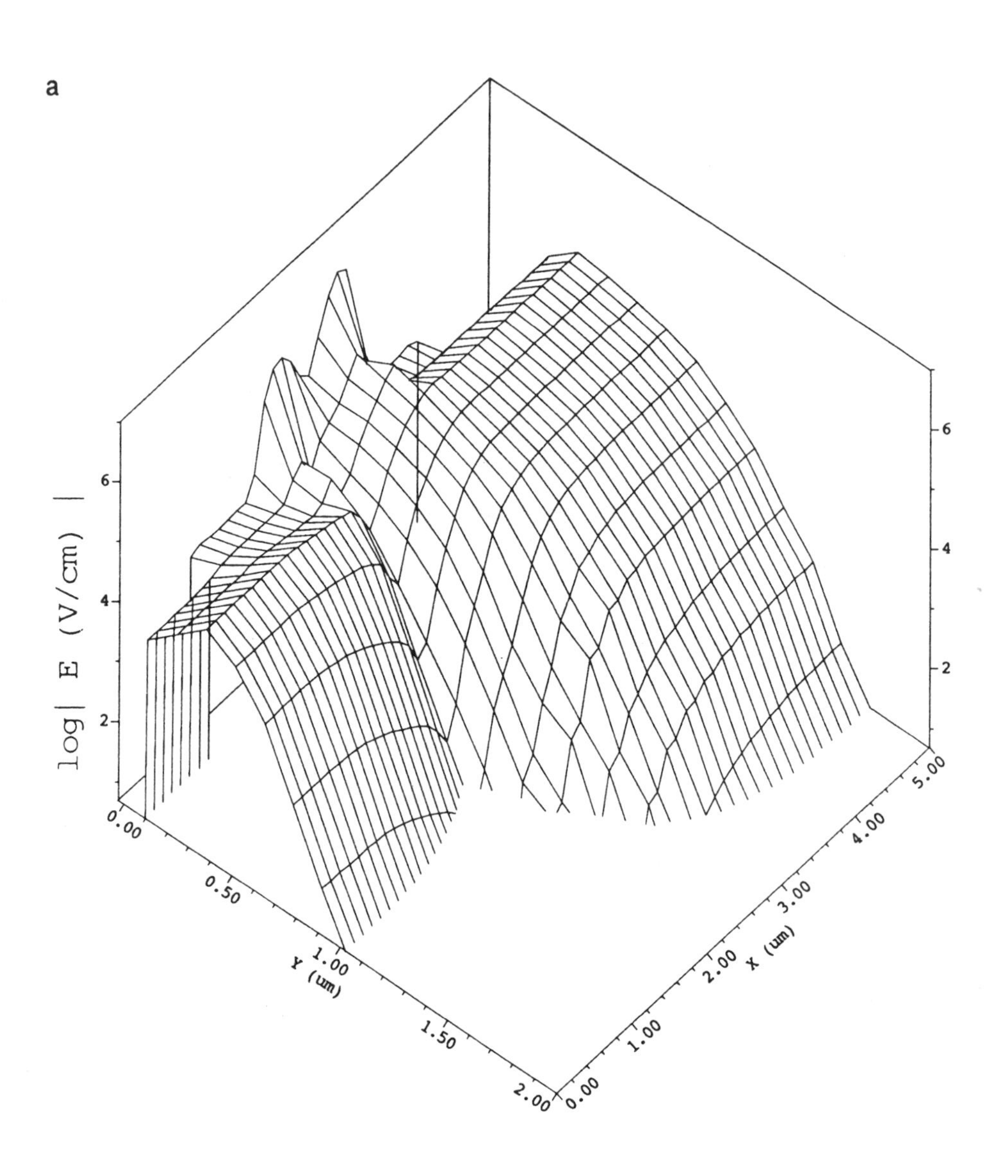

FIGURE 5.21. Electron field contours simulated at $V_D = 5$ V at (a) $t = 0.1$ ps, and (b) $t = 2$ ps.

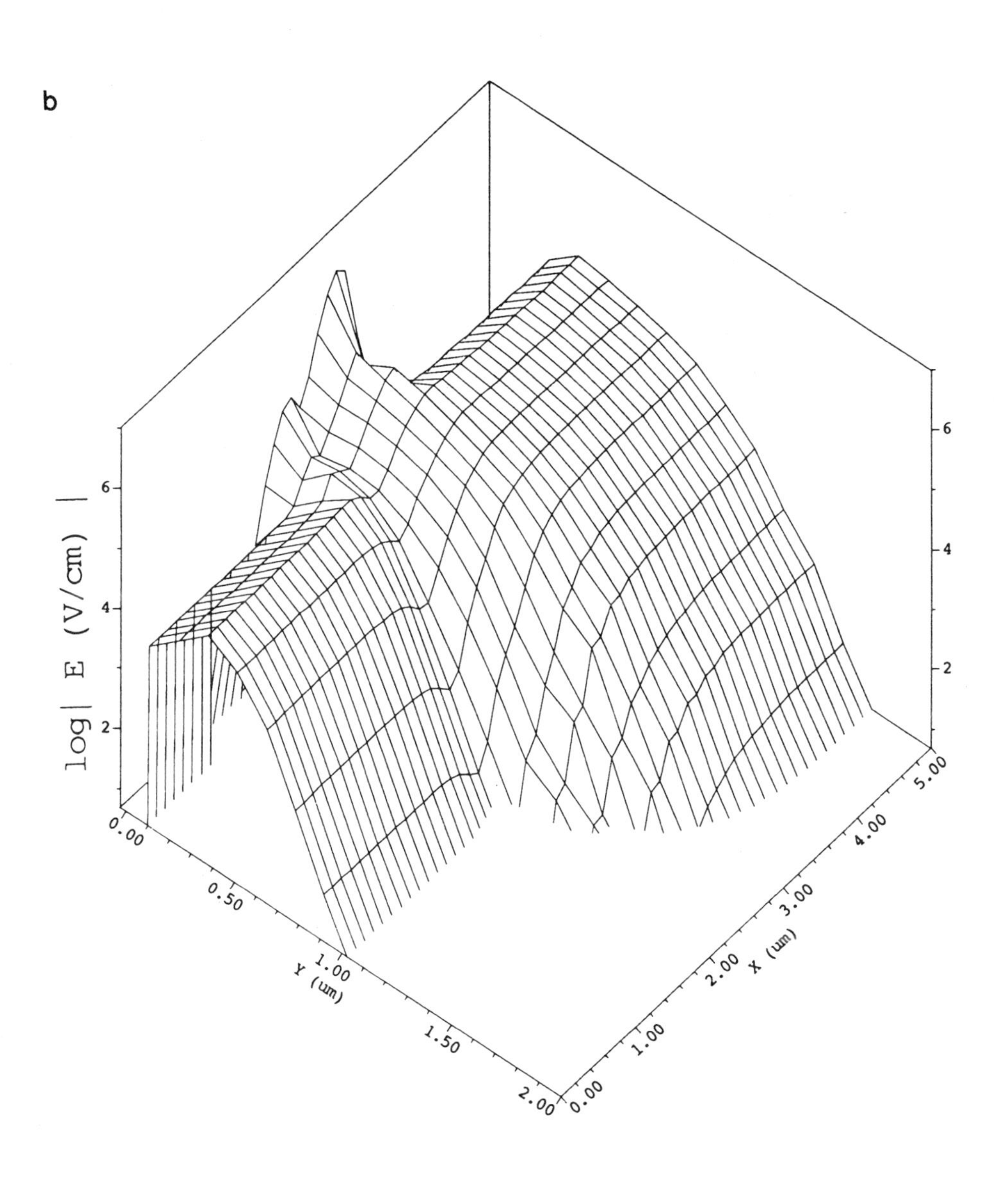

FIGURE 5.21. (Continued)

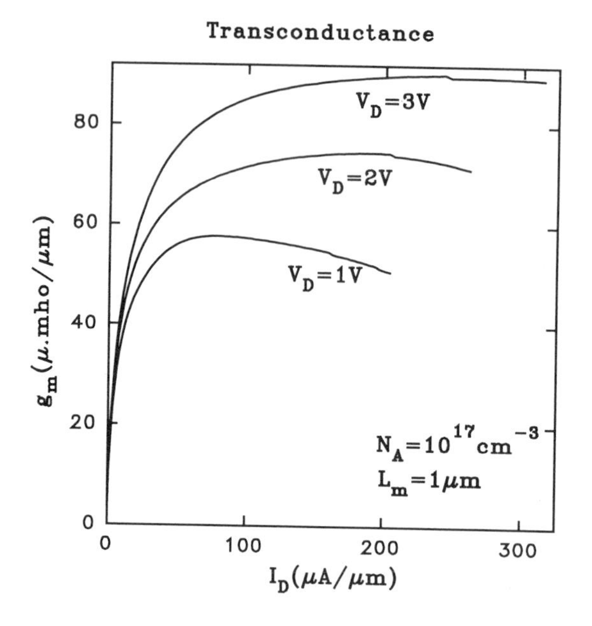

FIGURE 5.22. Transconductance versus I_D characteristics simulated for three V_D.

and then decreases as I_D is increased. Since the output conductance is affected strongly by the number of free carriers in the channel, g_o increases quickly if the gate voltage is increased.

Small-signal capacitances are also important parameters for characterizing MOSFET ac performance. Figure 5.24 shows the capacitances simulated with $V_G + dV_G$ taken

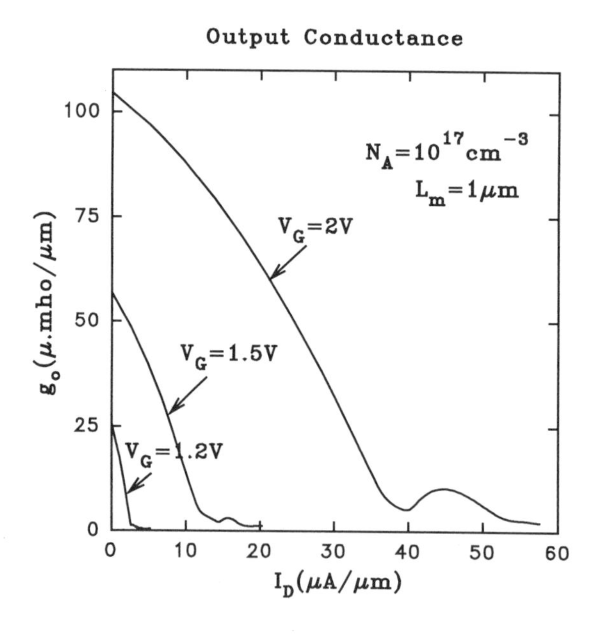

FIGURE 5.23. Output conductance versus I_D characteristics simulated for three V_G.

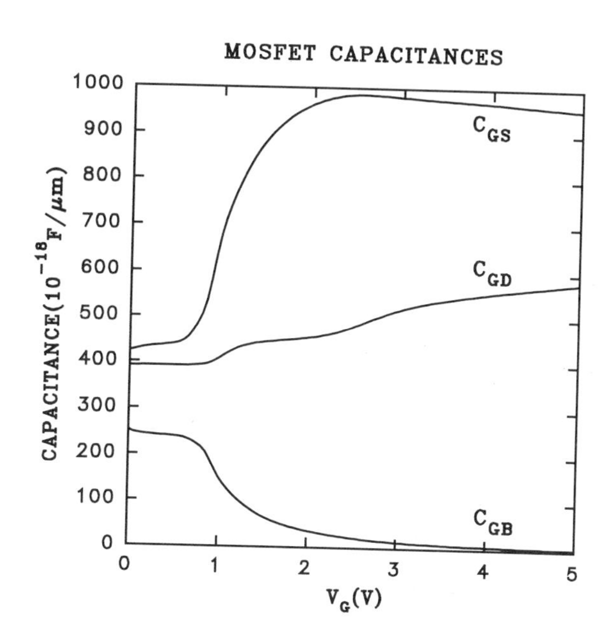

FIGURE 5.24. Small-signal capacitance versus V_G for a MOSFET with $N_D = 10^{17}$ cm^{-3} and $L_m = 1$ μm.

between the gate and source (C_{GS}), gate and drain (C_{GD}), and gate and body (C_{GB}). These capacitances describe the change of free-carrier charge storage between the indicated terminals when a small signal is superimposed on the dc gate voltage. Such parameters are needed in the MOSFET small-signal model for SPICE circuit simulations (Massobrio and Antognetti, 1993).

5.3. HOT ELECTRON EFFECT

The downscaling of transistor dimensions improves performance and packing density for VLSI circuits, but it degrades their reliability and functionality. The injection of carriers in the gate oxide of MOSFETs during operation has been acknowledged to be a severe reliability problem for scaled-down technologies. For instance, the stronger electric field in the scaled transistors accelerates many electrons and holes to high energy levels. These energetic (hot) carriers can tunnel through the barrier into the oxide near the drain region. Such hot carriers can become trapped in the oxide (Chung *et al.*, 1991), where they change the threshold voltage, *I–V* characteristics, and lifetime of the device.

The injected carriers trapped in the oxide can be hot electrons or hot holes. The hole injection level is orders of magnitude smaller than the electron injection level. However, injected holes generate three to four orders of magnitude more fast interface traps per injected carrier than electrons (Woltjer *et al.*, 1993). Injection of holes results in interface trap generation at the Si–SiO_2 interface and in hole trapping in the gate oxide. After channel hot hole stress, the positive charge of the trapped holes compensates for the negative charge in the generated interface traps. The net effect can be a small degradation

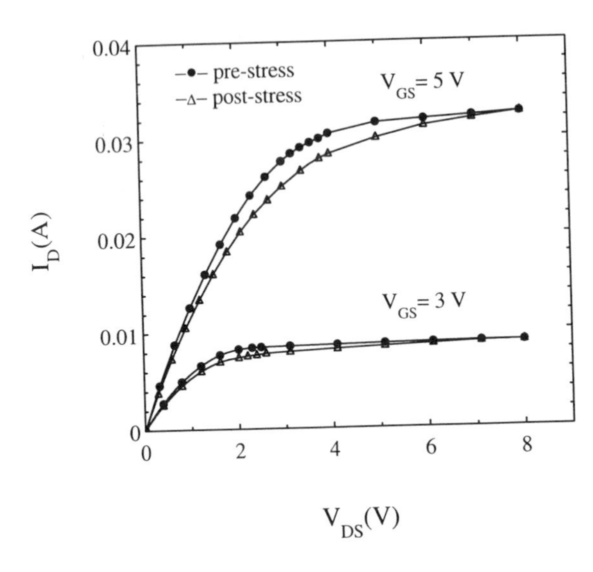

FIGURE 5.25. Drain current versus drain–source voltage before and after stress.

of drain current characteristics or a small net increase of the current, depending on whether the density of generated negatively charged fast interface traps is larger or smaller than the density of trapped holes. By a subsequent brief electron injection, the trapped holes are neutralized, and the fast interface traps generated during the hot hole stress provoke a degradation.

Impact-ionized holes injecting into the *p* substrate contribute to the substrate current. If the ohmic drop in the substrate due to hole injection in that region is significant, the source–substrate *n–p* junction becomes forward biased. Electrons are injected from the source to the *p* region. An *n–p–n* bipolar transistor action of an *n*-MOSFET can result within the source–channel–drain configuration and prevent gate control of the current.

Figure 5.25 shows the drain current versus drain–source voltage before and after hot carrier stress. The degradation in the linear region is stronger than that in the saturation region. This is because in the saturation region the effective channel length is reduced and the impact of hot electron damage is decreased.

Hot carrier effects can be reduced by reducing the doping in the source and drain regions so that the junction fields are smaller. However, lightly doped source and drain regions are incompatible with small-geometry devices because of contact resistance and other problems. A compromise design called the lightly doped drain (LDD) uses two doping levels, with heavy doping over most of the source and drain areas but with light doping in a region adjacent to the channel. The LDD structure decreases the field between the drain and channel regions, thereby reducing injection into the oxide, impact ionization, and other hot carrier effects.

The gate oxide thickness affects hot electron degradation (Toyoshima *et al.*; Doyle *et al.*, 1992). Figure 5.26 shows the substrate current versus gate–source voltage for oxide thicknesses of 7, 10, and 15 nm. The substrate current initially increases with increasing V_{GS} due to an increase of I_{DS}. A further increase in V_{GS} eventually results in a decrease in the substrate current due to an increase of V_{Dsat}, which in turn reduces the channel electric

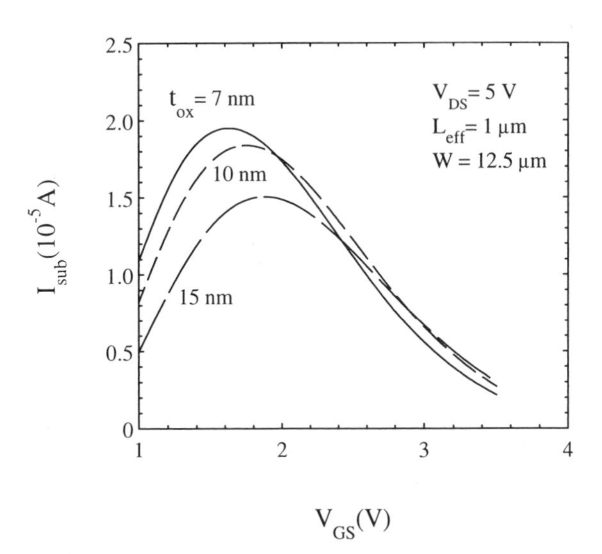

FIGURE 5.26. Substrate current versus gate–source voltage for different oxide thicknesses.

field. As the gate oxide thickness is decreased, the substrate current increases because of an enhanced vertical electric field and a large number of electron–hole pairs due to impact ionization. A thinner gate oxide device gives a higher substrate current but reduced hot electron effect. Figure 5.27 shows the drain current degradation normalized to its drain current before stress versus gate oxide thickness. This is because the thinner-gate-oxide device has smaller mobility and threshold voltage degradation due to a shift of damaged interface region toward the drain contact. Once the damaged interface region is moved toward the drain region, the effective hot electron effect is reduced.

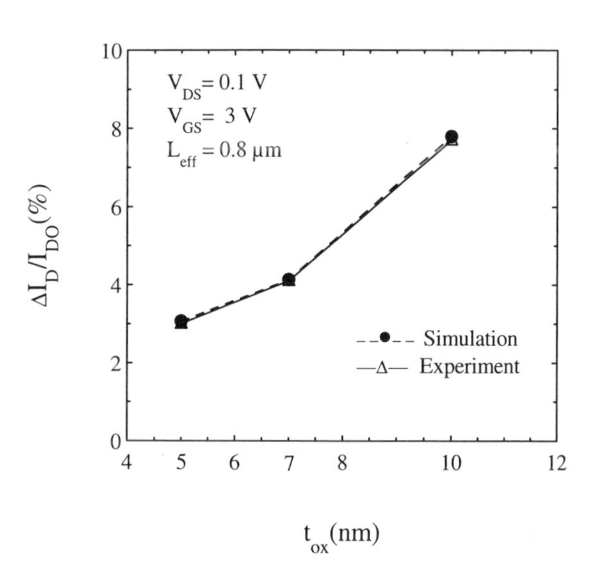

FIGURE 5.27. $\Delta I_D/I_D$ versus oxide thickness.

EXAMPLE 5.1

```
  COMMENT  One-micro n-MOSFET transient simulation
  COMMENT  Specify a rectangular mesh
     MESH  SMOOTH=1
   X.MESH  WIDTH=5 H1=0.125
   Y.MESH  N=1 L=-0.025
   Y.MESH  N=3 L=0.
   Y.MESH  DEPTH=1.0 H1=0.125
   Y.MESH  DEPTH=1.0 H1=0.25
  COMMENT  Eliminate some unnecessary substrate nodes
   ELIMIN  COLUMNS Y.MIN=1.1
  COMMENT  Specify oxide and silicon regions
   REGION  NUM=1 SILICON
   REGION  NUM=2 OXIDE IY.MAX=3
  COMMENT  Electrodes: #1=gate #2=Substrate #3=Source #4=Drain
   ELECTR  NUM=1 X.MIN=2 X.MAX=3 top
   ELECTR  NUM=2 BOTTOM
   ELECTR  NUM=3 X.MAX=1. IY.MAX=3
   ELECTR  NUM=4 X.MIN=4 IY.MAX=3
  COMMENT  Specify impurity profiles and fixed oxide charge
  PROFILE  P-TYPE N.PEAK=1E16 UNIFORM OUT.FILE=PMOSS1
  PROFILE  N-TYPE N.PEAK=1E20 Y.JUNC=0.34 X.MIN=0.0 WIDTH=1.95
        +  XY.RAT=0.75
  PROFILE  N-TYPE N.PEAK=1E20 Y.JUNC=0.34 X.MIN=3.05 WIDTH=1.95
        +  XY.RAT=0.75
INTERFACE  QF=1E10
  COMMENT  Regrid on doping
   REGRID  DOPING LOG IGNORE=2 RATIO=2 SMOOTH=1 IN.FILE=PMOSS1
        +  OUT.FILE=PMOSS2
  PLOT.2D  Boundary TITLE="DOPING REGRID" FILL SCALE
     FILL  C.OXIDE=2 C.ELECTR=4
  COMMENT  Specify contact parameters
  CONTACT  NUM=1 n.POLY
  COMMENT  Specify models to use
   MODELS  CONMOB FLDMOB SRFMOB2
  COMMENT  Symbolic factorization
     SYMB  CARRIERS=0
   METHOD  ICCG DAMPED
    SOLVE  OUT.FILE=ZERSOL
  COMMENT  Impurity profile plots
  PLOT.1D  DOPING X.START=0.25 X.END=0.25 Y.START=0.0 Y.END=2 Y.LOG
        +  POINTS BOT=1E15 TOP=1E21 COLOR=2
  PLOT.1D  DOPING X.START=1.50 X.END=1.50 Y.START=0.0 Y.END=2 Y.LOG
        +  POINTS BOT=1E15 TOP=1E18 COLOR=2
  CONTOUR  DOPING LOG MIN=16 MAX=20 DEL=.5 COLOR=2
  CONTOUR  DOPING LOG MIN=-16 MAX=-15 DEL=.5 COLOR=1 LINE=2
  COMMENT  Calculate drain characteristics
     LOAD  IN.FILE=ZERSOL
```

```
    SYMB   NEWTON CARRIERS=1 electrons
 COMMENT   Setup log file for I-V data
     LOG   IVFILE=PMOSd1
 COMMENT   Solve for Vgs= and then ramp drain
 COMMENT   Electrodes: #1=gate #2=Substrate #3=Source #4=Drain
   SOLVE   V4=1 ELEC=4 VSTEP=0.5 NSTEP=6
   SOLVE   V1=3 TSTEP=2E-12 TSTOP=10E-12 OUT.FILE=ON1
 COMMENT   Plot Id and Vg versus time
CALCULAT   name=Id A=I4
CALCULAT   name=Vg A=V1
 PLOT.1D   Y.AXIS=Id X.AXIS=TIME POINTS COLOR=2 OUT.FILE=ON4
 PLOT.1D   Y.AXIS=Vg X.AXIS=TIME POINTS COLOR=2
 COMMENT   Plot 3D electron contour and field distribution
 PLOT.3D   ELECTRON X.MIN=0 X.MAX=5.0 LOGARITH
 PLOT.3D   E.FIELD X.MIN=0 X.MAX=5.0 Y.MIN=0.25
```

REFERENCES

Chung, J. E., P.-K. Ko, and C. Hu (1991), *IEEE Trans. Electron Devices* **ED-38**, 1362.

Doyle, M. S., K. R. Mistry, and C. L. Huang (1992), *IEEE Trans. Electron Devices* **ED-39**, 1223.

Heil, O. (1935), British Patent 439,457.

Hofstein, S. R. and F. P. Heiman (1963), *Proc. IEEE* **50**, 1190.

Ihantola, H. K. J. (1961), Stanford Electron., Lab. Tech. Rep., No. 1661-1.

Ihantola, H. K. J. and J. L. Moll (1964), *Solid-State Electron.* **7**, 423.

Kahng, D. and M. M. Atalla (1960), *IRE Solid-State Device Res. Conf.* Pittsburgh, PA.

Lilienfeld, J. E. (1930), U.S. Patent 1,745,175.

Liou, J. J. (1994), *Advanced Semiconductor Device Physics and Modeling* (Artech House: Boston).

Massobrio, G. and P. Antognetti (1993), *Semiconductor Device Modeling with SPICE*, 2nd ed. (McGraw-Hill, New York).

Muller, R. S. and T. I. Kamins (1986), *Device Electronics for Integrated Circuits*, 2nd ed. (Wiley, New York).

Narayanan, R., A. Ortiz-Conde, J. J. Liou, F. J. Garic Sanchez, and A. Parthasarathy (1995), *Solid-State Electron.* **38**, 1155.

Nicollian, E. H. and J. R. Brews (1982), *MOS Physics and Technology* (Wiley, New York).

Pao, H. C. and C. T. Sah (1966), *Solid-State Electron.* **10**, 927.

Pierret, R. F. (1983), *Field Effect Devices* (Addison-Wesley, Reading, MA).

Sah, C. T. (1964), *IEEE Trans. Electron Devices* **ED-11**, 324.

Shockley, W. and G. L. Pearson (1948), *Phys. Rev.* **74**, 232.

Toyoshima, Y., H. Iwai, F. Matsuoka, H. Hayashida, K. Maeguchi, and K. Kanzaki (1990), *IEEE Trans. Electron Devices* **ED-37**, 1496.

Troutman, R. R. (1974), *IEEE J. Solid State Circuits* **SC-9**, 55.

Tsividis, Y. P. (1987), *Operation and Modeling of the MOS Transistor* (McGraw Hill, New York).

Woltjer, R., A. Hamada, and E. Takeda (1993), *IEEE Trans. Electron Devices* **ED-40**, 392.

6

BiCMOS Devices

Attempts to combine bipolar and MOS transistors on a common integrated circuit date to the late sixties (Lin *et al.*, 1969). RCA was an early leader in the introduction of BiCMOS operational amplifiers in the mid 1970s (Polinsky *et al.*, 1973). The next major trend was high-voltage BiCMOS pioneered at Stanford (Plummer and Meindl, 1976) and commercialized by Texas Instruments (Davis, 1979). This BIDFET technology combined CMOS, bipolar, and high-voltage lateral DMOS transistors. Most of these initial applications were in analog circuits. "Smart" power applications have evolved to include extremely high current (>20 A) and voltage (>500 V) levels. A third wave of BiCMOS applications developed in the mid 1980s. Motivated by the power dissipation constraints of bipolar circuits and the speed limitations of MOS transistors, Hitachi, Toshiba, and Motorola developed the 5-V digital BiCMOS technologies (Alvarez *et al.*, 1984a; Higuchi *et al.*, 1984; Miyamoto *et al.*, 1984). Subsequent BiCMOS technologies resulted in extremely high performance microprocessors and memories such as SRAMs (Hotta *et al.*, 1986; Bastani *et al.*, 1987; Tamba *et al.*, 1988).

6.1. COMPARISONS OF BiCMOS, CMOS, AND BJT

BiCMOS technology combines bipolar and CMOS transistors in a single integrated circuit. BiCMOS technology retains the benefits of bipolar and CMOS and is able to achieve VLSI circuits with high switching speed, low power dissipation, and high package densities. CMOS technology has advantages over bipolar in power dissipation, noise margins, package densities, and ability to integrate large complex functions with high yields. Bipolar technology has advantages over CMOS in switching speed, current drive per unit area, noise performance, analog capability, and I/O speed. BiCMOS technology offers the compromised advantages over bipolar and CMOS technologies such as (Alvarez and Schucker, 1988)

1. Improved speed over CMOS (but slower than bipolar)
2. Lower power dissipation than bipolar (but higher than CMOS)
3. Flexible I/Os (TTL, CMOS, or ECL)

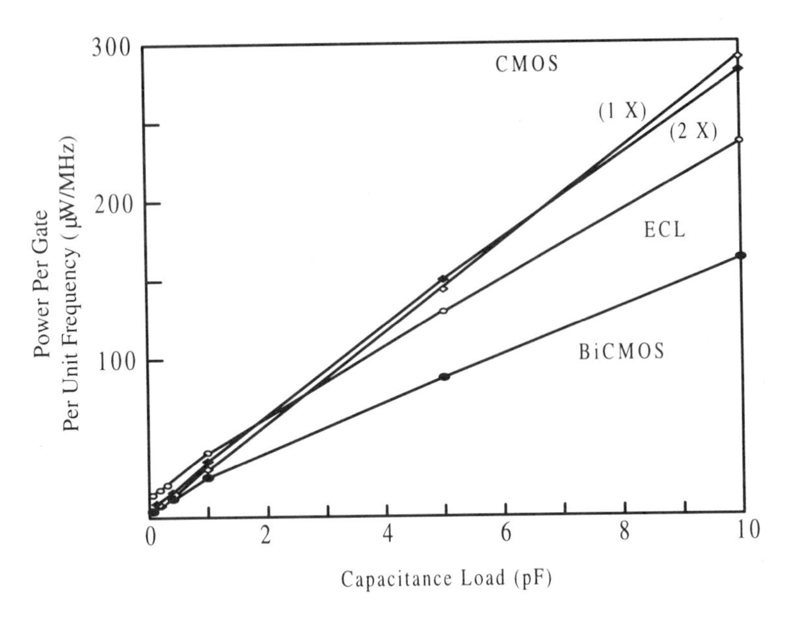

FIGURE 6.1. Delay versus capacitance (BiCMOS and CMOS).

4. High-performance analog
5. Latch-up immunity

BiCMOS technology has less dependence on capacitive loading and on process and temperature variations as compared to CMOS. The main drawbacks to BiCMOS are higher costs, lower yield, and longer fabrication cycle time compared to CMOS. As the CMOS process continues to become more complex, these drawbacks will become less significant. Compared to CMOS, a BiCMOS buffer dissipates less power because of the lower signal swing and the reduction in time spent in the transition region between the gates' on and off states (Alvarez *et al.*, 1984b).

The speed advantage of the BiCMOS buffer compared to a CMOS buffer is shown in Fig. 6.1. The BiCMOS is much faster than CMOS buffer provided the load capacitance is large.

6.2. PRINCIPLES OF BiCMOS OPERATION

BiCMOS inverters are used for driving large capacitances on-chip as well as off-chip. A circuit diagram of a BiCMOS buffer is shown in Fig. 6.2. The inverter consists of one *p*-MOSFET, three *n*-MOSFETs, and two *n–p–n* bipolar transistors. The BiCMOS inverter has a

1. High input impedance (Z_{in}) provided by the MOSFET
2. Low output impedance (Z_{out}) provided by the bipolar transistors

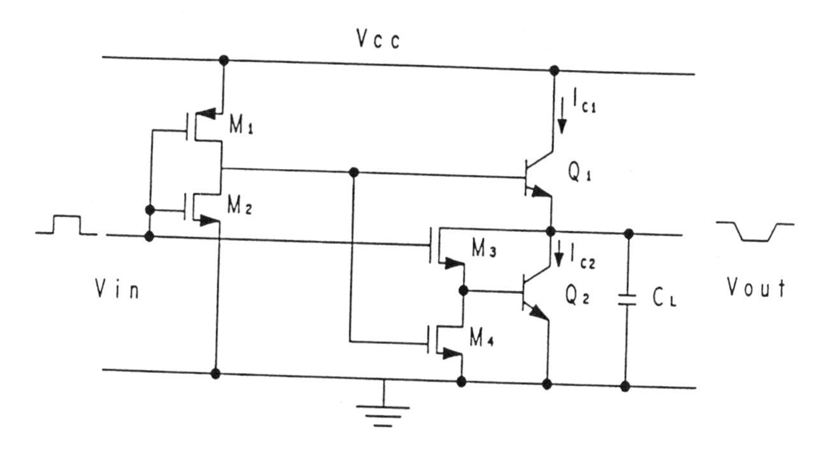

FIGURE 6.2. Schematic of a BiCMOS inverter.

3. Transient drive with no dc power component

The switching response of a BiCMOS gate consists of the rise delay, due to the charging of C_L through M_1 and Q_1, and the fall delay, due to the discharging of C_L through M_3 and Q_2. The N-channel transistors M_2 and M_4 provide a current discharging path for turning off the bipolar transistors. This reduces the power dissipation and saturation storage time. As the input applied voltage V_{in} goes from zero to V_{CC}, MOSFETs M_2 and M_3 are turned on. Consequently, the pull-up bipolar transistor Q_1 is turned off and the pull-down bipolar transistor Q_2 conducts. Then I_{C2}, the collector current of Q_2,

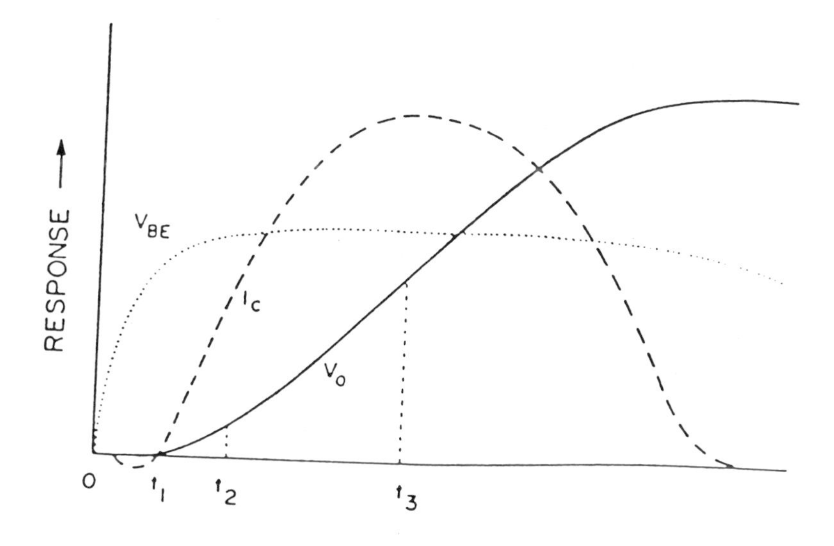

FIGURE 6.3. BiCMOS internal and output waveforms.

discharges the load capacitor C_L, and the output logic swing is constrained from going to zero by the *n–p–n* collector–emitter junction providing the low-level output voltage $V_{OL} = V_{CE2sat}$, the collector–emitter saturation voltage of Q_2. When V_{in} goes from V_{CC} to zero, M_1 conducts to turn on Q_1. So I_{C1}, the collector current of Q_1, charges C_L, and the high-level output voltage V_{OH} is $V_{CC} - V_{CE1sat}$, the collector–emitter saturation voltage of Q_1. Figure 6.3 shows the base–emitter voltage and collector current of Q_1 and the output waveforms V_o.

6.3. BiCMOS SWITCHING DELAY

For BiCMOS device operation, one needs to analyze the device and circuit interaction between the BJTs and MOSFETs, not just devices alone. The analysis involves switching of the MOSFETs, which are voltage-controlled devices, and the bipolar devices, which are charge controlled. In order to predict the gate delay analytically, careful attention must be paid to the surplus base charge that controls the collector current. During the BiCMOS switching response, the bipolar transistor experiences high-injection and high-current effects that severely affect BiCMOS performance.

The switching response of the BiCMOS logic is composed of a pull-up delay (a BiPMOS gate), due to the charging of C_L through M_1 and Q_1, and a pull-down delay (a BiNMOS gate), due to the discharging of C_L through M_3 and Q_2. In both instances the bipolar transistors act as emitter followers with respect to the charge and discharge paths of C_L. The delay of the pull-up transient is defined as the time for the output voltage to increase from zero to 50% of power supply voltage V_{CC}. The delay time of the pull-down transient is defined as the time for the output voltage to fall from power supply voltage to 50% V_{CC}. The gate delay is an average of the pull-up and pull-down delays.

The operation of the BiCMOS pull-up transient can be divided into three time intervals (Fig. 6.3) (Rosseel and Dutton, 1989; Greeneich and McLaughlin, 1988). A similar analysis can be applied to the BiCMOS pull-down transient:

1. During the initial delay interval $0 \le t \le t_1$, MOSFET M_1 switches on quickly due to its light load, but the *n–p–n* bipolar transistor does not conduct until its base–emitter voltage reaches about 0.65 V. Therefore there is a delay of time t_1 before the output voltage of the inverter begins to rise.
2. During the interval $t_1 \le t \le t_2$, M_1 still operates in the saturation region provided the absolute threshold voltage $|V_{TP}|$ of M_1 is greater than the base–emitter turn-on voltage V_{BE}(on) of Q_1 and the bipolar transistor Q_1 is on.
3. During the interval $t_2 \le t \le t_3$, where t_3 is the time point at 50% V_{CC}, M_1 enters the triode region and Q_1 operates at a high level of injection.

Once the three operating regions are identified, transient analysis can be carried out and the delay expression for BiCMOS pull-up is equal to $t_1 + t_2 + t_3$.

During the first interval, the bipolar transistor Q_1 is off and M_1 is in saturation. The *p*-MOS transistor M_1 can be replaced by a current source. The time at which the internal

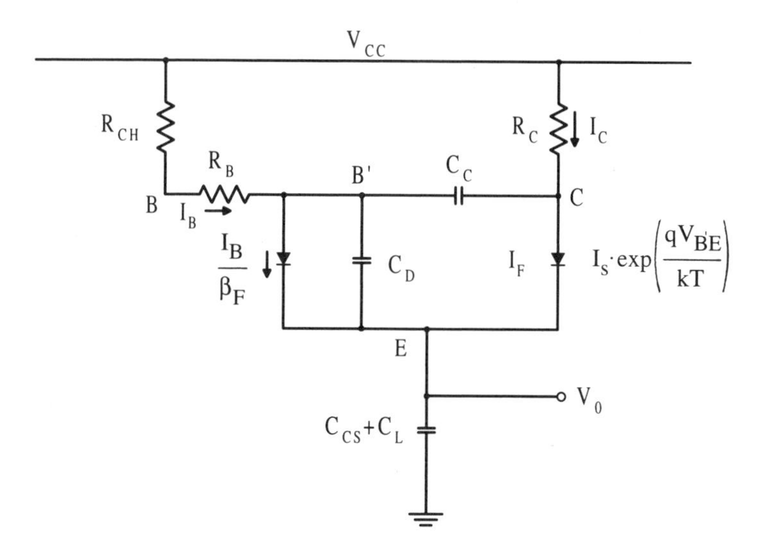

FIGURE 6.4. Equivalent circuit for the second time interval.

base of Q_1 reaches its turn-on voltage can be calculated as (Greeneich and McLaughlin, 1988)

$$t_1 = \frac{R_{CH}(C_E + C_C)V_{BE}(on)}{V_{CC} - |V_{TP}|} - \frac{R_C C_C^2}{C_E + C_C} \tag{6.3.1}$$

where R_{CH} is the equivalent dc channel resistance:

$$R_{CH} = \frac{2L}{W_1 \mu_p C_{ox}(V_{CC} - |V_{TP}|)} \tag{6.3.2}$$

During the second interval, Q_1 is on and M_1 is in saturation. Assuming the base–emitter voltage is relatively constant, C_E can be neglected. The resulting equivalent circuit for this interval is shown in Fig. 6.4. The time t_2 can be solved from the following equation (Greeneich and McLaughlin, 1988):

$$t_2 = \sqrt{\frac{2R_{CH} C_L^* \tau_F^* (|V_{TP}| - V_{BE}(on))}{V_{CC} - |V_{TP}|}} \tag{6.3.3}$$

where $C_L^* = C_L + C_S + C_C$ and $\tau_F^* = \tau_F + R_C C_C$.

During the third interval, Q_1 is on and M_1 is in the triode region. The MOSFET is modeled by an equivalent channel resistance. The resulting equivalent circuit for this interval is shown in Fig. 6.5. From Fig. 6.5 the current through M_1 can be expressed as

$$I_B = \frac{V_{CC} - V_{BE} - V_o}{R_{CH} + R_B} \tag{6.3.4}$$

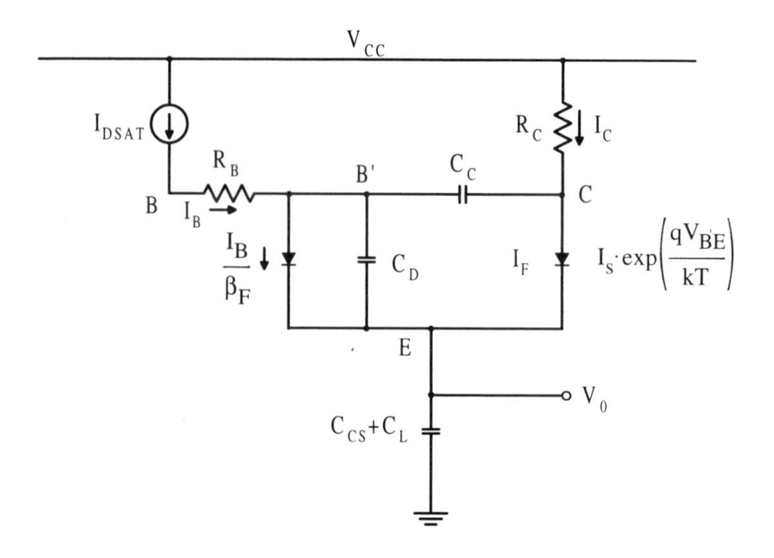

FIGURE 6.5. Equivalent circuit for the third time interval.

From the equivalent circuit, the base current can also be written as

$$I_B = \frac{I_C}{\beta_F} + C_D \frac{dV_{BE}}{dt} + C_C \frac{d(V_o + V_{BE} - V_{CC} + I_C R_C)}{dt} \tag{6.3.5}$$

where $I_C = I_S \exp(qV_{BE}/kT)$ and $C_D = q\tau_F I_C/kT$. Since V_{BE} is relatively constant compared to V_o, the term $C_C dV_{BE}/dt$ can be neglected. Taking advantage of the fact that

$$\frac{dI_C}{dt} \approx \frac{qI_C}{kT}\frac{dV_{BE}}{dt} \tag{6.3.6}$$

$$\frac{dV_o}{dt} \approx \frac{I_C}{C_L^*} \tag{6.3.7}$$

and substituting (6.3.6) into (6.3.5) gives (Fang *et al.*, 1992)

$$\tau_F^* \frac{d^2V_o}{dt^2} + \frac{1}{\beta_F^*}\frac{dV_o}{dt} + \frac{V_o}{(R_{\text{CH}} + R_B)C_L^*} = \frac{V_{CC} - V_{BE}(\text{on})}{(R_{\text{CH}} + R_B)C_L^*} \tag{6.3.8}$$

where

$$\frac{1}{\beta_F^*} = \frac{1}{\beta_F} + \frac{C_C}{C_L^*} \tag{6.3.9}$$

Equation (6.3.8) has the solution

$$V_o(t) = [C \sin(t/T) + D \cos(t/T)]e^{-t/2\beta_F^* \tau_F^*} \tag{6.3.10}$$

where

$$T = \frac{\sqrt{(R_{CH} + R_B)C_L^* \tau_F^*}}{1 - (T_0/2\beta_F^* \tau_F^*)^2} \tag{6.3.11}$$

The constants C and D are determined from the initial conditions $V_o(t = t_2)$ and $I_C(t = t_2) = C_L^* dV_o/dt$. The solution of the analytical equation, however, does not account for the high-current effect of the BJT, which is important in BiCMOS switching.

6.4. BiCMOS DEVICE SIMULATION

The performance of the BiPMOS device has been analyzed with PISCES (Kuo *et al.*, 1989). A merged BiPMOS device structure has been introduced to reduce the device size for BiCMOS VLSI. More physical insight into device operation can be obtained by examining total recombination rates for the device during transient conditions. A two-dimensional device simulation using PISCES to study the steady-state and turn-on transient behavior of a BiNMOS device has also been presented (Chen and Kuo, 1992). The turn-on transient performance of the BiNMOS device shows that, at 77 K, the switching time, which is determined by the load-related delay and the intrinsic delay of the bipolar device, increases about 45% from its 300 K value for an output load of 0.1 pF/μm.

The use of the MEDICI device simulator to study high-current phenomenon of the BJT in a BiCMOS device has been presented by Yuan (1992). A typical BiCMOS gate has been simulated using MEDICI. The BiPMOS structure in the MEDICI simulation is shown in Fig. 6.6. The MOSFET and the bipolar transistor are separated by a recessed

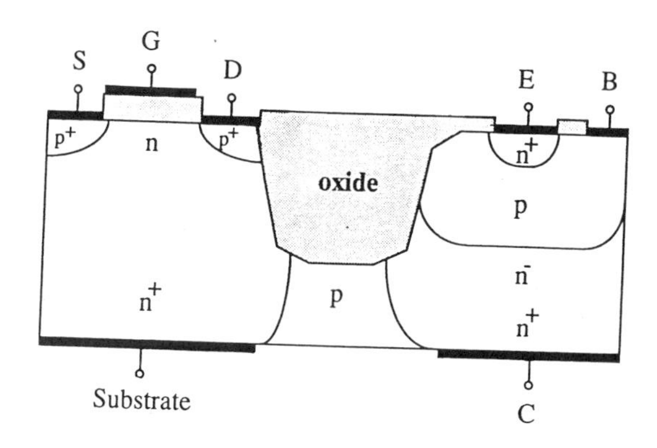

FIGURE 6.6. BiPMOS structure used for MEDICI simulation.

oxide. The 3-D doping profiles used in the BiPMOS structure are displayed in Fig. 6.7. The MOSFETs have threshold voltages of approximately 1 V, and the bipolar transistors have a maximum forward current gain of 80. Buried layers were used in the MOSFET and the bipolar transistor to ensure good contacts; they were separated by a *p*-doped channel stop under the recessed oxide. The polysilicon emitter of the BJT was simulated by specifying a surface recombination velocity of 10^4 cm/s at the polysilicon emitter interface. A transient analysis was performed by ramping the input voltage from 0 to V_{CC} or V_{CC} to 0 in time increments of 50 ps. The MEDICI input file is given in Example 6.1. The current vectors during the switching transient of the BiPMOS gate are shown in Fig. 6.8.

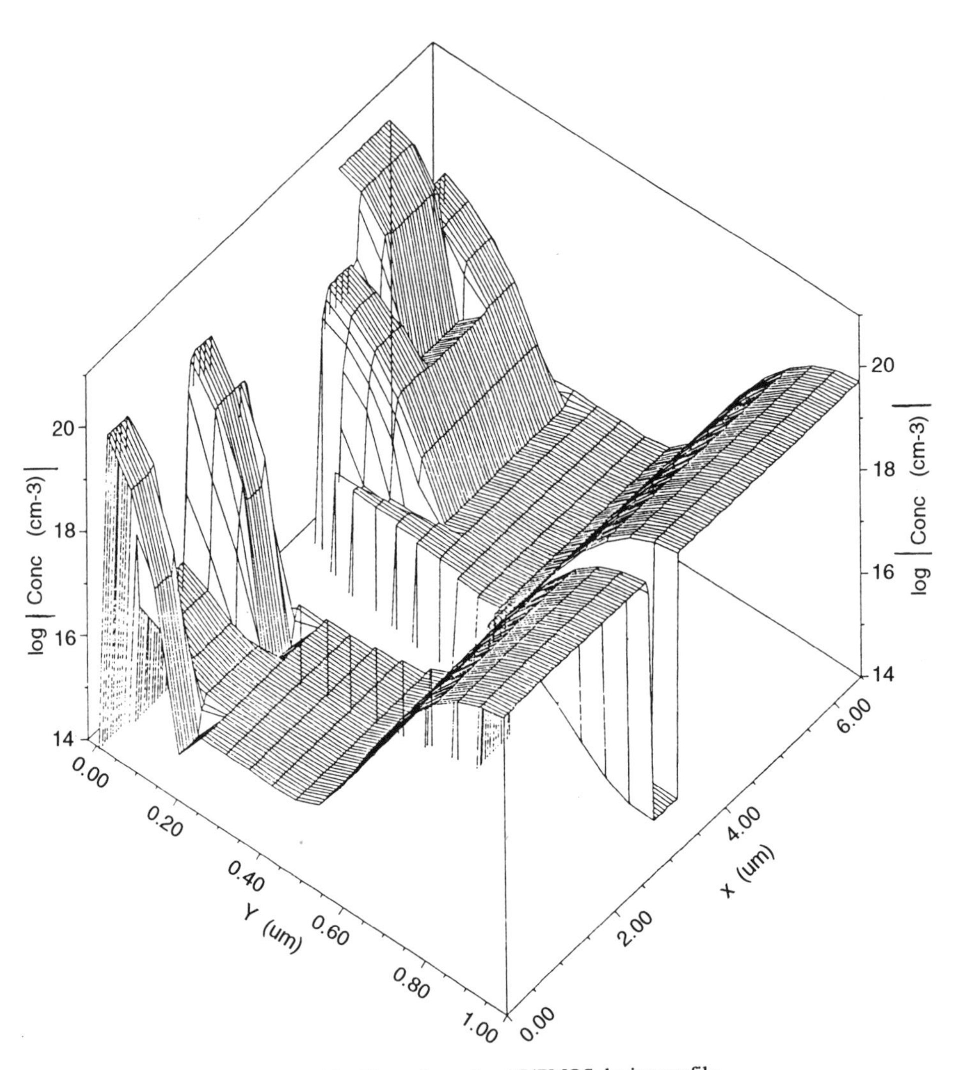

FIGURE 6.7. Three-dimensional BiPMOS doping profile.

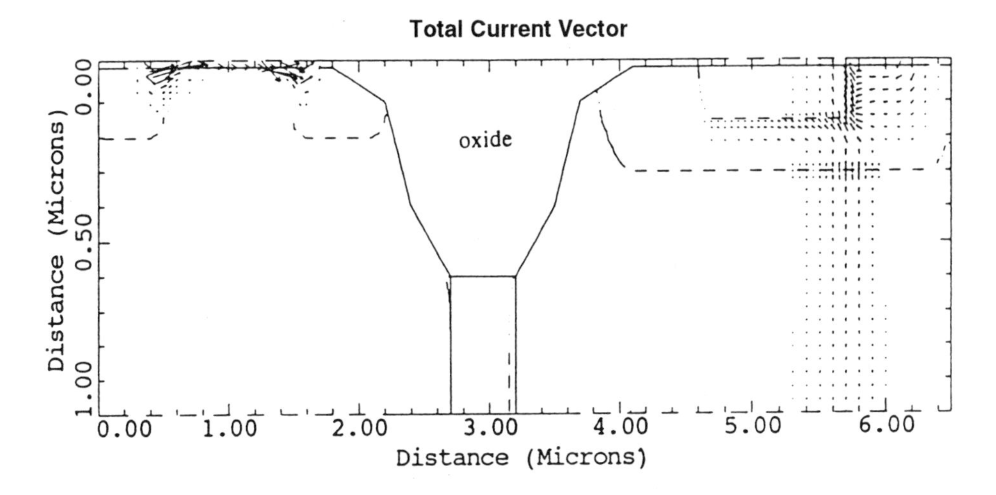

FIGURE 6.8. Current vectors of during switching transient.

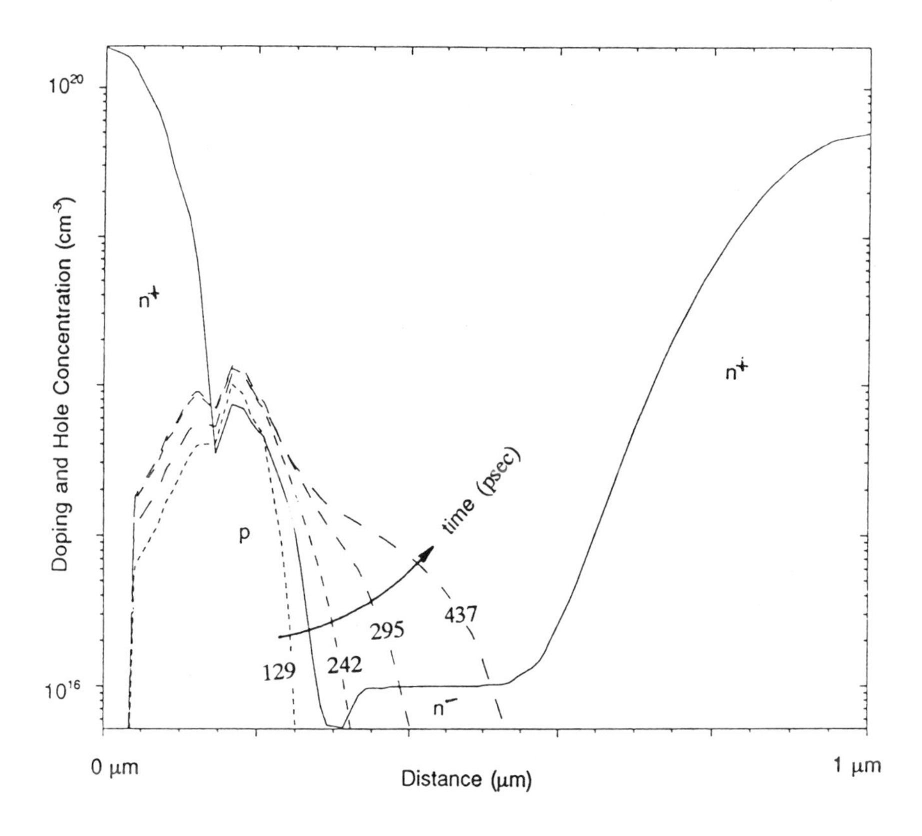

FIGURE 6.9. Hole concentration as a function of time.

With MEDICI the BiCMOS switching pull-up and pull-down delays have been analyzed. When the capacitive load increases from 0.1 to 0.5 pF/μm, the current required to drive the capacitive load increases. However, as the collector current increases, high-current degradation, such as high-injection effects and base pushout, increases, and, therefore, the current cannot scale up to the increase in the capacitive load. When the collector current of the bipolar transistor exceeds a critical value, the base extends into the epitaxial collector. This causes base widening and increases transit time through the base. The forward gain of the BJT degrades and, therefore, collector current decreases. Figure 6.9 shows the hole concentration in the BJT as a function of depth in the base and of time during the transient response of a BiPMOS inverter. One observes that base pushout takes place and the switching delay of the BiCMOS inverter is degraded. It is therefore important to account for base pushout in order to predict the BiCMOS switching delay accurately.

6.5. BiCMOS TRANSIENT ANALYSIS INCLUDING HIGH-CURRENT EFFECTS

The BiCMOS analysis presented in Sec. 6.3 assumes a constant V_{BE} in the second interval and voltage-independent base resistance, emitter–base capacitance, collector–base capacitance, and forward transit time. It neglects high-current effects, such as base widening, which severely affect BiCMOS performance. High-current effects also result in nonlinear forward transit time, base resistance, and base–emitter voltage. To accurately predict the transient response of the BiCMOS buffer, physical mechanisms of base-width modulation, base conductivity modulation, base pushout, bias-dependent base resistance, forward transit time, and emitter–base and collector–base junction capacitances must be accounted for. The BiCMOS switching response here is evaluated by using analytical and numerical approaches. In the analytical approach, a set of equations has been derived that predicts the pull-up 50% delay of a BiPMOS inverter. In the numerical approach, a system of equations describing a BiCMOS inverter is solved simultaneously by the Newton–Raphson method or Keller's shooting method. The circuit parameters at a given point in time are calculated in terms of the solution of the previous point in time.

6.5.1. *Analytical Approach*

During the switching transient the base–emitter voltage of the BJT changes with time. From the results of the device simulation, it is reasonable to consider V_{BE} as a linear profile with respect to time in the second and the third time intervals. Under this assumption, the following analytical models have been developed (Phanse, 1994). One can write the KCL equations at the base and the emitter of the BJT in the circuit as follows:

$$I_B = \frac{I_F}{\beta_F} + (C_{BE} + C_{BC})\frac{\partial V_{BE}}{\partial t} + C_{BC}R_C\frac{\partial I_C}{\partial t} + C_{BE}\frac{\partial V_o}{\partial t} \tag{6.5.1}$$

$$I_E = C_L\frac{\partial V_o}{\partial t} = I_F\left(1 + \frac{1}{\beta_F}\right) + C_{BC}\frac{\partial V_{BE}}{\partial t} \tag{6.5.2}$$

During the first interval, M_1 is in saturation and Q_1 is off. Equation (6.3.1) is still valid for calculating t_1. During the second interval ($t_1 \leq t \leq t_2$), M_1 is in saturation and Q_1 is on; V_{BE} is assumed to have a linear dependence with time. At time t_1 we have

$$V_{BE}(t_1) = V_{BE}\,(\text{on}), \quad V_o(t_1) = 0, \quad I_C(t_1) = 0 \tag{6.5.3}$$

$$V_{BE}(t) = V_{BE}\,(\text{on}) + S_1 \frac{t}{t_2 - t_1} \tag{6.5.4}$$

where $S_1 = V_{BE}(t_2) - V_{BE}(\text{on})/(t_2 - t_1)$.

At time t_2 the MOSFET enters the triode region of operation. Therefore, at t_2 we have $V_{GS} - V_{\text{TP}} = V_{DS}$, or

$$V_o(t_2) = |V_{\text{TP}}| - I_{D\text{sat}}\,R_B - V_{BE}(t_2) \tag{6.5.5}$$

By incorporating these equations into (6.5.2), and integrating between t_2 and t_3, we get

$$C_L \frac{\partial V_o}{\partial t} = I_{SS} e^{qV_{BE}(\text{on})/kT}\, e^{qS_1/kT} \left(1 + \frac{1}{\beta_F}\right) + C_{BE} \frac{\partial V_{BE}}{\partial t} \tag{6.5.6}$$

$$\begin{aligned} C_L[V_o(t_2) - V_o(t_1)] &= C_{BE}[V_{BE}\,(t_2) - V_{BE}\,(\text{on})] \\ &\quad + I_{SS} e^{qV_{BE}/kT} \left(e^{qS_1/kT} - e^{qS_s/kT}\right)\left(1 + \frac{1}{\beta_F}\right) \frac{kT}{qS_1} \end{aligned} \tag{6.5.7}$$

In this interval, the base current of Q_1 is the drain current of the MOSFET in saturation. Eliminating I_F from Eqs. (6.5.1) and (6.5.2) and integrating between t_1 and t_2, we get

$$\begin{aligned} I_{D\text{sat}}\,(t_2 - t_1) &= \left(\frac{\beta_F\, C_{BE}}{1 + \beta_F} + C_{BC}\right)[V_{BE}\,(t_2) - V_{BE}\,(t_1)] \\ &\quad + C_{BC} R_C [I_C\,(t_2) - I_C\,(t_1)] + \left(C_{BC} + \frac{C_L}{1 + \beta_F}\right)[V_o(t_2) - V_o(t_1)] \end{aligned} \tag{6.5.8}$$

At t_2, the dc current I_F is far greater than the charging current through C_{BC}. Therefore,

$$I_C\,(t_2) = \frac{I_{SS}}{q_b q_{BPO}}\, e^{qV_{BE}(t_2)/kT} \tag{6.5.9}$$

where q_{BPO} is the excess charge in the base due to base pushout. The development of q_{BPO} is discussed in detail in Sec. 6.5.3. Using Eqs. (6.5.7) and (6.5.8), we can solve for $V_{BE}(t_2)$ and t_2. Knowing these, we can calculate $V_o(t_2)$, $I_C(t_2)$, and other circuit parameters at t_2.

During the third interval ($t_2 \le t \le t_3$), M_1 is in the linear region and Q_1 is in the high-current region. In this interval a linear V_{BE} with respect to time is modeled as

$$V_{BE}(t) = V_{BE}(t_2) + S_2 \frac{t}{t_3 - t_2} \tag{6.5.10}$$

where $S_2 = [V_{BE}(t_3) - V_{BE}(t_2)]/(t_3 - t_2)$. Proceeding in a manner similar to the derivation of Eq. (6.5.7), we get

$$C_L[V_o(t_2) - V_o(t_1)] = C_{BE}[V_{BE}(t_2) - V_{BE}\,(\text{on})] + I_{SS}e^{qV_{BE}/kT}\left(e^{qS_1t_3/kT} - e^{qS_2t_2/kT}\right)\left(1 + \frac{1}{\beta_F}\right)\frac{kT}{qS_2} \tag{6.5.11}$$

where $V_o(t_3) = V_{CC}/2$ represents the 50% delay. Eliminating I_F from Eqs. (6.5.1) and (6.5.2) and integrating between t_2 and t_3, we obtain

$$\int_{t_2}^{t_3} I_B(t)\,dt = \left(\frac{\beta_F C_{BE}}{1 + \beta_F} + C_{BC}\right)[V_{BE}(t_2) - V_{BE}(t_1)] + C_{BC}R_C[I_C(t_3) - I_C(t_2)] + \left(C_{BC} + \frac{C_L}{1 + \beta_F}\right)[V_o(t_2) - V_o(t_1)] \tag{6.5.12}$$

During this interval, the base current of Q_1 is equal to the drain current of the MOSFET in the triode region. In the linear region, the MOSFET can be approximated as an equivalent channel resistance R_{CH}:

$$I_B(t) = I_{\text{Dlin}} = \frac{V_{CC} - V_{BE}(t) - V_o(t)}{R_{\text{CH}} + R_B} \tag{6.5.13}$$

$$\int_{t_2}^{t_3} I_B(t)\,dt = \frac{(t_3 - t_2)\{V_{CC} - 0.5[V_{BE}(t_3) + V_{BE}(t_2)] - 0.5[V_o(t_3) + V_o(t_2)]\}}{R_{\text{CH}} + R_B} \tag{6.5.14}$$

For the purpose of integrating the base current, the output voltage V_o can be considered to be linear with respect to time in this interval. From the MEDICI simulation results we observe that this assumption is reasonable. Moreover the overall error introduced in the calculation of t_3 is not significant. Once again, the dc collector current is far greater than the charging current through C_{BC}, so the collector current at t_3 is calculated from

$$I_C(t_3) \approx \frac{I_{SS}}{q_b q_{BPO}} e^{qV_{BE}(t_3)/kT} \tag{6.5.15}$$

Now using Eqs. (6.5.13) to (6.5.15) we can solve for the two unknowns: $V_{BE}(t_3)$ and t_3.

The analytical model for calculating the pull-up delay t_3 of the BiCMOS circuit is developed. However, an analytical approach cannot account for the time-dependent base-pushout, base-width and conductivity modulation, current-dependent base resistance, bias-dependent junction capacitance, and free-carrier capacitance. These factors significantly affect the BiCMOS switching delay, so it is necessary to solve for the transient response of the BiCMOS gate numerically.

6.5.2. *Numerical Approach*

In the numerical model, the circuit parameters at a given time point are used to calculate the circuit parameters of the next time point. The charging and discharging characteristics of all the capacitive elements of the circuit are carefully accounted for and a more accurate solution is obtained. The system of simultaneous equations describing the equivalent circuit of the BiCMOS inverter must be solved. Numerical algorithms such as the Keller shooting method and the Newton–Raphson method are examined. The Keller shooting method solves the nonlinear differential system by using a particular circuit constraint as a boundary condition at each time point. Given the value of V_{BE} at a time point, one must aim for the next time point in order to satisfy the boundary value (i.e., to make the error zero). One aims for the next time point and finds two points where the error is positive and negative, respectively. The zero error point may lie between these points. Convergence, which depends upon the solution, is arrived at by continuously finding an interval at whose boundaries the error is positive and negative, respectively. The algorithm for selecting the next target value is presented in Fig. 6.10. A new target value for V_{BE} is chosen to bring the "error" to zero (within specified error limits). The initial conditions used for the numerical model are $V_{BE}(t_1) = V_{BE}(\text{on})$, $V_{BC}(t_1) = V_{BE}(\text{on}) - V_{CC}$, $I_B(t_1) = I_C(t_1) = I_E(t_1) = 0$.

This approach did not converge when some of the nonlinear models describing the physical effects of the BJT and MOSFET were implemented. The Newton–Raphson method was then investigated. It solves the simultaneous equations

$$f_1(x_1, x_2, x_3, \ldots, x_n) = 0$$

$$f_2(x_1, x_2, x_3, \ldots, x_n) = 0$$

$$\vdots$$

$$f_n(x_1, x_2, x_3, \ldots, x_n) = 0 \tag{6.5.16}$$

which is symbolically written as $[f] = 0$. Performing a Taylor series expansion on the function f_i, dropping all terms higher than first order, and realizing that the desired result is $f_i = 0$, we get

$$f_i(x_{1,g}, x_{2,g}, \ldots, x_{n,g}) = \frac{\partial f(x_{1,g}, x_{2,g}, \ldots, x_{n,g})}{\partial x_1(x_{1,g} - x_{1,c})} + \frac{\partial f(x_{1,g}, x_{2,g}, \ldots, x_{n,g})}{\partial x_2(x_{2,g} - x_{2,c})}$$

$$+ \cdots + \frac{\partial f(x_{1,g}, x_{2,g}, \ldots, x_{n,g})}{\partial x_n(x_{n,g} - x_{n,c})} \tag{6.5.17}$$

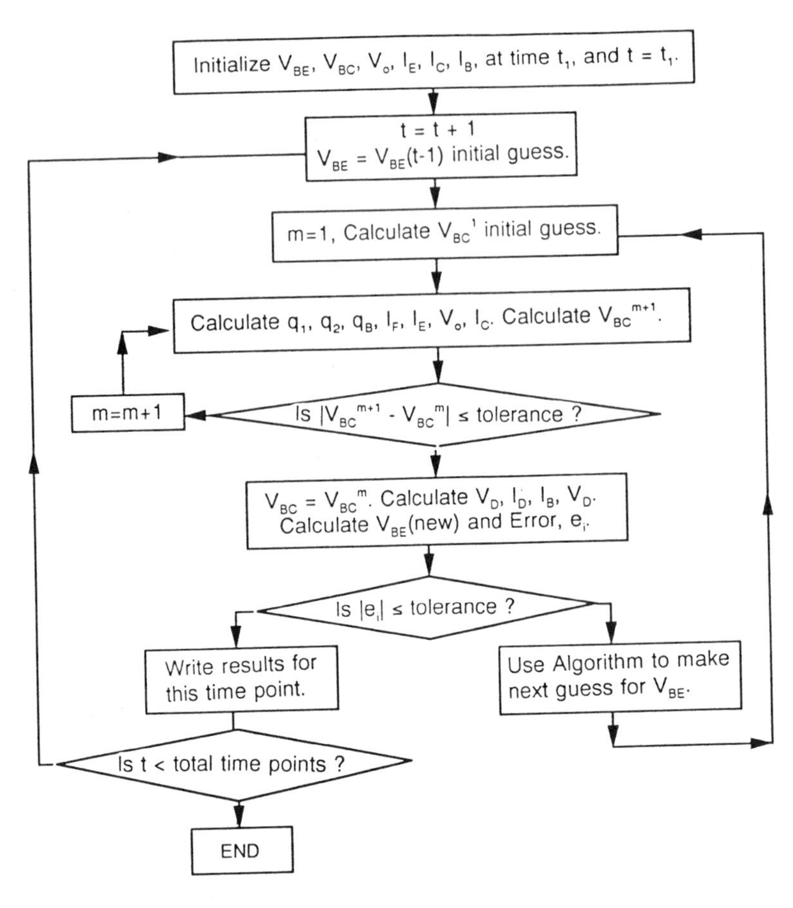

FIGURE 6.10. Flowchart for describing Keeler's shooting method.

where x_g's are the guess values and the x_c's are the values that give the correct value of zero for f_i. Equation (6.5.17) in matrix form is

$$\begin{bmatrix} \frac{\partial f_1}{\partial x_1} & \frac{\partial f_1}{\partial x_2} & \cdots & \frac{\partial f_1}{\partial x_n} \\ \frac{\partial f_2}{\partial x_1} & \frac{\partial f_2}{\partial x_2} & \cdots & \frac{\partial f_2}{\partial x_n} \\ \cdot & \cdot & \cdot & \cdot \\ \frac{\partial f_n}{\partial x_1} & \frac{\partial f_n}{\partial x_2} & \cdots & \frac{\partial f_n}{\partial x_n} \end{bmatrix} \begin{bmatrix} x_{1,g} - x_{1,c} \\ x_{2,g} - x_{x,c} \\ \cdots \\ x_{n,g} - x_{n,c} \end{bmatrix} = \begin{bmatrix} f_1 \\ f_2 \\ \vdots \\ f_n \end{bmatrix} \tag{6.5.18}$$

or, symbolically,

$$[J][\delta x] = [f] \tag{6.5.19}$$

where $[J]$ is the Jacobian matrix and $[\delta x]$ is the correction vector.

The partial derivatives are evaluated by a finite difference approximation. The Jacobian is inverted to solve for the x's required to move the values of the functions $f_{1,2,3,..,n}$ toward zero. Thus, $[\delta x] = [J]^{-1}[f]$. The next time point voltages are then updated by computing new x's as

$$x_i(\text{new}) = x_i(\text{old}) - \delta x \tag{6.5.20}$$

The iteration proceeds until all residual terms are within a specified tolerance. The flowchart for the algorithm is shown in Fig. 6.11.

The Newton–Raphson method is used for BiCMOS pull-up and pull-down transients. The equivalent circuit of the BiCMOS pull-up transient is shown in Fig. 6.12. The input voltage, at the gate of the PMOSFET, changes from V_{CC} to 0 V in the time t_{ramp}.

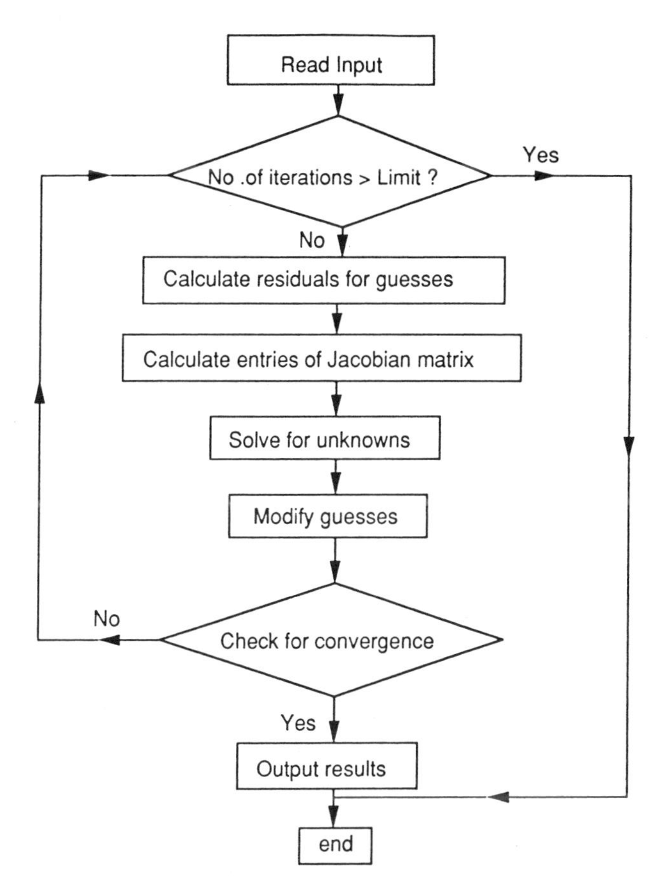

FIGURE 6.11. Flowchart for the Newton–Raphson method.

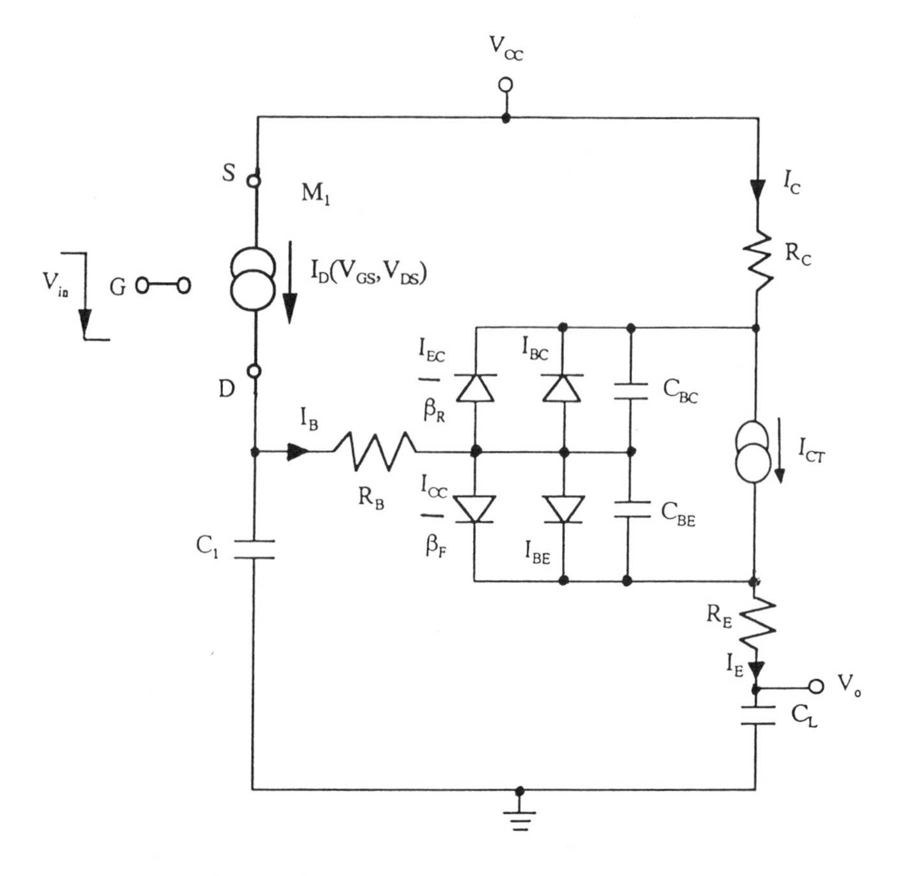

FIGURE 6.12. Equivalent circuit of BiCMOS pull-up.

The output voltage V_o, at the emitter of the BJT, changes from 0 V to V_{OH}. The equations describing the equivalent circuit are

$$V_o(t) = V_o(t - \Delta t) + \frac{[I_E(t) + I_E(t - \Delta t)]\Delta t}{2C_L} \tag{6.5.21}$$

$$V_D(t) = V_o(t) + I_E R_E + V_{BE}(t) + I_B(t - \Delta t)R_B \tag{6.5.22}$$

$$I_1 = \frac{C_1[V_D(t) - V_D(t - \Delta t)]}{\Delta t} \tag{6.5.23}$$

where I_1 the current through the parasitic capacitance C_1 and Δt is the time step of the transient simulation.

The following equations can be solved for every time point by using the Newton–Raphson method:

$$f_1(V_{BE}, V_{BC}) = I_D - I_1 - I_B = 0 \tag{6.5.24}$$

$$f_2(V_{BE}, V_{BC}) = V_{CC} - I_C R_C - V_{BC} + V_{BE} - I_E R_E - V_o = 0 \tag{6.5.25}$$

The simultaneous solution of these equations gives the correct internal junction voltages V_{BE} and V_{BC}, from which all circuit voltages and currents for this time point can be evaluated.

The equivalent circuit of the BiCMOS pull-down transient is shown in Fig. 6.13. The input voltage, at the gate of the n-MOSFET, changes from 0 to V_{CC} V in time t_{ramp}. The output voltage V_o, at the collector of the BJT, changes from V_{CC} to V_{OL}.

$$V_D(t) = I_E R_E + V_{BE}(t) - V_{BC}(t) + I_C R_C \tag{6.5.26}$$

$$V_S(t) = I_E R_E + V_{BE}(t) + I_B(t - \Delta t)\, R_B \tag{6.5.27}$$

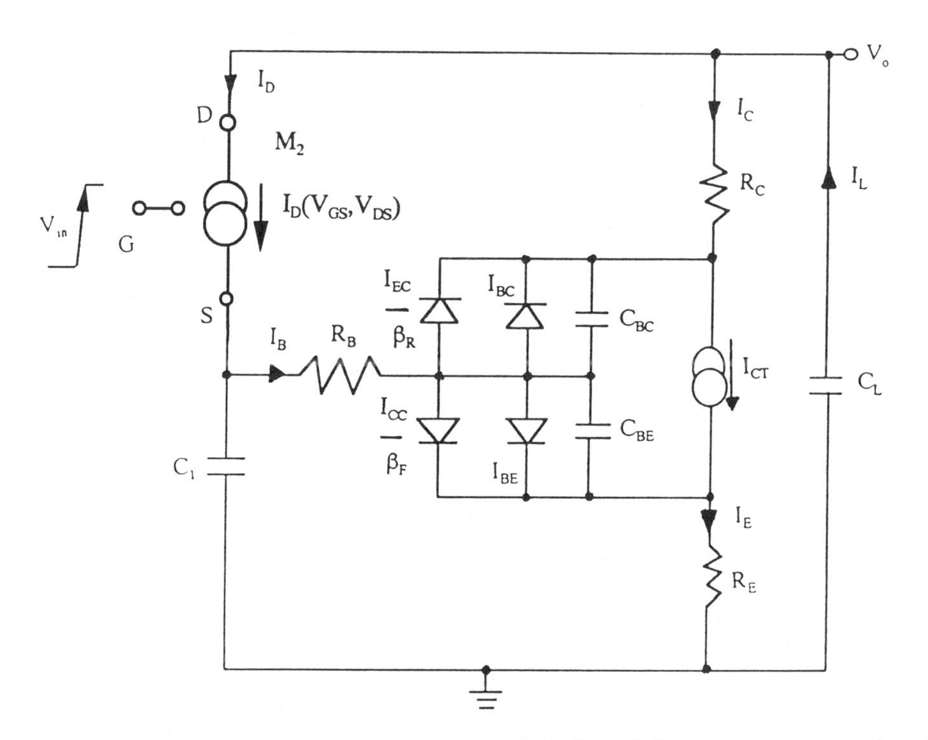

FIGURE 6.13. Equivalent circuit of BiCMOS pull-down.

$$I_L(t) = I_C + I_D \tag{6.5.28}$$

$$V_o(t) = V_o(t - \Delta t) - \frac{[I_L(t) + I_L(t + \Delta t)]\Delta t}{2C_L} \tag{6.5.29}$$

$$I_L(t) = \frac{C_1[V_S(t) - V_S(t - \Delta t)]}{\Delta t} \tag{6.5.30}$$

The following equations can be solved by using the Newton–Raphson method for time-dependent V_{BE} and V_{BC} during the pull-down transient of the BiCMOS gate:

$$f_1(V_{BE}, V_{BC}) = I_D - I_1 - I_B = 0 \tag{6.5.31}$$

$$f_2(V_{BE}, V_{BC}) = V_o - V_D = 0 \tag{6.5.32}$$

The Gummel–Poon model equations in Getren (1987) are used to describe the BJT. Base-width modulation and base conductivity modulation effects are accounted for by the normalized base charge q_b. The base pushout effect on the collector current is described by q_{BPO} (see Sec. 6.3.3). Writing the KCL equations at the emitter, collector, and base of the BJT, we get

$$I_E(t) = \frac{C_{BE}[V_{BE}(t) - V_{BE}(t - \Delta t)]}{\Delta t} + \frac{I_{CC} - I_{EC}}{q_b} + \frac{I_{CC}}{\beta_F} + I_{BE} \tag{6.5.33}$$

$$I_C(t) = \frac{I_{CC} - I_{EC}}{q_b} - \frac{C_{BC}[V_{BC}(t) - V_{BC}(t - \Delta t)]}{\Delta t} - \frac{I_{EC}}{\beta_R} - I_{BC} \tag{6.5.34}$$

$$I_B(t) = \frac{C_{BE}[V_{BE}(t) - V_{BE}(t - \Delta t)]}{\Delta t} + \frac{C_{BC}[V_{BC}(t) - V_{BC}(t - \Delta t)]}{\Delta t}$$

$$+ \frac{I_{CC}}{\beta_F} + \frac{I_{EC}}{\beta_R} + I_{BE} + I_{BC} \tag{6.5.35}$$

The parameters of the BJT and MOSFET models depend on their terminal voltages and the currents flowing through them. For simplicity the base resistance is modeled with a piecewise linear approximation: For $V_{BE} \geq V_{BH}$, $R_B = R_{BH}$; for $V_{BL} \leq V_{BE} \leq V_{BH}$, $R_B = R_{BL} + (R_{BH} - R_{BL})(V_{BH} - V_{BE})/(V_{BH} - V_{BL})$; for $V_{BE} \leq V_{BL}$, $R_B = R_{BL}$. The base–collector and base–emitter capacitances are sums of junction and diffusion capacitances. Under reverse bias the junction capacitance can be calculated from the depletion approximation. The diffusion capacitances are modeled as

$$C_{DE,C} = \frac{q\tau_{F,R}I_{CC,EC}}{kT} \tag{6.5.36}$$

where τ_F, the forward transit time through the base, is a function of the extended base width:

$$\tau_F = \tau_{F0}\left(\frac{X_B + X_{CIB}}{X_B}\right)^2 \tag{6.5.37}$$

where τ_{F0} is the forward transit time at low collector current, X_{CIB} is the extended base width due to base pushout (Sec. 6.5.3), and τ_R, the reverse transit time through the base, is assumed to be a constant.

In the linear region, the drain current of the MOSFET is given by (Yamaguchi, 1983)

$$I_D = g_{\text{surf}}\mu_0\sqrt{\frac{1 + E_{\text{perpend}}/E_{\text{critical}}}{1 + N_{\text{ch}}/(N_{\text{ch}}/S + N_r)}}\,\frac{C_{\text{ox}}W[(V_{GS} - V_{\text{TH}})V_{DS} - 0.5V_{DS}^2]}{L_{\text{eff}}} \tag{6.5.38}$$

and in the saturation region the drain current is

$$I_D = g_{\text{surf}}\mu_0\sqrt{\frac{1 + E_{\text{perpend}}/E_{\text{critical}}}{1 + N_{\text{ch}}/(N_{\text{ch}}/S + N_r)}}\,\frac{C_{\text{ox}}W(V_{GS} - V_{\text{TH}})^2(1 + \lambda V_{DS})}{L_{\text{eff}}} \tag{6.5.39}$$

where g_{surf} is the surface degradation factor; N_{ch} is the channel doping concentration; $S = 350$, $N_r = 3 \times 10^{16}\ \text{cm}^{-3}$, and $\mu_0 = 1400\ \text{cm}^2/\text{V-s}$ for electrons; $S = 81$, $N_r = 4 \times 10^{16}\ \text{cm}^{-3}$, and $\mu_0 = 480\ \text{cm}^2/\text{V-s}$ for holes; E_{perpend} is the effective perpendicular field in the inversion channel; $E_{\text{critical}} = 0.09$ MV/cm is the rate at which mobility degrades with increase in the perpendicular field; C_{ox} is the oxide capacitance per unit area; W is the channel width; L_{eff} is the effective channel length; and λ is the channel-length modulation factor.

6.5.3. Normalized Base Pushout Factor

When the collector current exceeds a critical value, injection of majority carriers (holes in the case of an *n–p–n* BJT) into the lightly doped epitaxial collector occurs. The epitaxial collector region is very lightly doped compared to the base; therefore, the entire base–collector depletion region lies in the epitaxial collector. The electric field in the epitaxial layer is

$$E_{\text{epi}} = \frac{V_{CB}}{X_{\text{epi}} - X_{CIB}} \tag{6.5.40}$$

The critical electric field value representing the junction between high and low field conditions is 10^5 V/cm. The critical current density J_0 is required to create the space-charge-limited condition and, hence, induce base pushout effects.

When $E_{epi} < 10^5$ V/cm (Whittier and Tremere, 1969),

$$J_0 = qv_{lim}\left(N_{epi} + \frac{2\varepsilon|V_{CB}|}{qX_{epi}^2}\right) \tag{6.5.41}$$

$$X_{CIB} = X_{epi}\left[1 - \sqrt{\frac{J_0 - qv_{lim}N_{epi}}{J_C - qv_{lim}N_{epi}}}\right], \quad \text{where } J_C > J_0 \tag{6.5.42}$$

When $E_{epi} > 10^5$ V/cm,

$$J_0 = q\mu_n N_{epi}|V_{CB}|/X_{epi} \tag{6.5.43}$$

$$X_{CIB} = (1 - J_0/J_C)X_{epi} \tag{6.5.44}$$

Now the extended base width is $X_B' = X_B + X_{CIB}$. The forward current of the bipolar transistor of the Gummel–Poon model is

$$I_{CT} = \frac{I_{SS}\int_{X_{E0}}^{X_{C0}} N_A(x)\,dx\left(e^{qV_{BE}/kT} - e^{qV_{BC}/kT}\right)}{\int_0^{X_B'} p(x)\,dx} \tag{6.5.45}$$

$$I_{CT} = \frac{I_{SS}\int_{X_{E0}}^{X_{C0}} N_A(x)\,dx\int_0^{X_B} p_1(x)\,dx\left(e^{qV_{BE}/kT} - e^{qV_{BC}/kT}\right)}{\int_0^{X_B} p_1(x)\,dx\int_0^{X_B'} p_2(x)\,dx} \tag{6.5.46}$$

where $p_2(x)$ is the actual hole profile in the base, including extended base effects, and $p_1(x)$ is the hole profile, assuming that base pushout does not occur as shown in Fig. 6.14. The above equation can be expressed as

$$I_{CT} = \frac{I_{SS}(e^{qV_{BE}/kT} - e^{qV_{BC}/kT})}{q_b q_{BPO}} \tag{6.5.47}$$

while q_{BPO} represents the excess charge in the base due to base pushout:

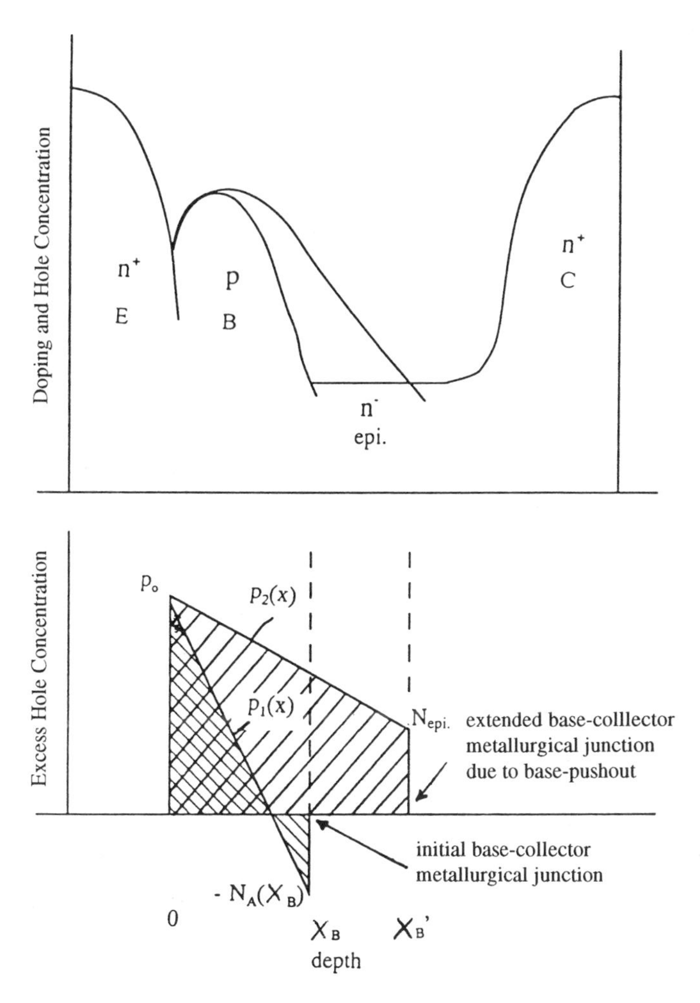

FIGURE 6.14. Hole concentration distribution including base pushout.

$$q_{BPO} = \frac{\int_0^{X_B'} p_2(x)\,dx}{\int_0^{X_B} p_1(x)\,dx} \tag{6.5.48}$$

At the emitter–base junction electrons are injected into the base. To maintain space-charge neutrality, the excess holes in the base must equal the injected electrons. Assuming base pushout does not occur, the hole concentration at the edge of the depletion width goes to zero. Therefore, the excess hole concentration at the edge of the depletion layer, $\Delta p_1(X_B)$, equals $-N_A(X_B)$. A linear profile assumption can be made for the excess hole concentration in the short base. Considering the base pushout effects, holes are injected across the initial base–collector junction to form the extended base. At the extended base–collector junction, all the holes are compensated for by the electrons supplied by N_{epi} doping. There is no base doping in the extended base, so $\Delta p_2(X_B')$ equals N_{epi}. After base pushout a linear profile for the excess hole concentration has been assumed as shown in Fig. 6.15.

$$\Delta p_1(0) = \Delta p_2(0) = \frac{n_i^2 e^{qV_{BE}/kT}}{N_A(0)} = p_0 \tag{6.5.49}$$

$$\Delta p_1(x) = p_0 - \frac{[p_0 + N_B(X_B)]x}{X_A'} \tag{6.5.50}$$

$$p_1(x) = \Delta p_1(x) + N_A(x) \tag{6.5.51}$$

$$\Delta p_2(x) = p_0 - \frac{(p_0 - N_{\text{epi}})x}{X_B'} \tag{6.5.52}$$

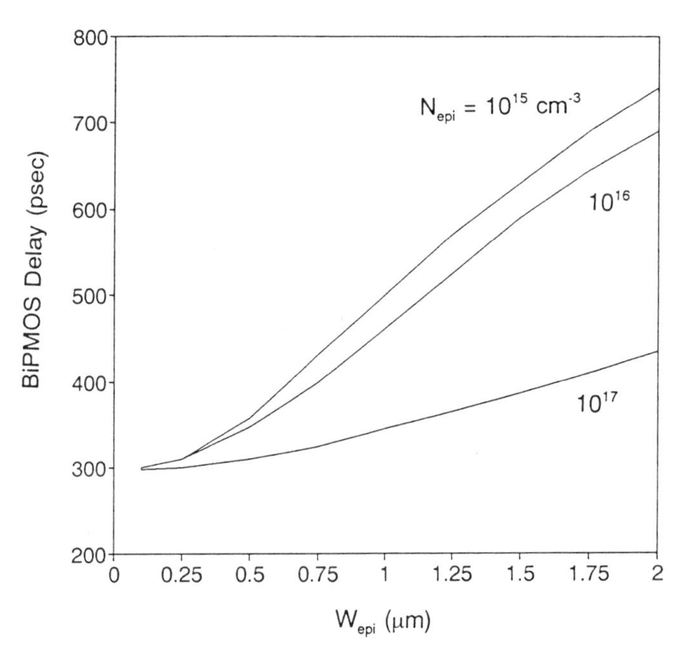

FIGURE 6.15. BiPMOS delay as a function of epitaxial layer thickness and concentration.

$$p_2(x) = \Delta p_2(x) + N_A(x) \tag{6.5.53}$$

$$q_{BPO} = \frac{Q_B + 0.5qX'_B(p_0 + N_{\text{epi}})}{Q_B + 0.5qX_B[p_0 - N_A(X_B)]} \tag{6.5.54}$$

where $Q_B = q\int N_A(x)\,dx$. At $x = X_B$, we have the base–collector metallurgical junction, so $N_A(X_B) = N_{\text{epi}}$ and

$$q_{BPO} = 1 + \frac{2qX_BN_{\text{epi}} + qX_{CIB}(p_0 + N_{\text{epi}})}{2Q_B + qX_B(p_0 - N_{\text{epi}})} \tag{6.5.55}$$

The base pushout factor q_{BPO} is used to calculate I_{CT} in the Gummel–Poon model for the BJT.

The current-induced base width is a function of the doping concentration of the epitaxial collector. The higher the epitaxial collector doping, the smaller is the width of the current-induced base; consequently, the BiCMOS switching response is faster. However, increasing the collector doping results in a decrease in the collector–base breakdown voltage, which is undesirable. The epitaxial collector doping concentration was varied between 10^{15} and 10^{17} cm^{-3}, while the epitaxial layer thickness was maintained at 0.25 μm. Both device simulations and model predictions indicate that there is no significant change in the collector current and the switching delay. When the epitaxial layer width is varied between 0.25 and 1.0 μm while the epitaxial layer doping concentration is kept at 10^{16} cm^{-3}, significant base pushout effect on the switching response is observed. Figure 6.15 shows predictions of the effect of base pushout on BiPMOS switching as a function of both epitaxial layer thickness and doping concentration. When epitaxial layer thickness is increased, base pushout is enhanced, collector current degrades, and, hence, delay increases. Decreasing the epitaxial collector doping concentration or increasing the epitaxial layer width enhances base pushout effects; therefore, the BiCMOS switching response degrades. At small epitaxial layer thicknesses, BiCMOS delay is relatively insensitive to variations in epitaxial layer doping concentration.

6.6. RADIATION EFFECT ON BiCMOS PERFORMANCE

In space environments electronic circuits are exposed to gamma energetic electrons and protons. These ionizing radiations cause defects that result in the appearance of traps in the forbidden energy band gap. Ionizing radiation also causes defects, such as neutral electron traps and neutral hole traps. Neutral traps do not significantly contribute to carrier scattering or the change of electron or hole concentration in the semiconductor, so they have been neglected. Ionizing radiation also causes interface-state generation (N_{it}) at the Si–SiO_2 interface and trapping of fixed positive charge in the oxide (N_{ot}). Interface states act as recombination centers and therefore increase the surface recombination velocity. The charge trapped in the oxide and in the interface states scatters the carriers (coulombic scattering) near the surface and hence results in mobility degradation of carriers near the interface. The radiation-induced buildup of interface traps and oxide charge generally occurs over time (Boesch, 1988). Here, total dose effects have been considered; i.e., the system has been evaluated when the interface trap with respect to time has stabilized. The interface-state density and the oxide-trapped charge density are assumed to change linearly with radiation dose. This assumption is valid when the radiation dose is less than 1 Mrad(Si) (Wilson and Blue, 1984; Baze *et al.*, 1989). Radiation-induced bipolar gain degradation occurs in linear bipolar and digital BiCMOS technologies (Enlow *et al.*, 1991). A cross section of a BiCMOS device (Pease *et al.*, 1992) shows a buried-layer (BL)-to-BL leakage path. The leakage current cross section in recessed field oxide (Pease *et al.*, 1985). To evaluate radiation effects on BiCMOS performance, one examines key degradation mechanisms of the BJT and MOSFET and radiation-induced leakage paths in the BiCMOS circuit.

6.6.1. Radiation Effects on the BJT in a BiCMOS Device

Radiation causes interface-state generation at the Si–SiO_2 interface, which increases the surface recombination velocity. It has been experimentally proven that the surface recombination velocity is proportional to the density of surface states (Grove, 1979); i.e., $S_n = \sigma_n v_{th} N_{it}$, where σ_n is the recombination probability (cross section) per carrier, v_{th} is the thermal velocity of the carrier, and N_{it} is the surface trap density. The oxide-trapped charge, N_{ot}, repels holes in the subsurface region of the *p* base of the *n–p–n* BJT and generates a depletion layer at the surface. As the oxide charge density increases, the width of the depletion layer increases until it reaches a maximum when the surface becomes strongly inverted ($\rho_s = 2\phi_F$). The surface begins to be inverted when $\rho_s = \phi_F$, and interface states charge negatively as they trap the inversion carriers (i.e., electrons). Shockley–Read–Hall recombination occurs in the depletion region at the interface. Depletion and subsequent inversion of the *p* base can occur under the spacer and field oxides in the extrinsic base (the region annotated $A_{S_{BE}}$ in Fig. 6.16), thereby causing recombination in

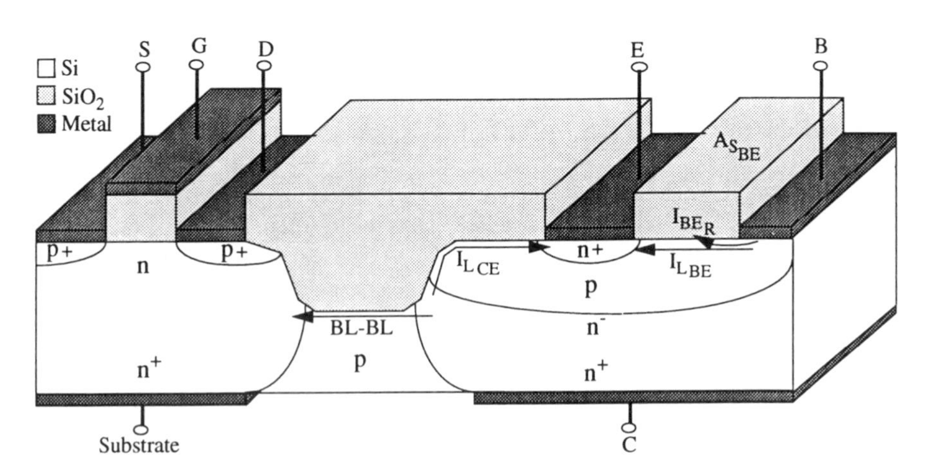

FIGURE 6.16. BiCMOS leakage path.

the semiconductor between the base and emitter contacts (I_{BE_R}), and channeling along the recessed oxide sidewall between the emitter and collector ($I_{L_{CE}}$). As the oxide charge density increases, the surface depletion region increases and more interface traps become effective recombination centers. Interface states lying over depleted surfaces act as recombination centers. As the interface-state density increases, the surface recombination velocity increases, resulting in more recombination current. Performing an analysis similar to Grove (1979), we derive the recombination current at the surface, and in the oxide-charge-induced depletion region between the emitter and the base contact of the BJT:

$$I_{BE_R} = 0.5qn_i\left(\frac{x_d}{\tau_{OXJ}} + S_n\right)A_{S_{BE}}\left(e^{qV_{BE}/n_{EL_R}kT} - 1\right) \tag{6.6.1}$$

where τ_{OXJ} is the recombination lifetime in the oxide-charge-induced junction, $A_{S_{BE}}$ is the area of the extrinsic base between the emitter and base contacts that is exposed to oxide, and n_{EL_R} is the base current ideality factor.

Equation (6.6.1) is valid when the surface in the base is depleted; i.e., $N_{ot} - N_{it} \leq N_A x_{dmax}$. Once the surface in the base is inverted, the crossover condition ($n \approx p$) is no longer met and, therefore, effective recombination does not take place at the surface. The base recombination current after inversion is derived as

$$I_{BE_R} = \frac{qA_{S_{BE}} n_i x_{\text{dmax}}}{2\tau_{OXJ}}\left(e^{qV_{BE}/n_{EL_R}kT} - 1\right) \tag{6.6.2}$$

I_{BC_R} is defined similarly. The radiation-induced base recombination currents, I_{BE_R} and I_{BC_R} are now added to the Gummel–Poon architecture.

6.6.2. *Radiation Effects on the MOSFET in a BiCMOS device*

The MOSFET is a surface transport device and, therefore, seriously affected by the $Si–SiO_2$ interface degradation. Carrier scattering due to interface trapped charge, oxide trapped charge, and increased recombination at the surface results in the degradation of the inversion mobility. An empirical model for inversion mobility as a function of interface-state density and oxide charge density (Zupac *et al.*, 1992) has been used in this analysis:

$$\mu(N_{it}, N_{ot}) = \frac{\mu_0}{1 + \alpha_{it}N_{it} + \alpha_{ot}N_{ot}} \tag{6.6.3}$$

where α_{it} and α_{ot} are process-dependent parameters. The mobility due to a given spatial distribution of oxide charge, $\mu(N_{ot})$, has been derived (Phanse *et al.*, 1993) as

$$\mu(N_{ot}) = \frac{32\sqrt{\pi}\varepsilon^2 (2kT)^{3/2} Z_{\text{inv}}}{C_o n_I^{-1/4} q^3 \sqrt{m^*} \int_0^{t_{ox}} N_{ot}(z)g(z)\,dz} \tag{6.6.4}$$

where Z_{inv} is the inversion-layer thickness, C_0 is a proportionality constant, n_I is the carrier density in the inversion layer, m^* is the effective mass of the carrier in the inversion layer, $N(z)$ is the spatial oxide charge per cm^{-3}, $g(z)$ is a function of the distance from the Si–SiO_2 interface, z, and t_{ox} is the thickness of the oxide. If the spatial distribution of the charge in the oxide is known, radiation-induced inversion mobility can be accurately modeled as $1/\mu = 1/\mu(N_{it}) + 1/\mu(N_{ot})$, where $\mu(N_{it}) = \mu_0/(1 + \alpha_{it}N_{it})$ (Galloway *et al.*, 1984). Mobility degradation results in a decrease in drain current, which drives the base of the BJT, in a BiCMOS inverter, and thus an increase in the time required for the BJT to turn on. When the MOSFET is in strong inversion, all the interface states are assumed to be filled by the inversion carriers. The charge trapped in the oxide and at the interface induces charge at the surface and therefore alters the threshold voltage of the MOSFET. The postradiation threshold voltage is expressed as $V_{\text{TP}} = V_{\text{TP0}} - Q_{ot}^p/C_{\text{ox}} - Q_{it}^p/C_{\text{ox}}$, and $V_{TN} = V_{\text{TN0}} - Q_{ot}^n/C_{\text{ox}} + Q_{it}^n/C_{\text{ox}}$. Q_{it}^p, Q_{ot}^p, Q_{it}^n, and Q_{ot}^n are the interface trap charge and oxide trap charge due to radiation effects for n- and p-channel transistors, respectively. The BiCMOS switching response comprises the pull-up (BiPMOS) transient and pull-down (BiNMOS) transient. In the BiPMOS structure the positive fixed oxide charge and the positively charged interface states tend to increase the threshold voltage of the MOSFET and thus decrease the drain current, resulting in increased switching delay. In the BiNMOS structure the positive fixed oxide charge and the negatively charged interface states produce opposite shifts in V_{TN}. Compensation of threshold voltage and the higher electron mobility make the BiNMOS gate more resistant to radiation.

6.6.3. Leakage Paths in BiCMOS

The principal failure modes in modern BiCMOS technologies are associated with the recessed field oxide isolation, edge leakage in n-channel MOSFETs and BL-to-BL leakage in the bipolar transistor (Pease *et al.*, 1992). Modern CMOS technologies utilize lateral oxide isolation and eliminate p^+ guard bands around the n-channel transistors. This results in a failure mode called edge leakage, which is the result of the turn-on of a parasitic parallel transistor near the "bird's beak" region of the lateral oxide isolation, and is due to trapped oxide positive charge in the thick isolation oxide or in the oxide spacer next to the polysilicon gate. In digital bipolar technologies the principal ionizing-radiation-induced failure mechanisms are

a. BL–BL leakage caused by inversion of p^+ channel stop under the recessed field oxide
b. Collector–emitter (C–E) edge leakage in walled emitter technologies
c. Base–emitter (B–E) edge leakage due to inversion of field oxide

In BiCMOS gates we can also have BL–BL leakage between the MOSFET and the bipolar structures. The cross section of BiCMOS technology showing the location of BL–BL and C–E leakage is displayed in Fig. 6.17. Radiation-induced trapped holes repel electrons and invert the underlying semiconductor. After inversion there is a leakage path in the semiconductor along the surface of the oxide. The inversion and depletion charges balance the net positive charge at the surface. Therefore the inversion channel charge is

$$Q_n = Q_{ot} - Q_{it} - qN_{\text{Asurface}}\, x_{d\max} \tag{6.6.5}$$

The channel conductance is $G_{\text{leak}} = \mu_n(N_{it}, N_{ot})Q_n W_1/L_1$, where W_1 and L_1 are the width and length of the leakage path due to the inverted semiconductor, and μ_n is the average electron mobility in the channel. A resistance $R_{\text{leak}} = 1/G_{\text{leak}}$ can represent the radiation-induced inverted channel in the equivalent circuit when the condition for inversion is met. Base–emitter, emitter–collector, and BL–BL leakage paths in BiCMOS circuits are modeled in this manner.

6.6.4. *MEDICI Simulation Including Radiation Effects*

The 2-D device simulator MEDICI was used to simulate the BiCMOS gate. BiPMOS and BiNMOS gates were simulated to obtain the pull-up and pull-down switching responses, respectively. The boundary statement was used to define silicon, oxide, and electron (contact) regions to form the structure. Typical MOSFET and bipolar doping profiles of present-day technologies were used. The MOSFET channel length is 0.6 μm, and the emitter area of the BJT is 0.65 μm^2. The MOSFET has a threshold voltage of 1 V, and the *n–p–n* BJT has a maximum forward current gain of 80. The MOSFET and the BJT are separated by a recessed field oxide. Under the recessed field oxide there is a *p*-doped channel stop. Buried layers were introduced in the MOSFET substrate and bipolar collector.

The preradiation surface degradation factor for inversion mobility, $g_{\text{surf}} = 0.9$, was used to simulate surface scattering of the inversion carriers due to broken bonds at the interface. Radiation-induced interface states and oxide charge degrade inversion mobility. The value $g_{\text{surf}} = 0.9(1 + \alpha_{it}N_{it} + \alpha_{ot}N_{ot})^{-1}$ was used to account for radiation-induced degradation of inversion mobility. The values used were $\alpha_{it} = 1 \times 10^{-12}$ cm^{-2} and $\alpha_{ot} = 1 \times 10^{-13}$ cm^{-2} for PMOS, and $\alpha_{it} = 2 \times 10^{-12}$ cm^{-2} and $\alpha_{ot} = 2 \times 10^{-13}$ cm^{-2} for NMOS.

In strong inversion nearly all interface states trap inversion carriers and acquire charge. The positive charge in the bulk of the oxide can be represented as an equivalent charge, N_{ot}, at the Si–SiO_2 interface. The radiation-induced charge was simulated in MEDICI by placing a charge $QF = Q_{ot} + Q_{it}$ for PMOS, and $QF = Q_{ot} - Q_{it}$ for NMOS, at the interface. If the condition for inversion is satisfied, QF induces a leakage path along the SiO_2–*p*-Si interface. This effect is modeled by including radiation-induced leakage resistance, R_{leak}, in the equivalent BiCMOS circuit. The surface recombination velocity is proportional to the number of interface states. Therefore, radiation causes greater recombination current near the surface. The postradiation surface recombination velocity, $S_{nR} = S_{n0}N_{it}/N_{it0}$ was used during the simulation at all Si–SiO_2 interfaces except the interface above the MOSFET inversion layer (increase in surface recombination velocity is accounted for in inversion mobility degradation). The term S_{n0} is the preradiation surface recombination velocity, and $N_{it0} = 10^{10}$ cm^{-2} is a typical preradiation interface-state density.

Radiation-induced charge perturbs the local electron and hole concentration in the semiconductor and causes recombination current. We have included the effects of Shockley–Read–Hall recombination (including concentration-dependent lifetimes),

Auger recombination, surface recombination, band-gap narrowing, concentration, and perpendicular field-dependent mobility during the simulation to represent the physical mechanisms altered by radiation.

Leakage currents were observed between the buried layers, the collector and the emitter, and the base and emitter contacts. During the transient response of the BiCMOS gate, electron injection from the base into the epitaxial collector (base pushout) was observed.

6.6.5. *The Model Including Effects of Base Pushout and Radiation*

The BiCMOS switching operation is analyzed by solving a system of simultaneous equations describing the equivalent circuit of the BiCMOS inverter including radiation effects. The Newton–Raphson method is used to obtain a solution at every time point, using the solution of the previous time point.

For the BiCMOS pull-up and pull-down transient including radiation effects, the following current equations are modified for the equivalent circuit in Fig. 6.17:

$$I_E(t) = \frac{C_{BE}[V_{BE}(t) - V_{BE}(t-\Delta t)]}{\Delta t} + I_{CT} + \frac{I_{CC}}{\beta_F} + I_{BE} + I_{BE_R} + \frac{V_{BE}(t) - V_{BC}(t)}{R_{LCE}} + I_{L_{BE}} \tag{6.6.6}$$

$$I_C(t) = I_{CT} - \frac{C_{BC}[V_{BC}(t) - V_{BC}(t-\Delta t)]}{\Delta t} - \frac{I_{EC}}{\beta_R} - I_{BC} - I_{BC_R} + \frac{V_{BE}(t) - V_{BC}(t)}{R_{L_{CE}}} \tag{6.6.7}$$

$$I_B(t) = \frac{C_{BE}[V_{BE}(t) - V_{BE}(t-\Delta t)]}{\Delta t} + \frac{C_{BC}[V_{BC}(t) - V_{BC}(t-\Delta t)]}{\Delta t} + \frac{I_{CC}}{\beta_F} + \frac{I_{EC}}{\beta_R} + I_{BE} + I_{BC} + I_{BE_R} + I_{BC_R} + I_{L_{BE}} \tag{6.6.8}$$

Here $I_{L_{BE}}$ is the base–emitter leakage current through the leakage path resistance $R_{L_{BE}}$. The sidewall leakage resistance between the collector and the emitter is $R_{L_{CE}}$, and I_{BE_R} is the radiation-induced recombination current in the extrinsic base between the base and the emitter contacts that is exposed to oxide and is given by Eqs. (6.6.1) and (6.6.2). The current I_{BC_R} is defined similarly.

An increase in radiation dose produces an increase in interface-state generation and oxide-trapped charge and, consequently, an increase in the radiation-induced surface

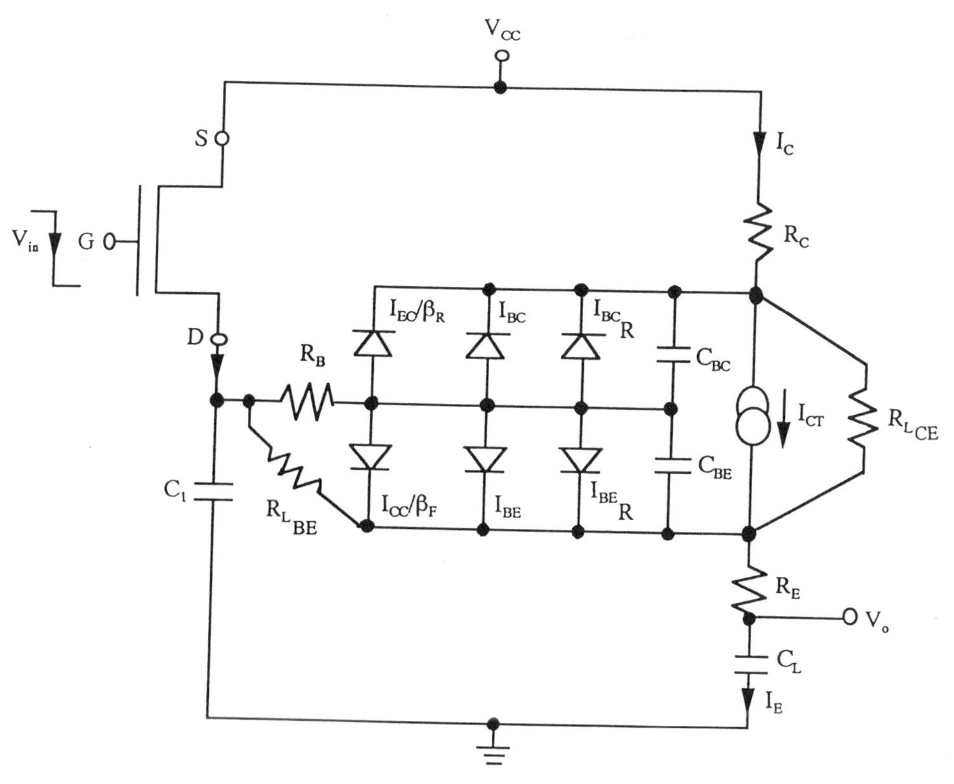

FIGURE 6.17. BiCMOS equivalent circuit including effect of radiation.

recombination current in the base. This leads to a decrease in the forward current gain of the bipolar transistor. The forward current gain of the bipolar transistor can be represented as

$$\beta = \frac{I_{CT} - I_{EC}/\beta_R - I_{BC} - I_{BC_R} + V_{CE}/R_{L_{CE}}}{I_{CC}/\beta_F + I_{EC}/\beta_R + I_{BC} + I_{BE} + I_{BC_R} + I_{BE_R} + V_{BE}/R_{L_{BE}}} \tag{6.6.9}$$

The normalized forward current gain of the bipolar transistor versus the base–emitter voltage, as a function of radiation dose (krad), is plotted in Fig. 6.18. The term β_{pk} is the peak preradiation current gain. Comparison of the model (symbols) with experimental data (solid lines) (Nowlin *et al.*, 1992) is shown for radiation doses of 0, 20, 100, and 500 krad. The dependencies $N_{ot} = 2.9 \times 10^8$ states/cm^2/rad(Si) and $N_{it} = 5.5 \times 10^7$ states/cm^2/rad(Si) (Pease *et al.*, 1983), $A_{S_{BE}} = 0.4\ \mu m^2$, $n_{EL_R} = 1.4$, and $\tau_{OXJ} = 1$ ns are used.

To verify the model for leakage resistance, the test structure in Fig. 6.19 was simulated by MEDICI. The substrate has a uniform doping of 10^{17} cm^{-3}. Using the condition of inversion derived earlier, the net positive charge at the Si–SiO$_2$ interface

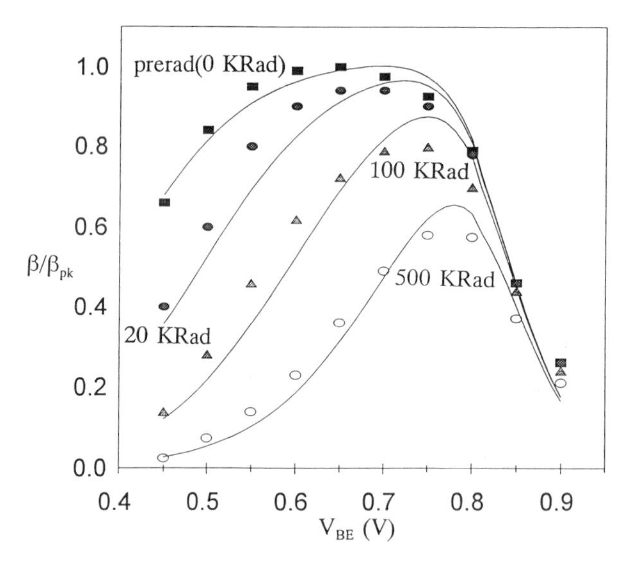

FIGURE 6.18. Normalized current gain versus base–emitter voltage.

required for inverting the surface is calculated as 10^{12} cm^{-2}. Different quantities of charge (10^{12}, 10^{13}, 10^{14}, 10^{15}) were placed at the interface, and the model predictions (solid lines) of the *I–V* characteristics of the leakage path were compared with MEDICI simulations (symbols), as shown in Fig. 6.20. The *I–V* characteristics are linear, indicating a pure resistive behavior. From MEDICI simulations we found that the net positive charge at the interface required for inversion is 10^{12}. This agrees with the model's prediction of the condition for inversion.

Technologies that use walled emitters are subject to channeling along the recessed field oxide sidewall due to inversion of the *p*-type base region. The experimental data (solid line) of the collector–emitter leakage current as a function of the radiation dose

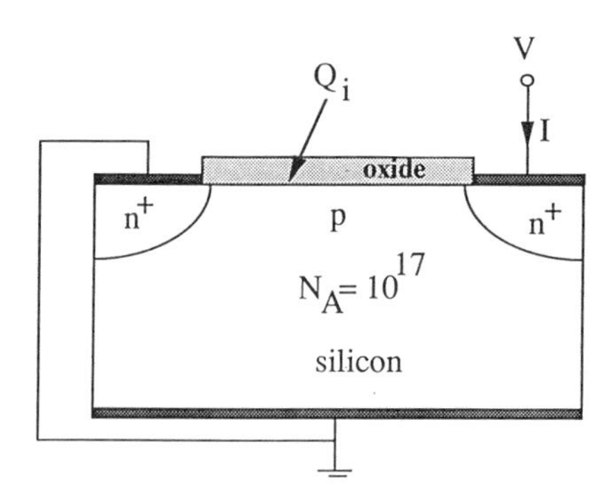

FIGURE 6.19. A test structure for leakage resistance.

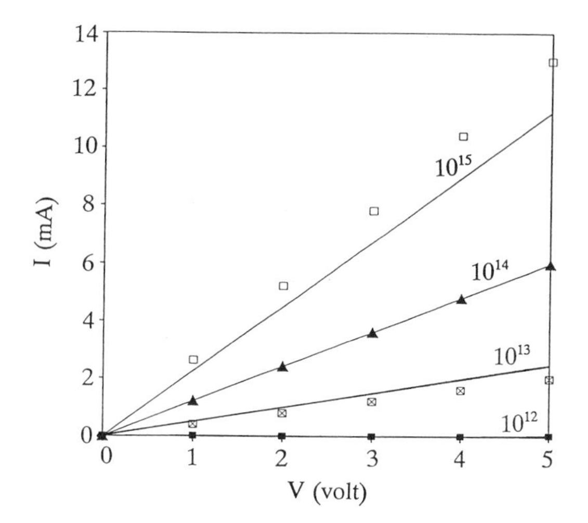

FIGURE 6.20. Current–voltage characteristic for leakage resistance.

was compared with the model predictions (symbols) and is shown in Fig. 6.21. Observe that at high radiation doses the collector–emitter leakage current curve flattens out because of the degraded mobility in the leakage path due to interface-state generation and oxide charge buildup.

The MEDICI simulation of the transient response of the BiCMOS gate is compared with model predictions for different N_{it} and N_{ot} in Figs. 6.22 and 6.23. The oxide charge density, N_{ot}, is kept as four times the interface-state density, N_{it}. In Fig. 6.22 the output voltage versus time is plotted during the pull-up transient. The model (solid lines) is compared with MEDICI simulation (symbols). Figure 6.23 shows the corresponding collector current during the BiCMOS pull-up transient. The output voltage of the

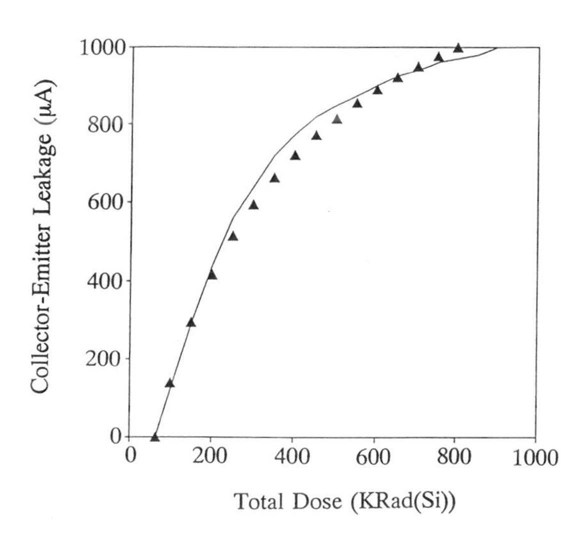

FIGURE 6.21. Collector–emitter leakage versus radiation dose.

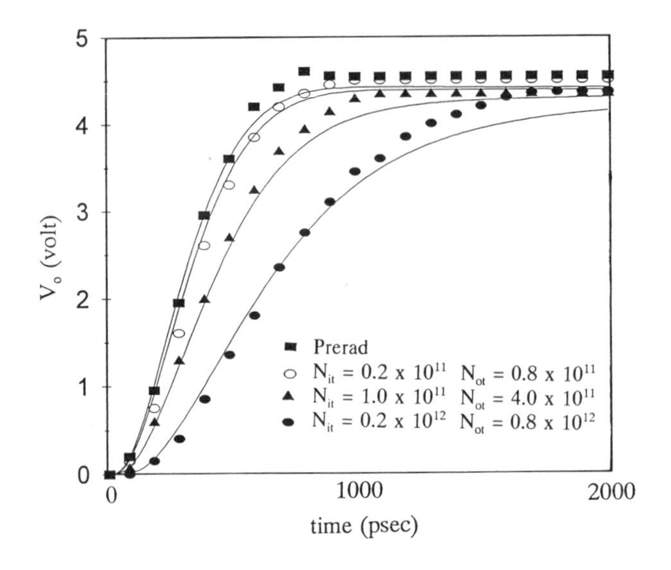

FIGURE 6.22. Output voltage versus radiation (BiCMOS pull-up).

BiCMOS gate during the pull-down transient is plotted in Fig. 6.24. The corresponding load current ($I_C + I_D$) during the pull-down transient is displayed in Fig. 6.25.

The pull-up and pull-down delays for different levels of radiation are displayed in Figs. 6.26 and 6.27. The model (solid line) is compared with MEDICI (symbol). The

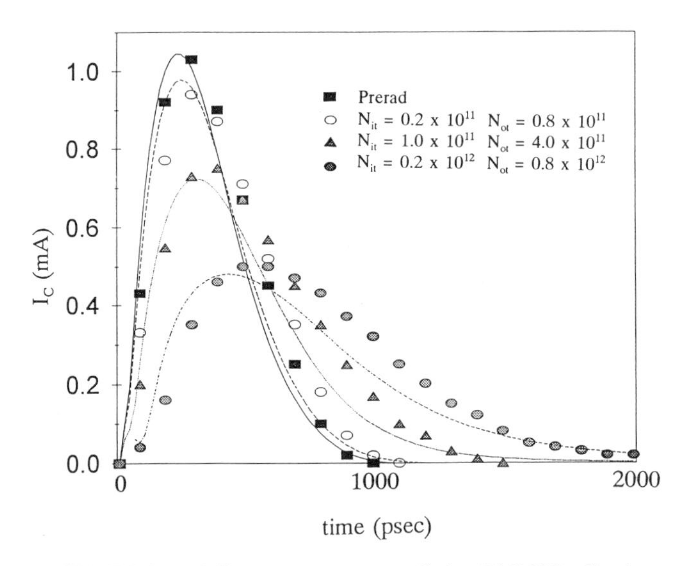

FIGURE 6.23. Collector current versus radiation (BiCMOS pull-up).

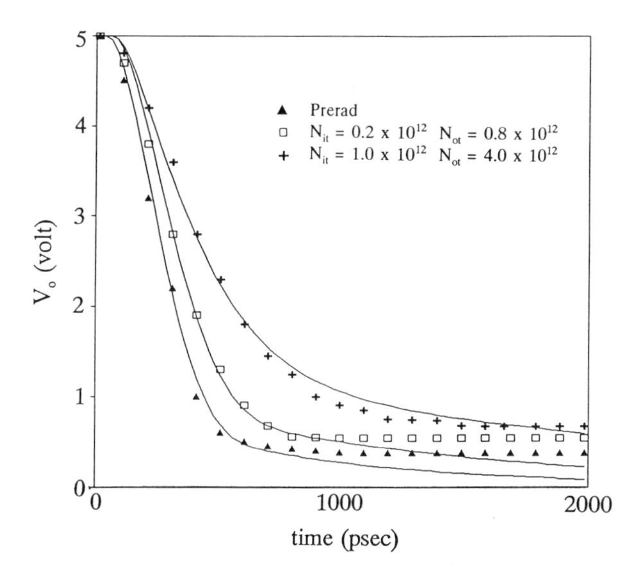

FIGURE 6.24. Output voltage versus radiation (BiCMOS pull-down).

increase in switching delay, over the preradiation delay, due to radiation effects on the MOSFET and on the BJT has been plotted. Both for BiCMOS pull-up and pull-down, one observes that at low levels of radiation, radiation effects on MOSFETs have a dominant effect on the overall switching response, while at higher levels of radiation,

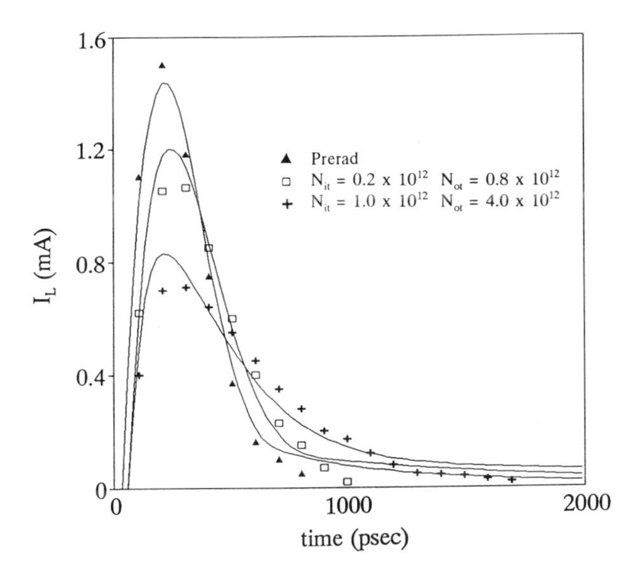

FIGURE 6.25. Load current versus radiation (BiCMOS pull-down).

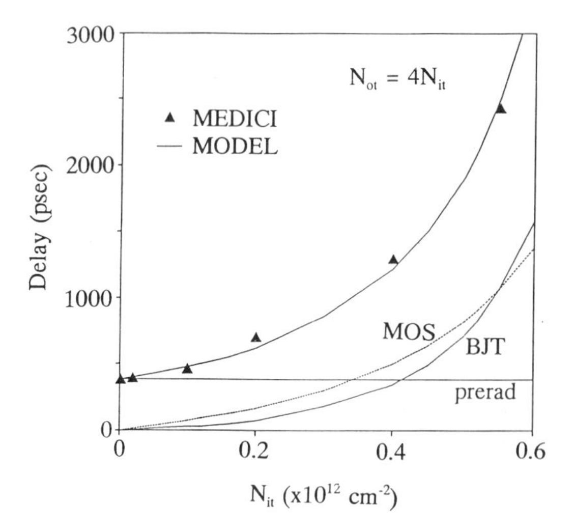

FIGURE 6.26. Pull-up delay versus radiation.

radiation effects on the BJT become more important. It is clear from these two figures that the BiCMOS pull-down is less susceptible to radiation effects. This is because N_{ot} and N_{it} produce opposite shifts in the threshold voltage, and due to the higher electron mobility the NMOSFET can drive sufficient current to turn on the bipolar transistor at higher radiation levels. At high radiation doses, a threshold voltage shift beyond the power supply voltage will make the BiCMOS gate nonoperational.

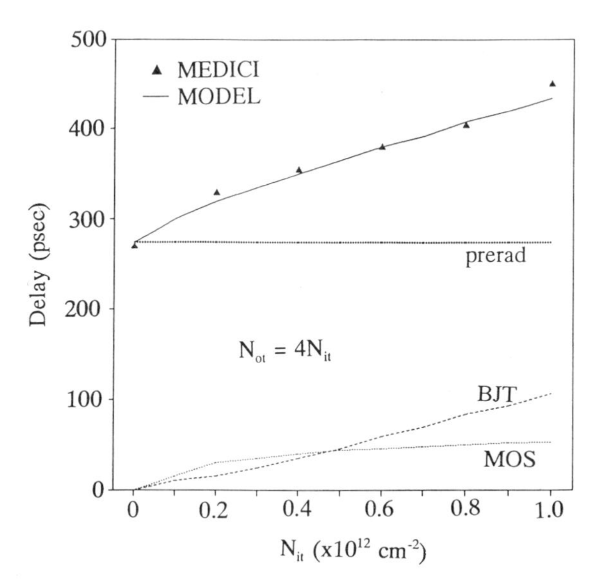

FIGURE 6.27. Pull-down delay versus radiation.

6.7. SCALING

Scaling in this section includes temperature scaling, geometrical scaling, and supply voltage scaling.

6.7.1. Temperature Scaling

When the temperature is increased, the MOSFET inversion mobility, and hence the drain current, decreases. The CMOS switching response degrades with temperature increase. Due to band-gap narrowing in the emitter of the BJT, the forward current gain of the BJT increases exponentially with temperature. Hence, the collector current of the BJT increases with temperature. Due to these opposing effects on the MOS transistor and the bipolar transistor, the BiCMOS switching delay is less sensitive to temperature variations compared to the CMOS switching delay.

To account for the effect at different temperatures in MEDICI simulation, temperature-dependent mobilities are used in the MEDICI simulation. Simulation results of the BiCMOS pull-up response and the BiCMOS pull-down response at –55, 25, and 150°C are given. Figures 6.28 and 6.29 show the BiNMOS pull-down response (output voltage and load current). A faster pull-down delay is obtained at a higher temperature. At higher temperatures, though the drain current decreases, the current gain and intrinsic carrier

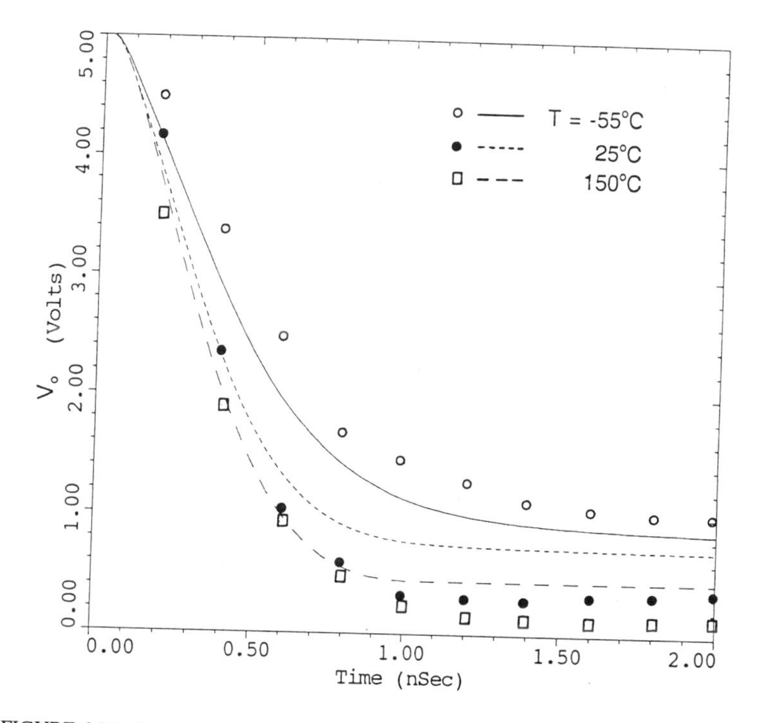

FIGURE 6.28. Output voltage waveforms at different temperatures (BiCMOS pull-down).

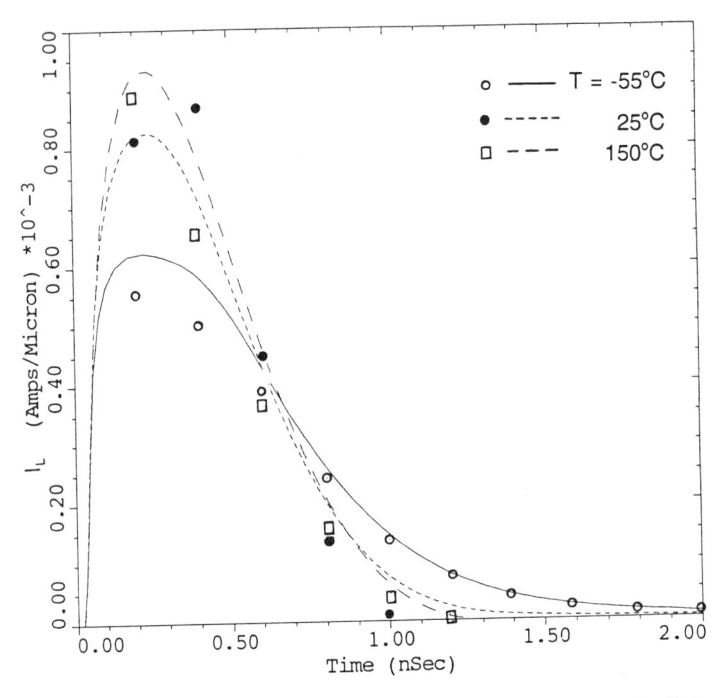

FIGURE 6.29. Output current waveforms at different temperatures (BiCMOS pull-down).

concentration of the BJT increase, which enhances the output current and decreases the propagation delay.

The band-gap narrowing in the emitter is a function of the emitter doping concentration. Decreasing emitter doping decreases band-gap narrowing in the emitter and, thus, decreases the temperature dependence of the forward current gain. For a BiNMOS driver, the load current that discharges the output capacitance consists of MOSFET drain current and BJT collector current. To make the BiNMOS delay insensitive to temperature, the load current should be made insensitive to temperature. As temperature increases, MOSFET drain current decreases (due to decreased MOSFET inversion mobility), and the collector current of the BJT increases (due to increased saturation current and forward current gain). Emitter doping can be adjusted to make the increase of collector current equal to the decrease of drain current, with increasing temperature.

Figure 6.30 shows the BiCMOS pull-down switching delay as a function of the emitter doping concentration for –55, 25, and 150°C. One observes that the BiCMOS pull-down switching delay becomes less dependent on temperature as emitter doping decreases, because reducing the emitter doping reduces the temperature dependence of the forward current gain of the BJT, and consequently reduces the temperature dependence of the switching delay. Such an approach can be used to design robust BiCMOS circuits that are insensitive to temperature variation.

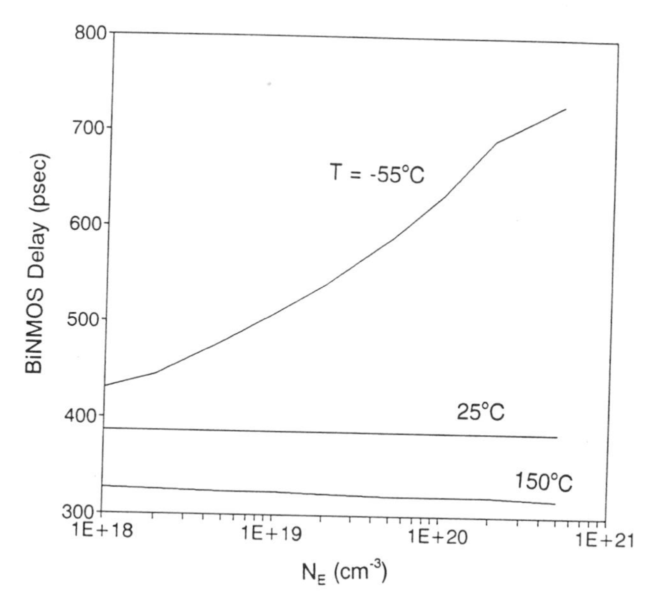

FIGURE 6.30. Pull-down delay versus emitter concentration at different temperatures.

6.7.2. *Geometrical Scaling*

The signal propagation delay due to large interconnect and off-chip capacitances is a major factor in determining the performance of VLSI circuits. In view of the driving capability of bipolar transistors, the BiCMOS buffer is usually adopted to drive heavy capacitive loads. It is apparent that the bigger the MOS transistors the higher the base and collector currents of the bipolar transistors, and thereby the shorter the delay. This will inevitably drive the bipolar transistors into the high-current regime where base pushout effect occurs.

In practice, however, the BiCMOS buffer is driven by an internal logic circuit. If the width of the MOS transistors increases, the gate capacitance increases and therefore the speed of the internal logic circuit decreases due to increased output capacitance. Similarly, if the BJT emitter length increases, the BJT collector current increases. This enhanced collector current charges the load capacitance faster to shorten the gate delay. However, the increased emitter length also increases the emitter–base junction and diffusion capacitances. The speed of the MOS transistors of the BiCMOS buffer decreases. In order to minimize the overall propagation delay, optimization of device dimensions for both the MOS and bipolar transistors is necessary.

Using the SPICE simulation, Fang *et al.* (1992) presented the overall BiCMOS delay as a function of sizes of bipolar and MOS transistors based on 2-μm emitter-width bipolar and 2-μm channel-length CMOS technologies (Fig. 6.31). It can be seen from Fig. 6.31 that increasing the width of MOS transistors reduces the delay of the BiCMOS buffer initially and then increases the delay at larger channel width. Increasing the length of

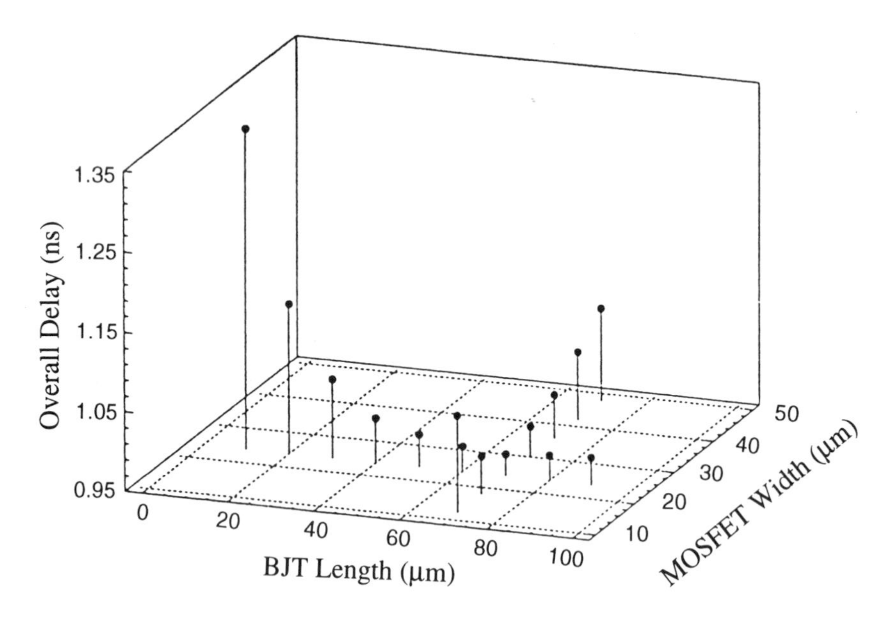

FIGURE 6.31. Delay as a function of BJT and MOS device sizes (Fang *et al.*, 1992).

bipolar transistors reduces the delay of the BiCMOS at first, but the delay increases when it passes a minimum. The optimized bipolar emitter length is about 70 μm, and the optimized MOS channel length is about 15 μm for an output load capacitance of 5 pF.

6.7.3. Supply Voltage Scaling

As the number of gates or transistors increases, power dissipation becomes a significant constraint. Furthermore, submicron MOS transistors encounter significant reliability problems such as hot electron degradation and device lifetime reduction. Thus, a supply voltage reduction, to 3.3 V for example, is necessary.

Experimental dependence of propagation delay times of BiCMOS gates on power supply voltage have been reported (Momose *et al.*, 1991). CMOS and BiCMOS ring oscillators with 41-state inverters were fabricated on the same wafers. Speed performance was evaluated and compared. Channel widths for NMOS and PMOS transistors are 9 and 18 μm, respectively. The channel length is 0.6 μm for NMOS and 0.8 μm for PMOS. The channel widths and lengths of NMOS transistors for sinking base charges are both 3 μm.

Figure 6.32 shows the comparison of propagation delay times of CMOS and BiCMOS ring oscillators. Both supply voltage and load capacitances were varied. At 5 V, the propagation delay times of the CMOS ring oscillators with and without a load of 1.3 pF were 110 and 550 ps, respectively. For the BiCMOS case, shown with closed circles, propagation delay times were 160 and 220 ps, respectively.

On the other hand, at lower supply voltages, a rapid degradation in speed was observed with a BiCMOS supply voltage of about 3.0 V, while CMOS maintains its speed

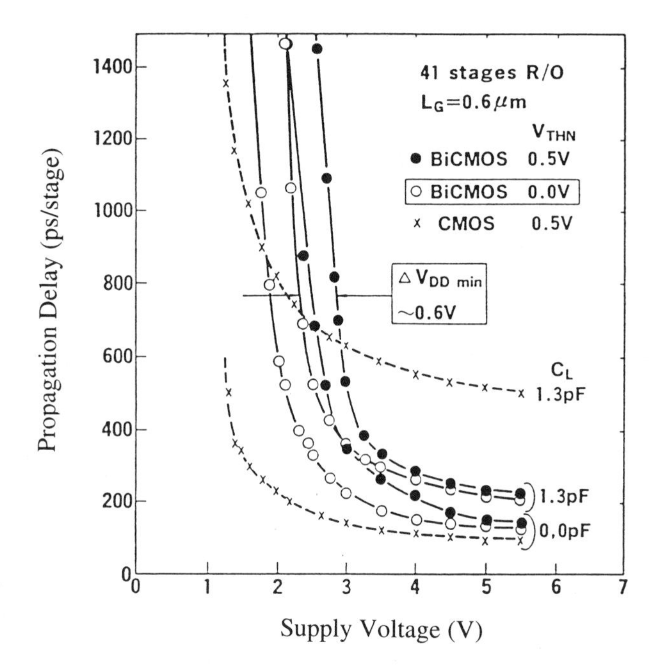

FIGURE 6.32. Delay versus supply voltage (Momose *et al.*, 1991).

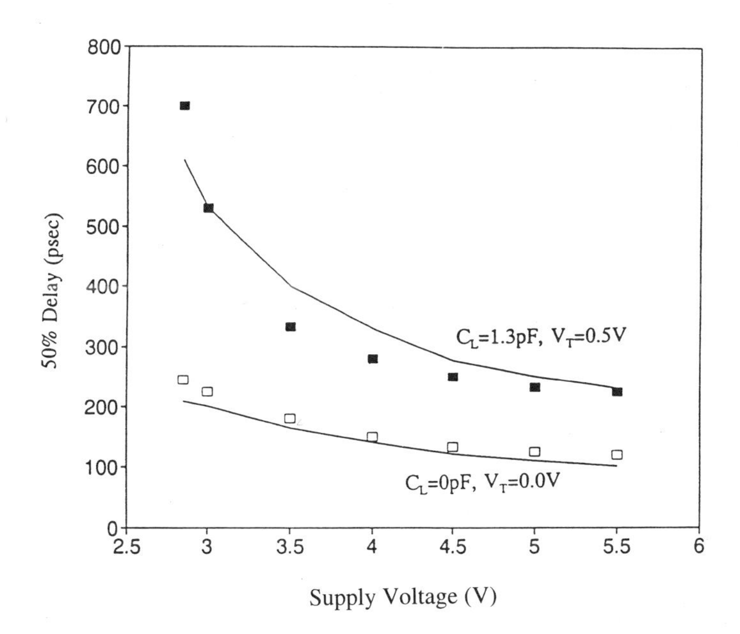

FIGURE 6.33. Comparisons of delay between the model predictions and experiments at different supply voltages.

at less than 2.5 V. It was also verified that this phenomenon in BiCMOS does not change appreciably by changing device dimensions, such as channel length and emitter length, except for the threshold voltage of the NMOS transistors. The minimum operating voltage at which the propagation delay time increases rapidly is approximately 3.0 V for a threshold voltage of 0.5 V, and 2.4 V for a threshold voltage of 0.0 V. Thus, a lower threshold voltage is the most effective way to improve speed at 3.3-V operation, if the leakage current is permissible.

MEDICI simulation is used to understand why the delay increases significantly at low supply voltages. As supply voltage decreases, the drain current supplied by the MOSFET, and thus the base current of the BJT, decreases. In addition, the quasi-saturation effect of the BJT increases when the supply voltage is decreased (or scaled). The time required to turn the BJT on and off increases, and the driving capability of the BJT decreases, which results in increased BiCMOS switching delay. The model's prediction for the BiCMOS switching delay when the power supply voltage is scaled from 5.5 to 2.75 V agrees with experimental data (Fig. 6.33).

6.8. HOT ELECTRON RELIABILITY OF BiCMOS DEVICES

When the input of a bipolar transistor in the output section of a typical BiCMOS gate circuit makes the transition from high to low, the output also goes from high to low after a certain delay. During this delay period, the emitter–base junction of the upstream bipolar transistor is reverse biased. The magnitude of this reverse bias is almost equal to the supply voltage, so junction breakdown occurs between the emitter and base, allowing a large reverse current to flow. In addition, hot carriers inject into the SiO_2 layer near the emitter–base junction. This increases surface states and changes the base current of the bipolar transistor. In addition, avalanche breakdown in the collector–base junction of the bipolar transistor generates excess carriers that reduce BJT base current. This decreases the logic swing and increases the standby current of the BiCMOS circuit (Lu and Chuang, 1992).

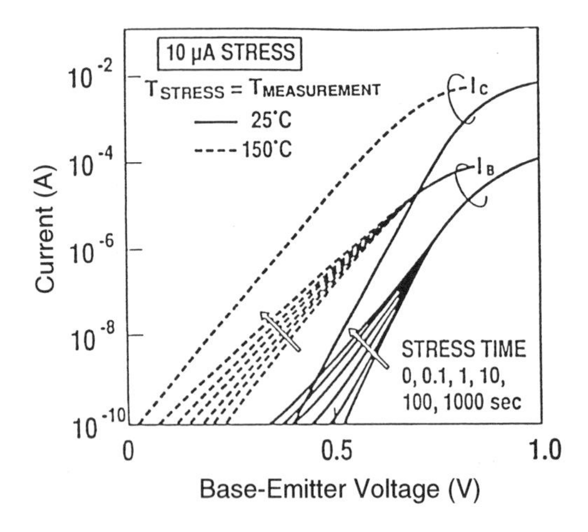

FIGURE 6.34. Base and collector currents versus base–emitter voltage under different stress conditions (Momose and Iwai, 1994).

Effects of hot electron stress on current and current gain of the bipolar transistor are presented. Figures 6.34 and 6.35 show the Gummel plot and current gain of the BJT at 25 and 150°C for different stress times under 10 μA stress, respectively. Hot electron stress increases the base current at low to moderate base–emitter voltages. This is because hot electrons increase the surface states, which increases the recombination current at the emitter–base surface. Since the collector current is determined by the minority transport in the quasi-neutral base, it is independent of hot electron stress. The increased base current degrades the current gain of the bipolar transistor significantly at low to moderate base–emitter bias. The temperature dependence of the base current degradation is plotted in Fig. 6.35. The stress and measurement temperatures are equal, and $\Delta I_B/I_B$ is evaluated 5000 s after stress. Note that the worst degradation occurs at around 50°C for both 10-μA and 0.1-μA stress cases. When either T_{stress} or T_{measure} is above 50°C, thermal annealing effects reduce the degradation (Momose and Iwai, 1994).

The comparison between hot electron stress and ionizing radiation stress has been studied (Kosier *et al.*, 1995). Figure 6.36 shows the normalized current gain versus base–emitter voltage for increasing levels of total ionizing dose, and Figure 6.37 displays the normalized current gain versus base–emitter voltage for increasing levels of hot carrier stress time. The current gain degrades substantially for both types of stress, and the degradation is most severe at lower values of base–emitter voltage. Note that the current gain degradation tends to saturate for large values of total ionizing doses, but shows no tendency to saturate for the stress times shown here. Although both types of stress lead to qualitatively similar changes in the current gain of the device, the physical mechanisms responsible for the degradation are quite different. In hot carrier stress, the damage is localized near the emitter–base junction, which causes the excess base current to have an ideality factor of 2. For ionizing radiation stress, the damage occurs along all

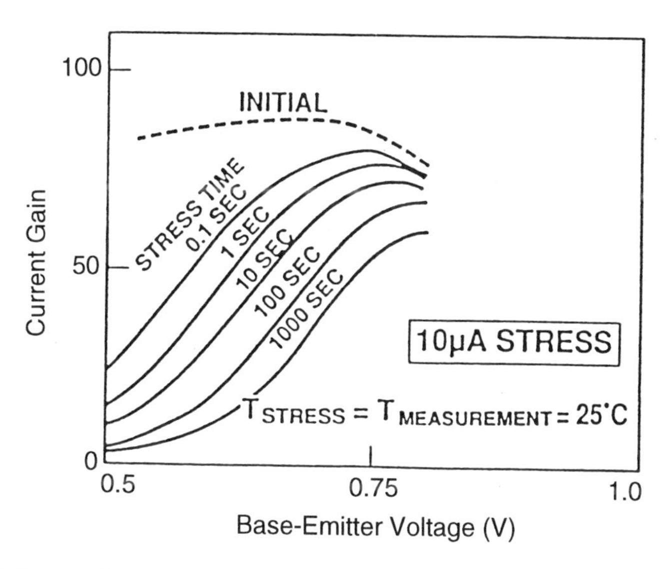

FIGURE 6.35. Current gain versus base–emitter voltage under different stress times (Momose and Iwai, 1994).

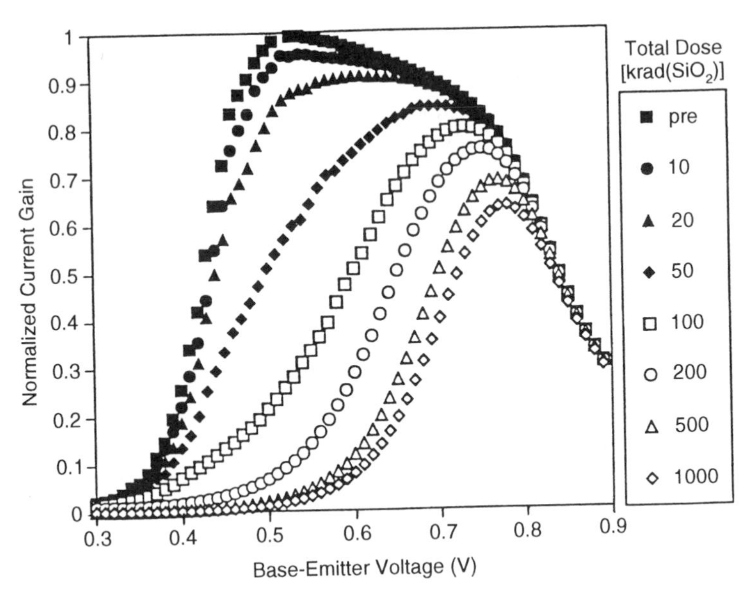

FIGURE 6.36. Current gain as a function of radiation (Kosier *et al.*, 1995).

oxide–silicon interfaces, which causes the excess base current to have an ideality factor between 1 and 2 for low total doses of ionizing radiation, but an ideality factor of 2 for large total doses. The different physical mechanisms that apply for each type of stress imply that improvement in resistance to one type of stress does not necessarily imply improvement in resistance to the other type of stress. For example, substantial improve-

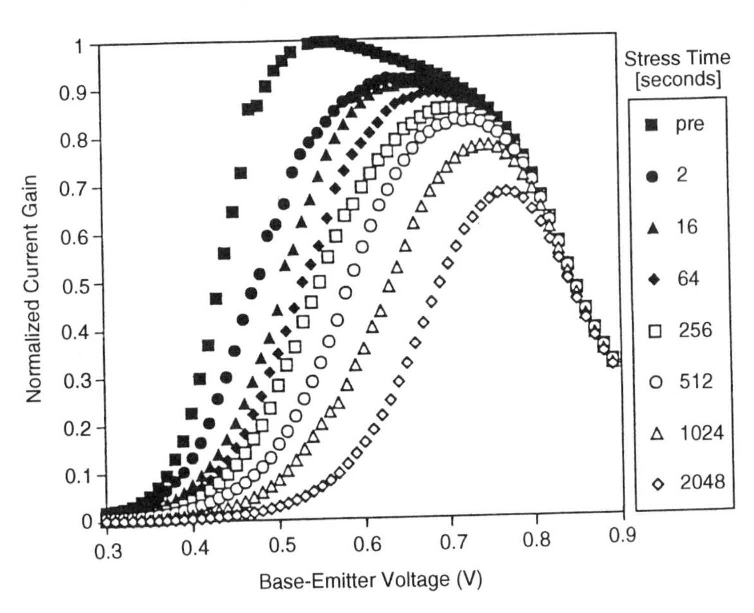

FIGURE 6.37. Current gain as a function of stress time (Kosier *et al.*, 1995).

ment in the radiation hardness of BJTs for surface recombination can be expected by increasing the surface doping of the intrinsic base. For hot carrier stress, the performance of the device depends on the details of the stress conditions. Increasing the doping in the intrinsic base leads to larger surface electric fields for a given reverse-bias voltage, which leads to worse performance under constant-current stress in the avalanche regime.

EXAMPLE 6.1

```
 COMMENT  BiPMOS transient simulation
 COMMENT  Specify mesh points
    MESH  SMOOTH=1
  X.MESH  WIDTH=1.5 H1=0.05
  X.MESH  WIDTH=0.2 H1=0.1
  X.MESH  WIDTH=1.2 H1=0.1
  Y.MESH  N=1 L=-0.02
  Y.MESH  N=6 L=0.0
  Y.MESH  DEPTH=1.0 H1=0.05
 COMMENT  Eliminate unnecessary mesh points
  ELIMIN  COLUMNS Y.MIN==0.5 X.MAX=1.6
  ELIMIN  ROWS Y.MIN=0.5 X.MAX=1.6
 COMMENT  Specify oxide and silicon regions
  REGION  NUM=1 X.MAX=1.6 Y.MIN=0.0 SILICON
  REGION  NUM=2 X.MIN=1.8 Y.MIN=0.0 SILICON
  REGION  NUM=3 X.MAX=1.6 Y.MAX=1.8 OXIDE
  REGION  NUM=4 X.MIN=1.6 X.MAX=0.0 OXIDE
  REGION  NUM=5 X.MIN=1.8 Y.MAX=0.0 OXIDE
  REGION  NUM=6 X.MAX=0.0 X.MIN=1.8 X.MAX=2.4 POLYSILI
 COMMENT  Electrodes: #1=Source #2=Gate #3=Drain #4=Emitter #5 Base #6=Collector
       +  #7=Substrate
  ELECTR  NUM=1 X.MAX=0.3 IY.LO=6 IY.HI=6
  ELECTR  NUM=2 X.MIN=0.45 Y.MIN=1.15 TOP
  ELECTR  NUM=3 X.MIN=1.3 X.MAX=1.6 IY.LO=6 IY.HI=6
  ELECTR  NUM=4 X.MIN=1.8 IY.MAX=2.4 IY.LO=6 IY.HI=6
  ELECTR  NUM=3 X.MIN=2.7 X.MAX=3.0 TOP
  ELECTR  NUM=6 X.MIN=1.8 BOTTOM
  ELECTR  NUM=1 X.MIN=1.6 BOTTOM
 COMMENT  Specify impurity profiles
 COMMENT  n-p-n BJT doping profile:
 PROFILE  REGION=2 N-TYPE N.PEAK=1E16 UNIFORM OUT.FILE=BPDOPF
 PROFILE  REGION=2 P-TYPE N.PEAK=1E19 Y.JUNC=0.273 X.PEAK=1.8
       +  WIDTH=1.0 XY.RAT=1
 PROFILE  REGION=2 P-TYPE N.PEAK=7E17 Y.PEAK=0.14 Y.CHAR=0.075
       +  X.PEAK=1.8 WIDTH=1.0 XY.RAT=1
 PROFILE  REGION=2 N-TYPE N.PEAK=2E20 Y.PEAK=0.0 Y.JUNC=0.15
       +  X.PEAK=1.8 WIDTH=0.6 XY.RAT=0.75
 PROFILE  REGION=2 N-TYPE N.PEAK=5E19 Y.PEAK=1.0 Y.CHAR=0.14
 COMMENT  p-MOSFET doping profile:
 PROFILE  REGION=1 N-TYPE N.PEAK=3E15 UNIFORM
```

```
   PROFILE  REGION=1 N-TYPE N.PEAK=2E16 Y.PEAK=0.0 Y.CHAR=0.1
   PROFILE  REGION=1 P-TYPE N.PEAK=2E20 Y.JUNC=0.2 X.PEAK=0.0
         +  WIDTH=0.3 XY.RAT=0.75
   PROFILE  REGION=1 P-TYPE N.PEAK=2E20 Y.JUNC=0.2 X.PEAK=1.3
         +  WIDTH=0.3 XY.RAT=0.75
 INTERFACE  QF=1E10
   COMMENT  Regrid on doping
    REGRID  DOPING LOG RATIO=3 SMOOTH=1 IN.FILE=BPDOPF
         +  OUT.FILE=BPMESH
  MOBILITY  REGION=6 CONC=7E19 HOLE FIRST LAST
  MATERIAL  REGION=6 taup0=8E-8
  MOBILITY  gsurfn=0.9 gsurfp=0.9
   COMMENT  Specify models to use
    MODELS  CONMOB CONSRH AUGER BGN PRPMOM FLDMOB
   COMMENT  Symbolic factorization
      SYMB  CARRIERS=0 NEWTON
    METHOD  ITLIMIT=30 X.NOR AUTONR
     SOLVE  INIT V1=5 V2=5 V6=5
   CONTACT  NUM=4 CAPACITANCE=1E-13
   CONTACT  NUM=3 CAPACITANCE=1E-16
     SOLVE  V2=0 TSTEP=3E-12 ENDRAMP=50E-12 TSTOP=1000E-12
   PLOT.1D  Y.AXIS=V4 X.AXIS=Time
```

REFERENCES

Alvarez, A. R., P. Meller, and B. Tien (1984a), *Int. Electron Devices Meeting*, 420.

Alvarez, A. R., P. Meller, and B. Tien (1984b), *Int. Electron Devices Meeting*, 420.

Alvarez, A. R. and D. W. Schucker (1988), *Cust. Int. Conf.*, 22.1.1.

Bastani, B., C. Lage, L. Wong, J. Small, R. Lahri, L. Bouknight, T. Bowman, J. Manoliu, and P. Tuntasood (1987), *VLSI Technol. Symp.*, 41.

Baze, M. P., R. E. Plaag, and A. H. Johnson (1989), *IEEE Trans. Nucl. Sci.* **NS-36**, 1858.

Boesch, H. Edwin, Jr. (1988), *IEEE Trans. Nucl. Sci.* **NS-35**, 1160.

Chen, Y.-W. and J. B. Kuo (1992), *IEEE Trans. Electron Devices* **ED-39**, 348.

Davis, S. (1979), *Electron. Design News*, 51.

Enlow, E. W., R. L. Pease, and W. Combs (1991), *IEEE Trans. Nucl. Sci.* **NS-38**, 1342.

Fang, W., A. Brunnschweiler, and P. Ashburn (1992), *IEEE J. Solid-State Circuits* **SC-27**, 191.

Galloway, K. F., M. Gaitan, and T. J. Russell (1984), *IEEE Trans. Nucl. Sci.* **NS-31**, 1497.

Getreu, I. E. (1987), *Modeling the Bipolar Transistor* (Elsevier, New York).

Greeneich, E. W. and K. L. McLaughlin (1989), *IEEE J. Solid-State Circuits* **SC-23**, 558.

Grove, A. S. (1979), *Physics and Technology of Semiconductor Devices* (Wiley, New York).

Higuchi, H., G. Kitsukawa, T. Ikeda, and Y. Nishio (1984), *Int. Electron Devices Meeting*, 694.

Hotta, T., I. Masuda, H. Maejima, and A. Hotta (1986), *Int. Solid-State Circuits Conf.*, 190.

Kosier, S. L., A. Wei, R. D. Schrimpf, D. M. Fleetwood, M. D. DeLaus, R. L. Pease, and W. E. Combs (1995), *IEEE Trans. Electron Devices* **ED-42**, 436.

Kuo, J. B., G. P. Rossel, and R. W. Dutton (1989), *IEEE Trans. Computer-Aided Design* **CAD-8**, 929.

Lin, H. G., J. C. Ho, R. R. Iyer, and K. Kwong (1969), *IEEE Trans. Electron Devices* **ED-16**, 945.

Lu, P. F. and C. T. Chuang (1992), *IEEE Trans. Electron Devices* **ED-39**, 1902.

Miyamoto, J., S. Saitoh, H. Momose, H. Shibata, K. Kanzake, and S. Kohyama (1984), *Int. Electron Devices Meeting*, 63.

Momose, H. S. and H. Iwai (1994), *IEEE Trans. Electron Devices* **ED-41**, 978.

Momose, H., Y. Unno, and T. Maeda (1991), *IEEE Trans. Electron Devices* **ED-38**, 566.
Nowlin, R. N., E. W. Enlow, R. D. Schrimpf, and W. E. Combs (1992), *IEEE Trans. Nucl. Sci.* **NS-39**, 2026.
Pease, R. L., W. Combs, and S. Clark (1992), *IEEE Trans. Nucl. Sci.* **NS-39**, 352.
Pease, R., D. Emily, and H. E. Boesch, Jr. (1985), *IEEE Trans. Nucl. Sci.* **NS-32**, 3946–3952.
Pease, R. L., R. M. Turfler, D. Platteter, D. Emily, and R. Blice (1983), *IEEE Trans. Nucl. Sci.* **NS-30**, 4216.
Phanse, A. (1994), *Analysis of BiCMOS Switching*, M.S. thesis (University of Central Florida, Orlando).
Phanse, A., D. Sharma, A. Mallik, and J. Vasi (1993), *J. Appl. Phys.* **74 (1)**, 757.
Plummer, J. D. and J. D. Meindl (1976), *J. Solid State Circuits* **SC-11**, 809.
Polinsky, M. A., O. H. Schade, and J. P. Keller (1973), *Int. Electron Devices Meeting*, 229.
Rosseel, G. P. and R. W. Dutton (1989), *IEEE J. Solid-State Circuits* **SC-24**, 90.
Tamba, N., S. Miyaoka, M. Odaka, M. Hirao, K. Ogiue, K. Tamada, T. Ikeda, H. Higuchi, and H. Uchida (1988), *Int. Solid-State Circuits Conf.*, 184.
Whittier, R. J. and D. A. Tremere (1969), *IEEE Trans. Electron Devices* **ED-16**, 39.
Wilson, C. L. and J. L. Blue (1984), *IEEE Trans. Nucl. Sci.* **NS-31**, 1448.
Yamaguchi, Ken (1983), *IEEE Trans. Electron Devices* **ED-30**, 659.
Yuan, J. S. (1992), *IEEE Trans. Electron Devices* **ED-39**, 587.
Zupac, D., K. F. Galloway, R. D. Schrimpf, and P. Augier (1992), *Appl. Phys. Lett.* **60(25)**, 3156.

7

Metal–Semiconductor Field-Effect Transistors

The concept of a metal–semiconductor junction field-effect transistor (MESFET) is very similar to that of a junction field-effect transistor (JFET) discussed in Chapter 4. The geometry of the conducting channel in a MESFET is confined by the top and bottom gate depletion regions, which are controlled by the gate and drain voltages. Thus, before the two depletion regions touch each other (channel pinch-off), the MESFET, in essence, acts like a variable resistance. Beyond the channel pinch-off, however, the drain current increases only slowly with increasing drain voltage, and the device is operated in the saturation region. Unlike the JFET, which is a bulk device, the MESFET is a surface device; i.e., one of the boundaries of the primary region (channel region) is the interface between the metal and semiconductor. Because many trapping states can be present at the metal–semiconductor interface, current transport in the MESFET is fluctuated by the capture and release of free carriers at these surface states. As a result, the MESFET has a higher noise level than the JFET. The major difference between the MESFET and the metal–oxide semiconductor field-effect transistor (MOSFET), which is also a surface device, is that a MOSFET is normally off until a voltage greater than the threshold is applied to the gate, whereas the MESFET is normally on unless a large reverse voltage is applied to the gate to cut off the conducting channel. If the channel layer (or epilayer) thickness is made very thin, however, the MESFET will be off unless a forward voltage is applied to the gate. Here, the normally on MESFET will be emphasized.

MESFETs have a great deal of potential for further advances due to the following reasons:

1. In addition to silicon, a variety of semiconductor materials (i.e., GaAs and InP) with majority carrier transport properties superior to silicon are applicable for MESFET fabrication.
2. Further miniaturization to submicron dimensions can be realized in most MESFET structures.
3. Monolithic integration of circuits on semi-insulting substrates enable device isolation with low parasitic capacitances, low-loss interconnections, and high packing density.

This chapter focuses on GaAs MESFETs, which has clearly emerged as the leading MESFET technology since the early 1980s. Since it is difficult to grow a stable native oxide with small interface trap density at the interface, and since high-quality aluminum Schottky-barrier gates can be formed on a GaAs surface, MESFETs fabricated with GaAs have been used widely in microwave and high-speed digital applications.

Because MESFETs utilize Schottky contacts to control the charge transport through the channel, the concept and modeling of such a contact are discussed first.

7.1. SCHOTTKY DIODE

7.1.1. Basic Concept

When putting a metal on top of a semiconductor, a Schottky contact or an ohmic contact is formed, depending on doping concentration, interface properties, and the difference between metal and semiconductor work functions. Theoretically, if the metal work function is larger than the semiconductor work function, a Schottky contact is formed; otherwise an ohmic contact is created. An ohmic contact can also be formed, even with the metal work function larger than the semiconductor work function, by inserting a heavily doped semiconductor layer adjacent to the junction. This heavily doped layer increases the junction barrier height and reduces the depletion region thickness, which then increases the free-carrier tunneling probability across the very thin potential barrier. As a result, the tunneling current from semiconductor to metal is significantly increased, and the contact resistance is greatly reduced (Shur, 1990; Sze, 1983).

Since the Schottky contact is used to form MESFETs, the basic concept of such a contact will be addressed. The Schottky diode is electrically similar to the abrupt one-sided *p–n* junction diode, but there is a major difference between the two: the Schottky diode, unlike the *p–n* junction diode, in most cases operates as a majority carrier device in which the minority carrier transport and storage are nearly absent. As a consequence, Schottky devices have a high switching speed and are suitable for high-speed applications such as digital switches and microwave detectors (Cowley and Sze, 1965).

Consider an ideal metal–semiconductor (*n*-type) Schottky junction at thermal equilibrium. When the metal and semiconductor are separated, they have different work functions (ϕ_m and ϕ_s) and, thus, have Fermi level (E_{Fm} and E_{Fs}) locations with respect to the local vacuum level. When the two materials contact each other, the Fermi levels must align. Therefore, we find that there is a work function difference $\phi_m - \phi_s$ across the interface (Fig. 7.1a). This potential difference can be regarded as the junction built-in potential for the semiconductor, and a band bending in the energy diagram is observed (Fig. 7.1a), where

$$\phi_i = \phi_m - \phi_s \tag{7.1.1}$$

is the junction built-in potential, X_d is the edge of the surface region (or the depletion region), and

a

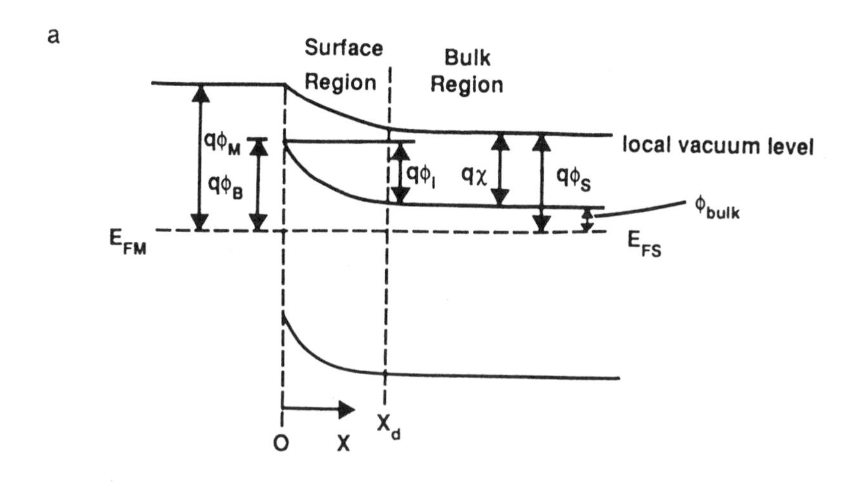

b

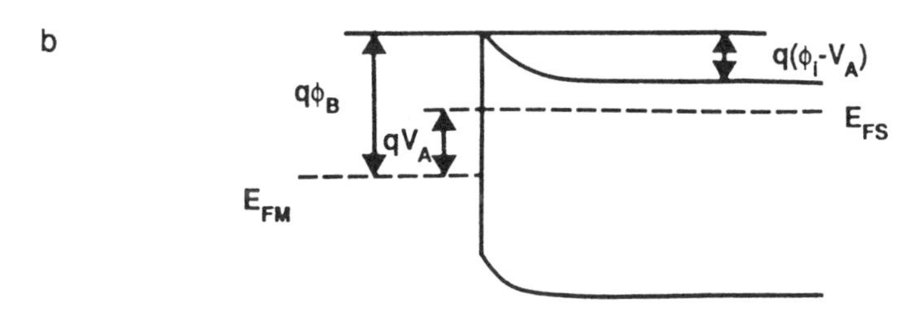

c

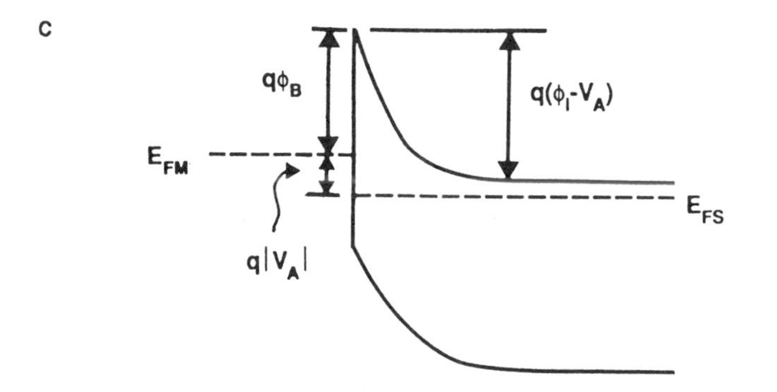

FIGURE 7.1. Schottky-contact energy-band diagram at (a) thermal equilibrium, (b) forward-bias condition, (c) reverse-bias condition.

$$\phi_B = \phi_m - \chi \tag{7.1.2}$$

is the barrier potential, where χ is the electron affinity in the semiconductor ($\chi = 4.05$ V for Si). Note that (Fig. 7.1a)

$$\phi_s = \chi + \phi_{\text{bulk}} \tag{7.1.3}$$

where $\phi_{\text{bulk}} = E_C - E_{Fs}$ is the bulk potential:

$$\phi_{\text{bulk}} = 0.5E_G - kT/q \ln(N_D/n_i) \tag{7.1.4}$$

Here E_G is the semiconductor band gap.

When a voltage V_A is applied to the metal contact, the two Fermi levels will then be separated by qV_A. The barrier potential in the semiconductor is also increased or decreased by this quantity. Figures 7.1b,c show the energy-band diagrams for the Schottky junction under forward- ($V_A > 0$) and reverse-bias ($V_A < 0$) conditions, respectively.

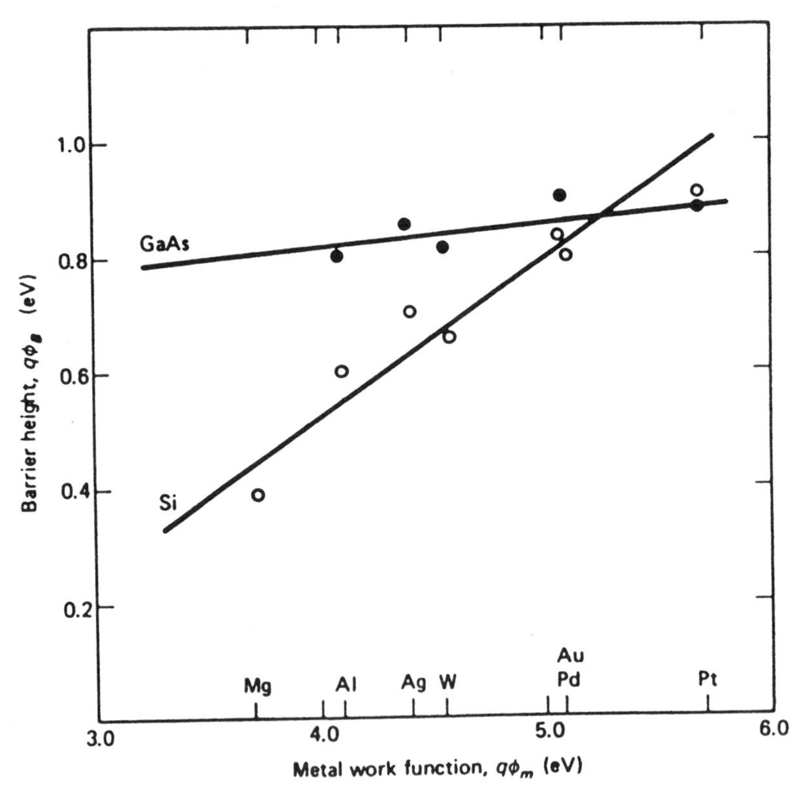

FIGURE 7.2. Schottky-barrier potential versus the metal work function for junctions formed on silicon and GaAs (Cowley and Sze, 1965).

One important implicit assumption is that the system is in quasi-equilibrium, that is, almost at equilibrium even though currents are flowing. The quasi-equilibrium condition was implicitly invoked in the energy-band diagrams in Figs. 7.1b,c, in which the Fermi energies are assumed to be flat in the metal–semiconductor contact under forward- and reverse-bias conditions. Often, the quasi-equilibrium approximation suffices under low-bias conditions and becomes less valid otherwise.

Figure 7.2 plots the Schottky-barrier potential as a function of metal work function for both GaAs and Si (Liao, 1985). From this, together with the doping concentration, one can calculate ϕ_i and ϕ_B from (7.1.1) and (7.1.2), respectively.

7.1.2. *Effect of Interface States*

Experimentally, it is found that the barrier height ϕ_B is a less sensitive function of ϕ_m than (7.1.2) would suggest. An explanation of this weak dependence on ϕ_m was put forward by Bardeen (1947), who suggested that the discrepancy may be due to the effect of interface states. Consider the case in which the metal and semiconductor are separated by a thin insulating layer (Fig. 7.3) and a continuous distribution of surface states is present at the interface, characterized by a neutral level ϕ_0. The number of surface states depends strongly on processing conditions as well as on the properties of the semiconductor surface before the metal is deposited.

If a large number of surface states are present near the metal–semiconductor interface, then the Fermi level is pinned at the interface (Fermi level pinning). This happens because, when the state density is large, a negligible movement of the Fermi level at the semiconductor surface would transfer sufficient charge to equalize the Fermi levels. When the Fermi level is pinned, the barrier height becomes (Crowell *et al.*, 1966)

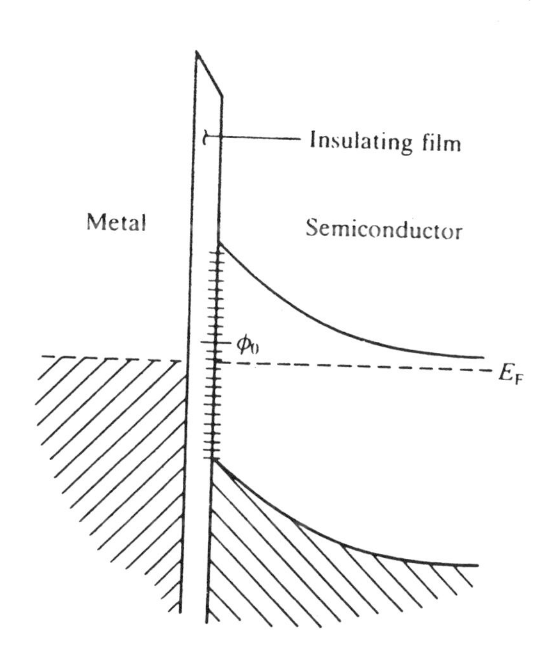

FIGURE 7.3. Schottky contact with an insulating interfacial layer and surface states.

$$q\phi_B = c_2 q(\phi_m - \chi) + (1 - c_2)(E_G - q\phi_0) \tag{7.1.5}$$

where

$$c_2 = \varepsilon_i/(\varepsilon_i + q^2 D_s \delta) \tag{7.1.6}$$

Here $q\phi_0$ is the distance between the semiconductor Fermi energy and the valence-band edge when the semiconductor is not covered by a metal, ε_i is the dielectric permittivity of the interfacial layer containing surface states, δ is the thickness of the interfacial layer, and D_s is the surface-state density. For silicon, as with germanium and gallium arsenide, the quantity $q\phi_0$ is found experimentally to be about (1/3) E_G. For a metal–semiconductor contact having a very large D_s, c_2 approaches zero and

$$q\phi_B \approx E_G - q\phi_0 \tag{7.1.7}$$

If the interface-state density is negligibly small, then (7.1.5) reduces to the simple Schottky theory:

$$\phi_B = \phi_m - \chi \tag{7.1.8}$$

If we assume that the occupation of the interface states is determined by the Fermi level in the metal, which is true for very thin interfacial layers, then the flat-band barrier height ϕ_{B0} (Fig. 7.4) is defined as

$$\phi_{B0} = \gamma(\phi_m - \phi_s) + (1 - \gamma)(E_G/q - \phi_0) \tag{7.1.9}$$

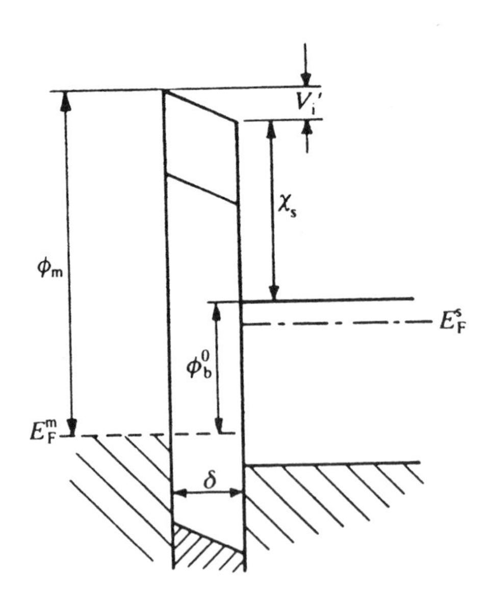

FIGURE 7.4. Schottky contact with insulating interfacial layer under flat-band condition.

$$\gamma = \varepsilon_i/(\varepsilon_i + q\delta D_s) \tag{7.1.10}$$

7.1.3. *Bias Dependence of the Barrier Height*

If there is no interfacial layer, the barrier height is independent of any electric field and bias that may exist inside the semiconductor. However, when there is an interfacial layer, the electric field in the interfacial layer modifies the barrier height. The charge Q_{ss} in the interface states is

$$Q_{ss} = qD_s(\phi_B + \phi_0 - E_G/q) \tag{7.1.11}$$

Gauss's theorem gives

$$\varepsilon_i E_i - \varepsilon E_{\max} = Q_{ss} \tag{7.1.12}$$

where $E_{\max}$ is the value of the semiconductor electric field at $x = 0$. This leads to (Rhoderick and Williams, 1988)

$$\phi_B = \phi_{B0} - \alpha E_{\max} \tag{7.1.13}$$

where

$$\alpha = \delta\varepsilon/(\varepsilon_i + q\delta D_s) \tag{7.1.14}$$

Since the Schottky junction can be approximated as a one-sided *p–n* junction, the space-charge-region thickness X_d can be modeled with the conventional depletion approximation (Mulle and Kamins, 1986) as

$$X_d = \sqrt{\frac{2\varepsilon(\phi_i - V_A)}{qN_D}} \tag{7.1.15}$$

The maximum electric field $E_{\max}$ at the metal–semiconductor interface ($x = 0$) and the electrostatic potential $\phi_{\text{scr}}(x)$ in the space-charge region can also be derived from Poisson's equation:

$$E(\mathrm{x} = 0) = E_{\max} = -qN_D X_d/\varepsilon \tag{7.1.16}$$

$$\phi_{\text{scr}}(\mathrm{x}) = (qN_D/\varepsilon)(X_d - 0.5x)x \tag{7.1.17}$$

The space-charge Q_s per unit area in the semiconductor is

$$Q_s = qN_D X_d = \sqrt{2q\varepsilon N_D(\phi_i - V_A)} \tag{7.1.18}$$

Under small-signal ac conditions, the junction shows a capacitive behavior that can be derived by using (7.1.18):

$$C_{\text{scr}} = \frac{\partial Q_S}{dV_A} = \sqrt{\frac{q\varepsilon N_D}{2(\phi_i - V_A)}} = \frac{\varepsilon}{X_d} \tag{7.1.19}$$

where C_{scr} is the metal–semiconductor space-charge-region capacitance.

7.1.4. Current–Voltage Characteristics

Two basic mechanisms have been proposed to describe current transport in the metal–semiconductor junction. Before electrons can be emitted over the barrier into the metal, they must first be transported from the bulk of the semiconductor to the interface. In traversing the semiconductor depletion region, their motion is governed by the usual mechanisms of diffusion and drift. On the other hand, when the electrons arrive at the interface, their emission into metal is determined by the rate of transfer of electrons across the interface. These two processes are effectively in series, and the current is determined predominantly by whichever causes the larger hindrance to the flow of electrons. According to the diffusion theory of Wagner (1931) and Schottky and Spenke (1939), the first of these two processes is the limiting factor, whereas according to the thermionic-emission theory of Bethe (1942), the second is more important.

The difference between the two theories can be clearly illustrated by the quasi-Fermi level for electrons (Fig. 7.5). According to diffusion theory, the concentration of electrons in the semiconductor adjacent to the interface is not altered by the applied bias. This is equivalent to assuming that at the interface the quasi-Fermi level in the semiconductor coincides with the Fermi level in the metal. In this case, the quasi-Fermi level decreases through the depletion region (Fig. 7.5). From the thermionic-emission point of view, the electrons emitted from the semiconductor into the metal are not in thermal equilibrium with the electrons in the metal. They can be described as "hot" electrons. As they penetrate the metal, the hot electrons lose energy by collisions with electrons in the metal. Their quasi-Fermi level falls until it ultimately coincides with the metal quasi-Fermi level. This

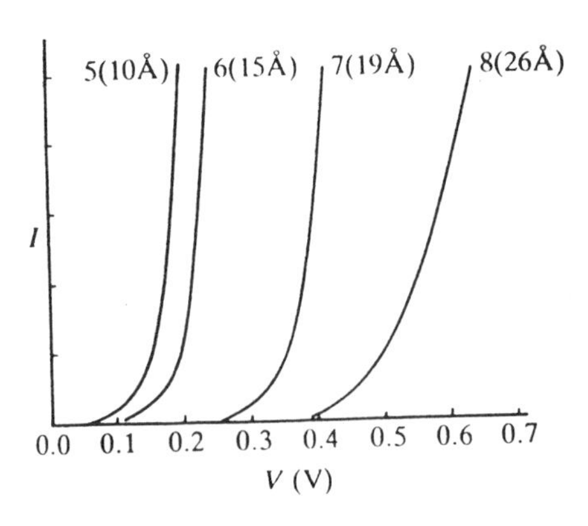

FIGURE 7.5. Forward current–voltage characteristics of Schottky diode with different oxide interfacial layer thicknesses (Rhoderick and Williams, 1988).

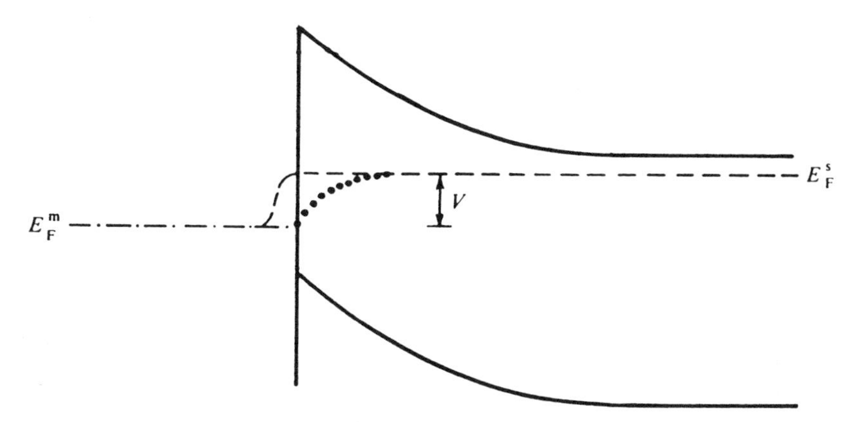

FIGURE 7.6. Electron quasi-Fermi level in a forward-biased Schottky junction according to diffusion theory (dotted lines) and thermionic-emission theory (dashed lines).

implies that the electron quasi-Fermi level at the interface does not have to coincide with the metal Fermi level, and the quasi-Fermi level remains flat through the depletion region (Fig. 7.6).

7.1.4.1. Diffusion Theory The dependence of current on applied voltage in a metal–semiconductor junction can be derived by integrating the carrier drift and diffusion in the depletion region from both sides of the contact. Assuming that the minority carrier current is negligibly small compared to the majority carrier current, the one-dimensional electron current density J is

$$J = q\mu_n nE + qD_n \frac{dn}{dx} = qD_n \left(\frac{-qn}{kT} \frac{d\phi}{dx} + \frac{dn}{dx} \right) \tag{7.1.20}$$

The electron concentration is defined by the electron quasi-Fermi level E_{Fn} as

$$n = N_C e^{-q(E_C - E_{Fn})/kT} \tag{7.1.21}$$

The current density can be written as

$$J = q\mu_n n \frac{dE_{Fn}}{dx} \tag{7.1.22}$$

Combining these two equations and integrating the resulting equation from $x = 0$ to $x = X_d$ gives (Rhoderick and Williams, 1988)

$$J = qN_C E_{\max} e^{-q\phi_B/kT} (e^{qV_A/kT} - 1) \tag{7.1.23}$$

7.1.4.2. Thermionic-Emission Theory In Bethe's thermionic-emission theory, the effect of drift and diffusion in the depletion region is assumed to be negligible. This, in turn, implies that the concentration of electrons on the semiconductor side of the interface is increased by a factor of $\exp(qV_A/kT)$ when a bias voltage is applied. Thus, in the semiconductor,

$$n = N_C e^{-q(\phi_B - V_A)/kT} \tag{7.1.24}$$

Since the number incident of electrons per second is given by elementary kinetic theory to be $nv_{th}/4$ (v_{th} is the average thermal velocity of electrons), the current density due to electrons passing from the semiconductor to the metal is

$$J_{sm} = \frac{qN_C v_{th}}{4} e^{-q(\phi_B - V_A)/kT} \tag{7.1.25}$$

There is also a flow of electrons from the metal into the semiconductor. Such a flow gives

$$J_{ms} = \frac{qN_C v_{th}}{4} e^{-q\phi_B/kT} \tag{7.1.26}$$

Hence,

$$J = J_{sm} - J_{ms} = A^* T^2 e^{-q\phi_B/kT}(e^{qV_A/kT} - 1) \tag{7.1.27}$$

where

$$A^* = 4\pi m^* qk^2/h^3 \tag{7.1.28}$$

Note that ϕ_B is, in general, affected by the bias. Equation (7.1.27) is often written in the form

$$J = J_S'(e^{qV_A/n_L kT} - 1) \tag{7.1.29}$$

$$J_S' = A^* T^2 e^{-q\phi_{B0}/kT} \tag{7.1.30}$$

where ϕ_{B0} is ϕ_B at zero bias and n_L is an empirical factor (called the ideality factor), which can be modeled as

$$\frac{1}{n_L} = 1 - \frac{d\phi_B}{dV_A} \tag{7.1.31}$$

Crowell and Beguwala (1971) have calculated the position of the quasi-Fermi level at the interface and have concluded that, for semiconductors with a fairly high mobility such as Ge, Si, and GaAs, the variation of the quasi-Fermi level through the depletion region is negligibly small. This justifies the fundamental assumption of thermionic-emission theory. Thus, it can be concluded that in Schottky diodes made from fairly high mobility semiconductors, the forward current is limited by thermionic emission.

7.1.5. *Effect of Interfacial Layer on Current Transport*

Schottky diodes nearly always have a thin insulating layer between metal and semiconductor. This layer has the following effects on the current transport:

1. Because of the potential drop in the layer, the zero-bias barrier height is lower than it would be in an ideal diode.
2. The electrons have to tunnel through the barrier associated with the insulator.
3. When a bias is applied, part of the bias voltage is dropped across the insulating layer so that the barrier height is a function of the bias voltage.

From Eq. (7.1.31) and the dependence of ϕ_B on V_A, an explicit expression for n as a function of the insulating layer thickness δ is

$$n_L = 1 + \frac{\delta\varepsilon}{X_d(q\delta D_s + \varepsilon_i)} \tag{7.1.32}$$

Figure 7.5 shows the I–V characteristics of Schottky diodes with different oxide interfacial layer thicknesses.

7.2. SIMPLE MESFET MODEL

Here we consider an n-channel MESFET (Fig. 7.7), but the treatment applies to a p-channel MESFET as well. In this device, the gate is used to modulate the channel geometry, and the current flows predominantly between the source and drain. Therefore, a Schottky contact under reverse-bias condition without current flow through the junction is desirable.

The approach used in modeling the JFET is still applicable here. The one-sided depletion layer thicknesses associated with the top and bottom gates are

$$X_{dT} = \sqrt{\frac{2\varepsilon[V(x) + \phi_{iT} - V_G]}{qN_D}} \tag{7.2.1}$$

$$X_{dB} = \sqrt{\frac{2\varepsilon[V(x) + \phi_{iB} - V_G]}{qN_D}} \tag{7.2.2}$$

where $V(x)$ is the potential drop along the channel, V_G is the applied gate voltage, N_D is the average doping concentration in the channel, and ϕ_{iT} and ϕ_{iB} are the top and bottom gate junction built-in potentials, respectively. For a MESFET with symmetrical top and bottom gates, the height of the conduction channel H is

$$H = h - X_{dT} - X_{dB} = h - 2\sqrt{\frac{2\varepsilon[V(x) + \phi_i - V_G]}{qN_D}} \tag{7.2.3}$$

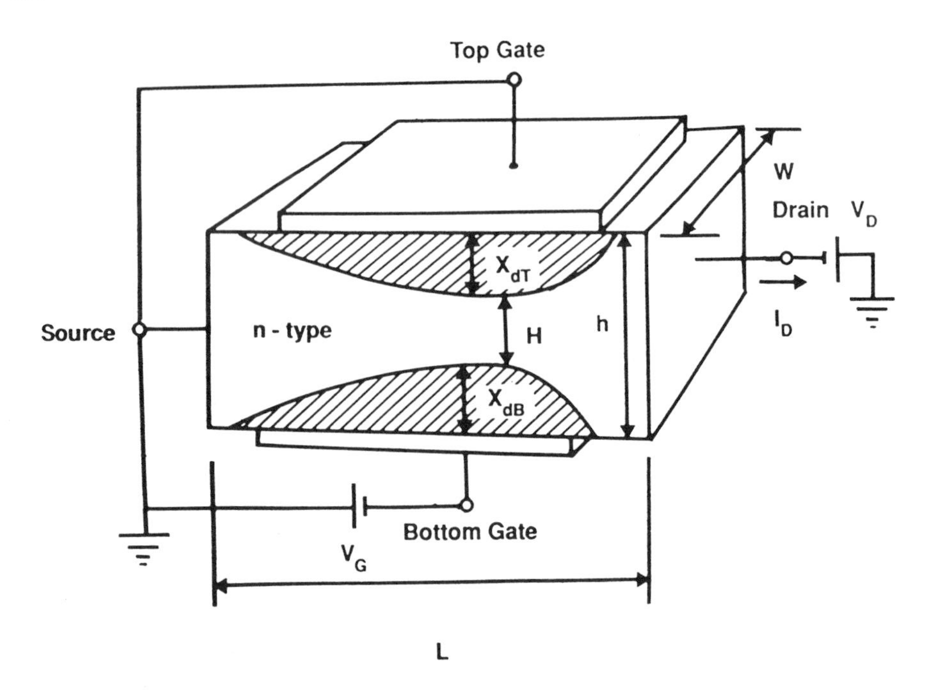

FIGURE 7.7. Simplified *p*-channel metal–semiconductor field-effect transistor structure, where the shaded regions are the depletion regions associated with the Schottky contacts.

Here h is the channel layer thickness and $\phi_i = \phi_{iT} = \phi_{iB}$. Assuming the drain current I_D is constant throughout the channel, the incremental change of the channel potential dV is related to the incremental change of the channel resistance dR as

$$dV = I_D dR = I_D dx/(q\mu_n N_D WH) \tag{7.2.4}$$

where μ_n is the average electron mobility in the channel and W is the width of the MESFET (Fig. 7.7). Putting (7.2.3) into (7.2.4), integrating the resulting equation from $x = 0$ to $x = L$ (L is the channel length), and using the boundary conditions $V(x = 0) = 0$ and $V(x = L) = V_D$, we derive I_D as (Shur, 1990)

$$I_D = \frac{q\mu_n N_D h^2}{2\varepsilon}\left[\frac{V_D - 2\,((V_D + \phi_i - V_G)^{1.5} - (\phi_i - V_G)^{1.5})}{3\sqrt{|V_P - \phi_i|}}\right] \tag{7.2.5}$$

where V_D is the drain voltage and $V_P = -0.5qN_D h^2/\varepsilon + \phi_i$ is the pinch-off voltage of the MESFET. Equation (7.2.5) is only applicable for such values of V_G and V_D when the conducting channel exists in the narrowest spot ($H > 0$ anywhere in the channel). This condition is violated when the two depletion regions touch ($H = 0$ and $h = X_{dT} + X_{dB}$). The drain voltage that causes this occurrence (called the channel pinch-off) is the drain saturation voltage V_{Dsat}:

$$V_{Dsat} = |V_P - V_G| \tag{7.2.6}$$

If V_D is replaced with V_{Dsat} in (7.2.5), the drain saturation current I_{Dsat} for $V_D \geq V_{Dsat}$ is then

$$I_{Dsat} = \frac{q\mu_n N_D h^2}{2\varepsilon}\left[\frac{|V_P - \phi_i|/3 - \phi_i + V_G + 2(\phi_i - V_G)^{1.5}}{3\sqrt{|V_P - \phi_i|}}\right] \tag{7.2.7}$$

The current–voltage characteristics of MESFETs have similar trends to those of JFETs.

7.3. MEDICI SIMULATION

The simple model developed in the previous section can be used to estimate the general trends of a MESFET, but it often fails to give an accurate description of the MESFETs behavior, particularly if the channel length is short (Cappy, 1980; Higgins and Pattanayak, 1982). This is because a high electric field can exist in the channel, leading to a highly nonlinear drift velocity (and mobility) versus field characteristic (Chang and Fetterman, 1986). The problem is further compounded by multidimensional effects, which are also important in submicron MESFETs (Chang and Day, 1989; Shih *et al.*, 1992; Himsworth, 1972).

We first consider a planar GaAs MESFET structure (Fig. 7.8). The device has an ion-implemented n channel with Gaussian profile and peak concentration of 10^{20} cm^{-3}

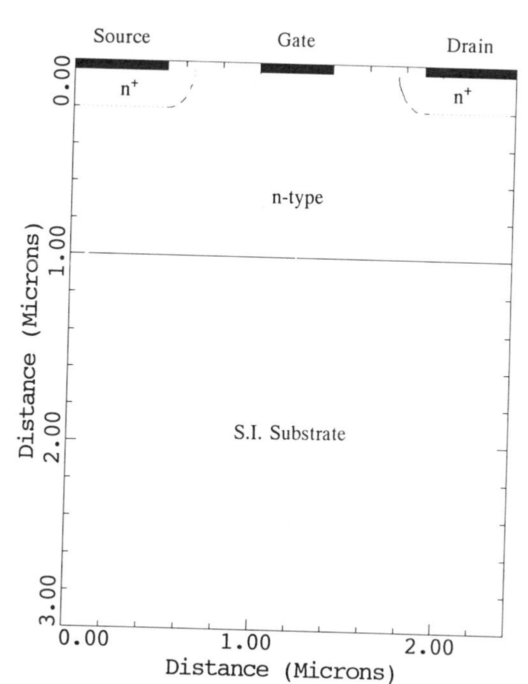

FIGURE 7.8. Structure of the GaAs MESFET used in simulation.

for the source and drain regions. A metal work function of 5.17 V (Au) is used for the gate contact. An advanced GaAs MESFET structure having a recess gate will be treated later. A MEDICI input file of a planar GaAs MESFET is given in Example 7.1 for reference.

7.3.1. *Steady-State Simulation*

Figure 7.9 illustrates the current–voltage characteristics of the GaAs MESFETs having four different gate lengths L biased at $V_G = -1$ V. Note that the simulated drain current has units of A/μm, where μm is the third dimension of the MESFET (i.e., 10^{-4} A/μm means 10^{-2} A for a 100-μm-wide MESFET). Since the current level in the MESFET is relatively low, the effect of lattice heating is not significant and is therefore negligible. This is demonstrated by the three-dimensional lattice temperature contours in Figs. 7.10a,b for MESFETs with $L = 1$ μm and $L = 0.4$ μm, respectively. Note that the lattice temperature is slightly higher in the $L = 0.4$-μm MESFET due to the higher drain current in such a device.

It is apparent from the results in Fig. 7.9 that the MESFET can generate a larger drain current if the channel length is reduced. Furthermore, for a given channel length when the gate is not placed at the center between the drain and source regions (asymmetrical case), the current is reduced at small V_D's but is increased as V_D is increased beyond 3 V.

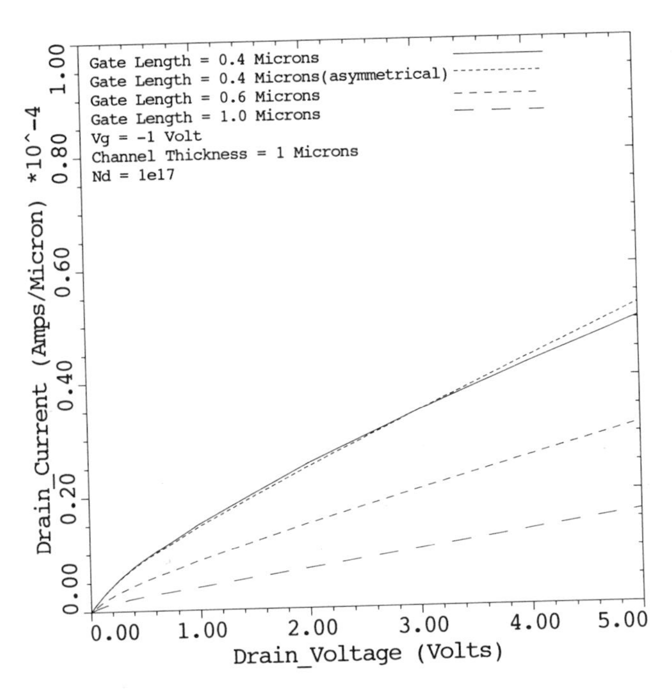

FIGURE 7.9. Drain current versus drain voltage characteristics simulated for different gate lengths.

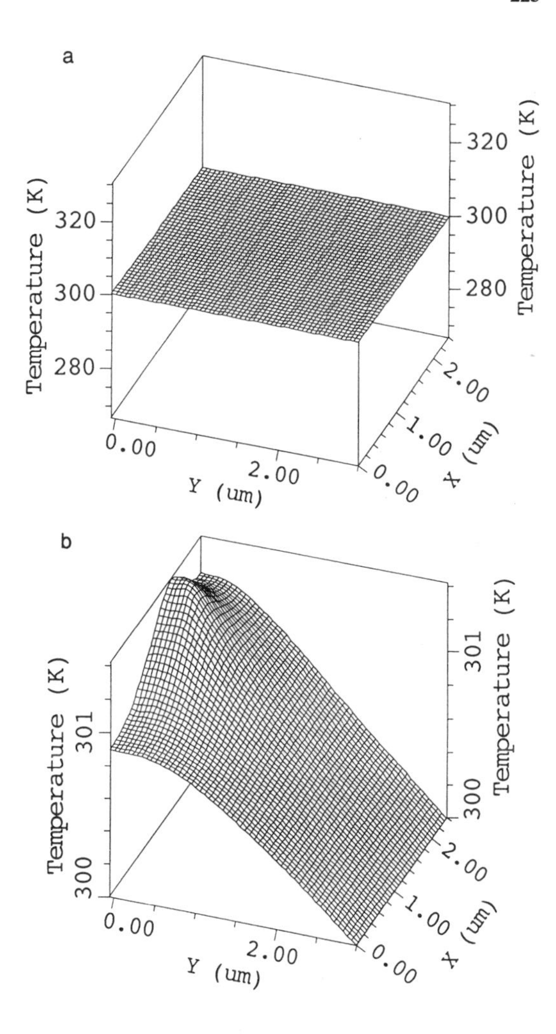

FIGURE 7.10. Lattice temperature contours of the (a) 1-μm and (b) 0.4-μm MESFETs at $V_D = 4$ V.

Figures 7.11a,b show the electric field contours in the 0.4-μm MESFET biased with $V_D = 1$ and 5 V, respectively. It can be seen that the field in the channel near the drain region is increased significantly as V_D is increased. Also note that a very high field exists at the region underneath the gate due to the reverse-biased Schottky contact there. The electric field vectors in the MESFET biased at $V_D = 1$ and 5 V are given in Figs. 7.12a,b, respectively. As expected, the electric field vectors near the drain region increase with V_D. Also, note that the electric field is nearly absent in a small region below the source. Figures 7.13a,b and 7.14a,b show the current vectors and 2-D electron contours, respectively, in the MESFET at two V_D's. If the free-carrier depletion is defined as $n < 10N_D$, then the edge of the space-charge region in the MESFET can be estimated by the 10^{16} cm^{-3} electron contour in Fig. 7.14.

The effects of the peak doping concentration N_D in the channel on the current–voltage characteristics are simulated as shown in Fig. 7.15. The results suggest that the drain current is strongly influenced by N_D; the larger N_D, the higher the drain current is.

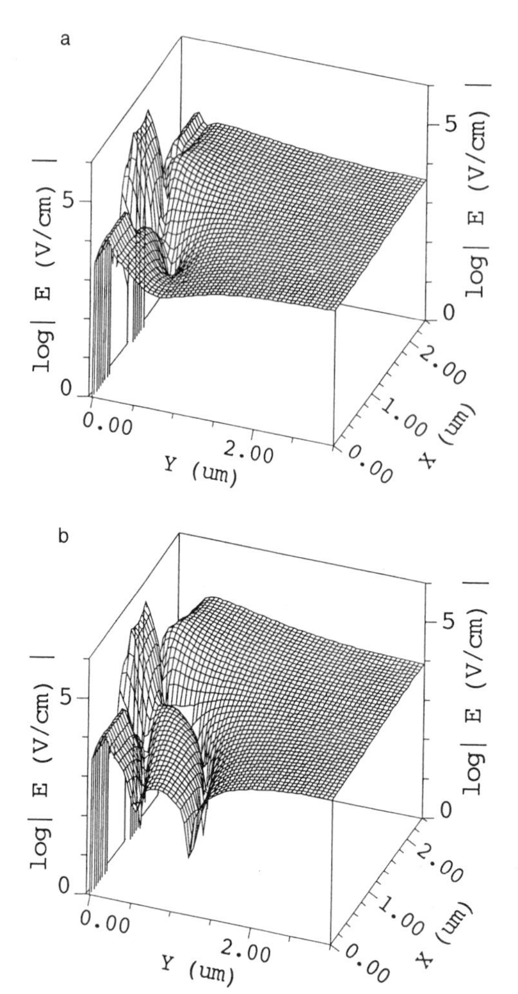

FIGURE 7.11. Three-dimensional electric field contours of the 0.4-μm MESFET simulated at $V_G = 1$ V and (a) $V_D = 1$ V and (b) $V_D = 5$ V.

Figure 7.16 shows the *I–V* results of MESFETs having three channel layer thicknesses. A higher current level is obtained when a thicker channel layer is used, because the height of the conducting channel is controlled by the depletion region associated with the gate as well as the depletion region associated with the epi channel and semi-insulating substrate. As a consequence, a thinner channel layer results in a smaller conducting channel height and thus a smaller drain current.

7.3.2. Transient Response

The transient response of the GaAs MESFET is studied. The simulation is done by applying a step-down gate voltage V_G of 0 to –5 V at $t = 0$. A fixed drain voltage of 5 V is also used. The *i–t* characteristics of MESFETs having four gate lengths are shown in

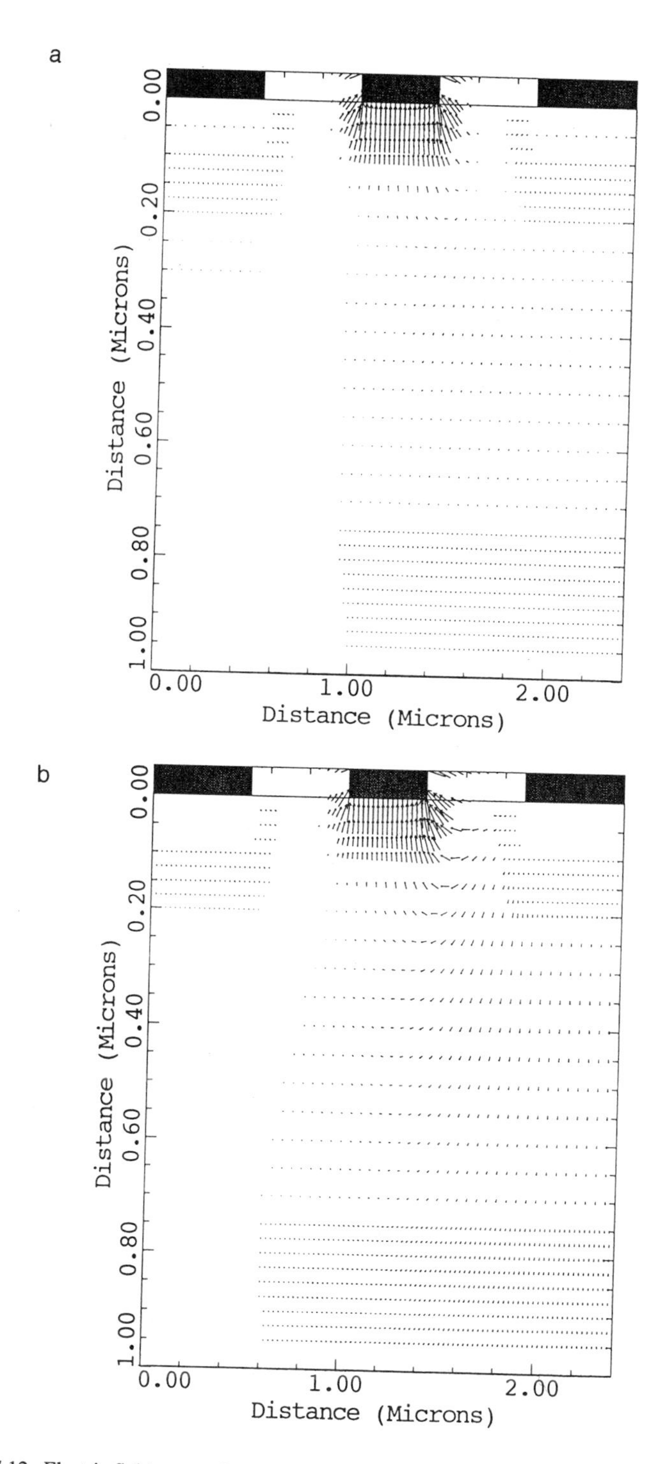

FIGURE 7.12. Electric field vectors in the MESFET biased at $V_G = 1$ V and (a) $V_D = 1$ V and (b) $V_D = 5$ V.

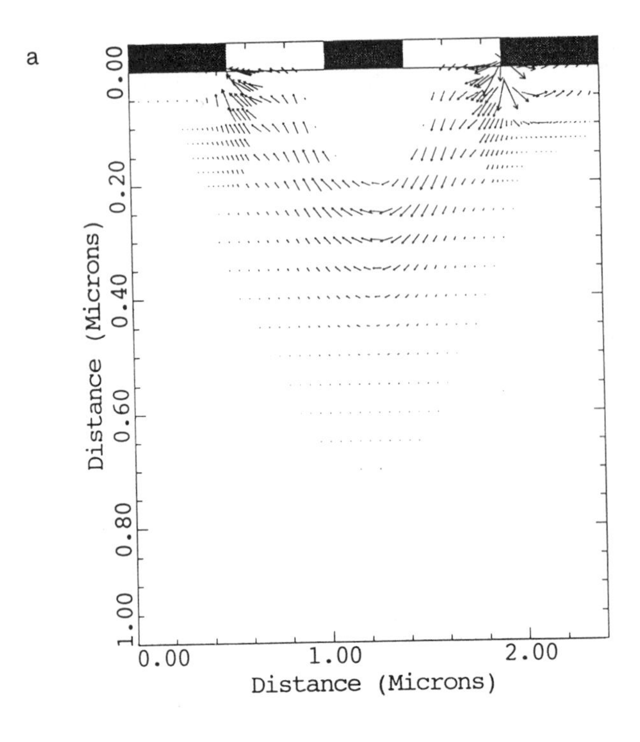

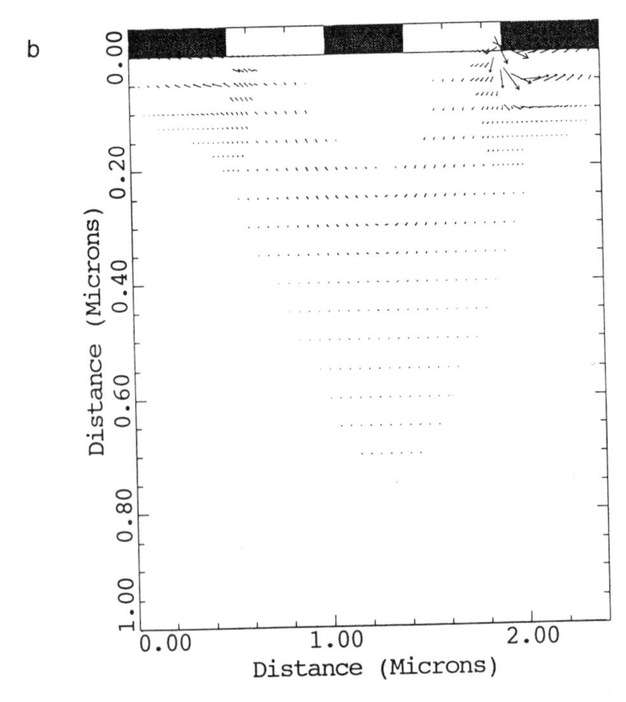

FIGURE 7.13. Current vectors in the MESFET biased at $V_G = 1$ V and (a) $V_D = 1$ V and (b) $V_D = 5$ V.

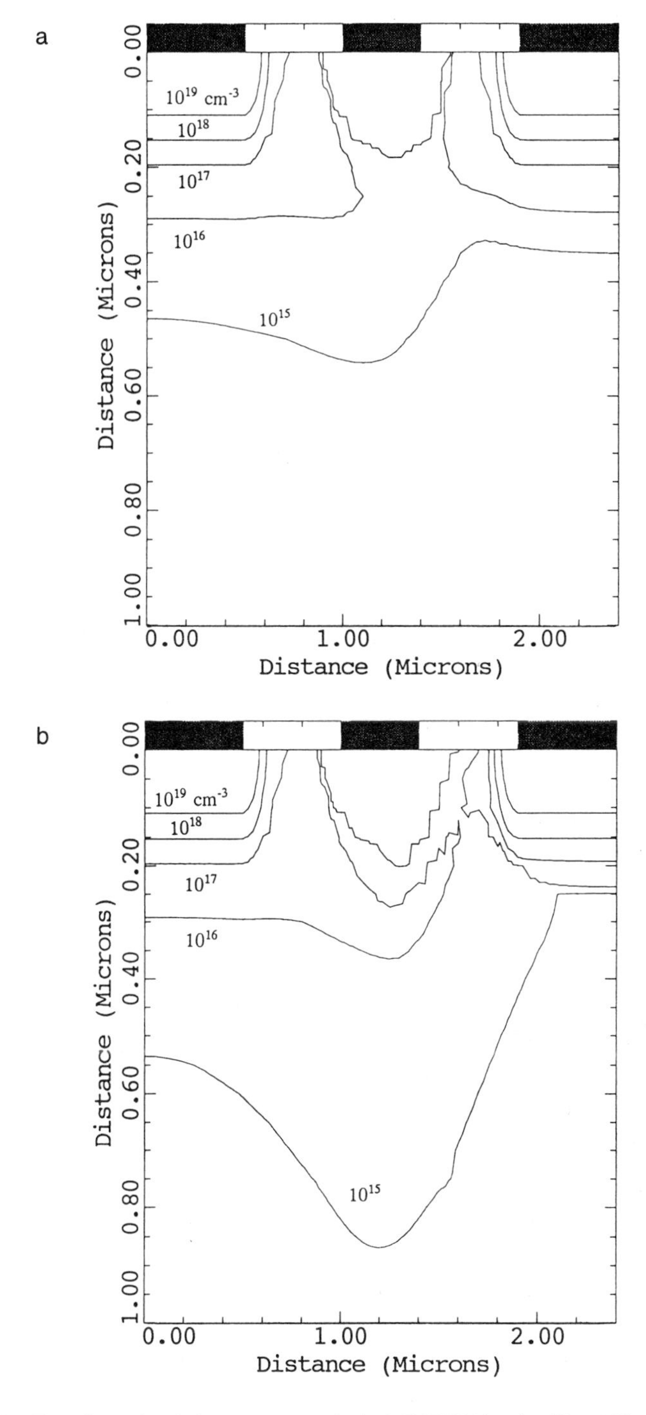

FIGURE 7.14. Two-dimensional electron contours in the MESFET biased at $V_G = 1$ V and (a) $V_D = 1$ V and (b) $V_D = 5$ V.

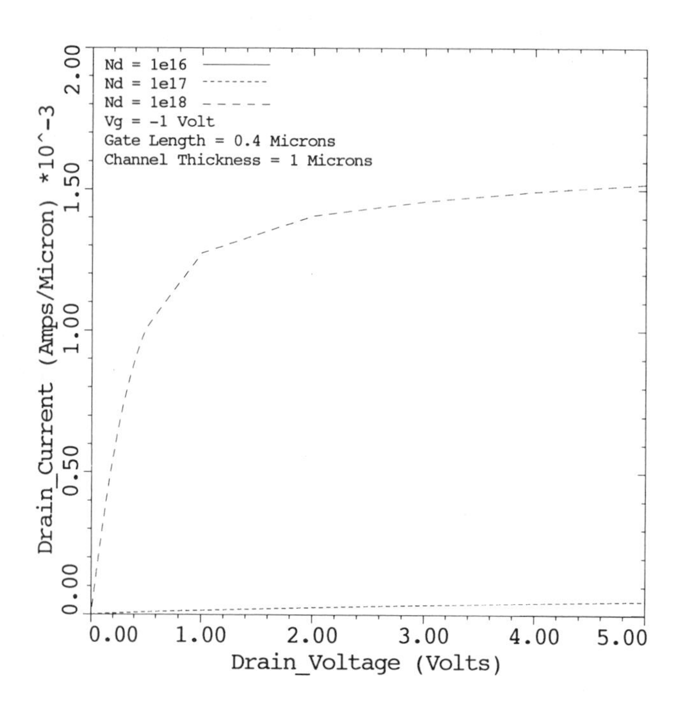

FIGURE 7.15. *I–V* characteristics of 0.4-μm MESFETs having different peak channel doping concentrations.

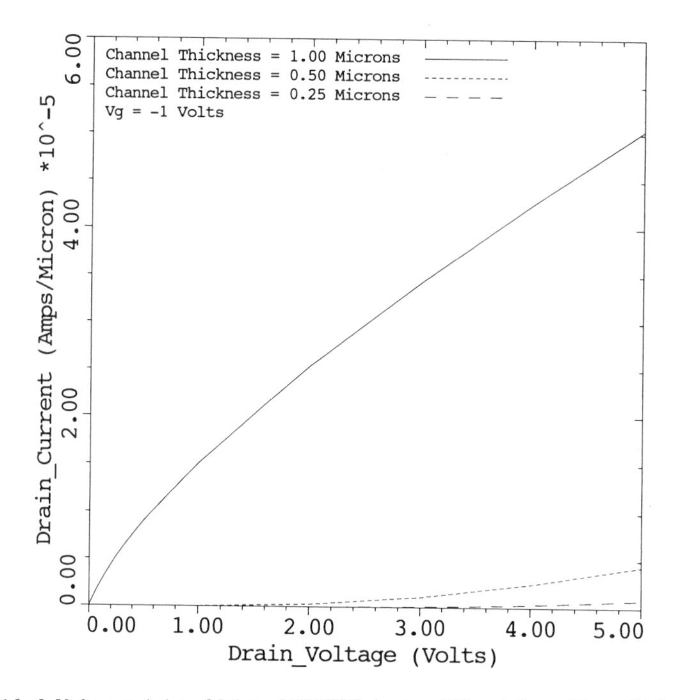

FIGURE 7.16. *I–V* characteristics of 0.4-μm MESFETs having different channel layer thicknesses.

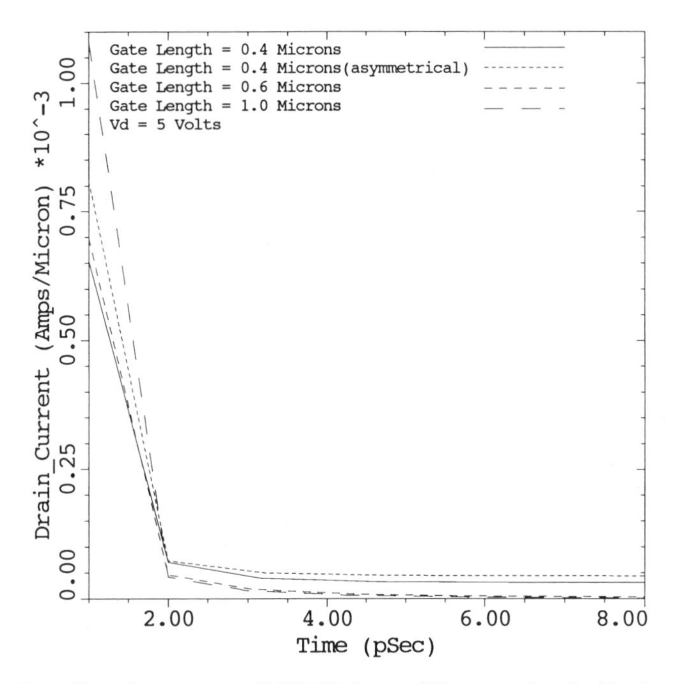

FIGURE 7.17. Turn-off transient response of MESFETs having different gate lengths. The devices are subject to a step-down V_G from −5 V to 0 at $t = 0$.

Fig. 7.17. The turn-off time (i.e., the time for the current to decay 90% of its initial value) seems to be insensitive to the channel length.

The effects of channel doping concentration on MESFET switching behavior are illustrated in Fig. 7.18. The switching speed appears to be insensitive to the doping concentration as well. The same statement also applies to the effects of channel layer thickness on the MESFET transient response, as evidenced by the results in Fig. 7.19.

Figures 7.20a,b plot the electron density contours in the MESFET at $t = 1$ and 5 ps. Clearly, the electron density decreases rapidly as time progresses, especially for those underneath the gate, due to the expansion of the depletion region. The flow of electrons during the transients is particularly interesting (Figs. 7.21a–c). Initially, a zero gate voltage is applied, and the channel is filled with electrons. At $t = 0$, V_G is switched to −5 V, and the expansion of the depletion region will push the electrons toward both sides of the channel (Fig. 7.21a). Since the source terminal is grounded and the drain terminal has a positive voltage, the electrons being pushed to the source region must flow to the drain region and exit there (Fig. 7.21b). After a few picoseconds, the depletion region under the gate reaches its steady-state position, and the electrons in the MESFET resume their normal flow from source to drain (Fig. 7.21c).

7.3.3. Small-Signal Analysis

Small-signal simulation is carried out in this section. In addition to the dc bias voltage, a sinusoidal voltage with a magnitude of 26 mV is applied to the gate terminal.

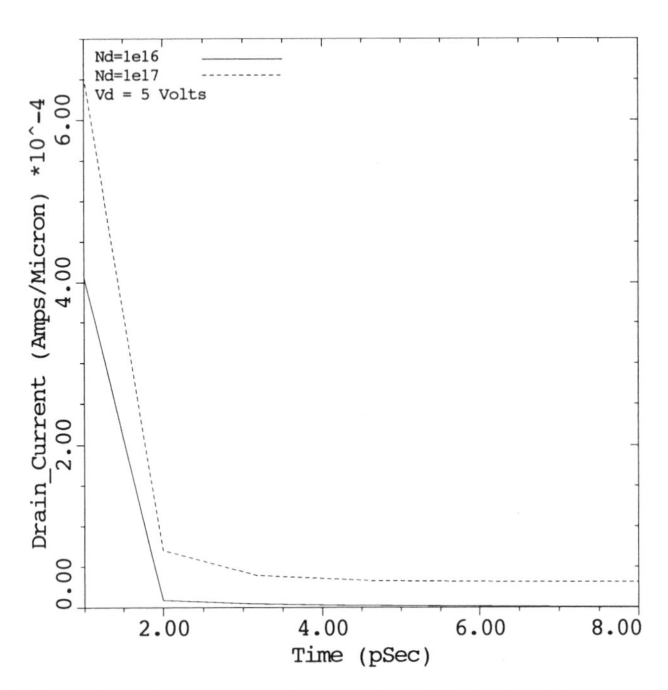

FIGURE 7.18. Turn-off transient response of 0.4-μm MESFETs having different peak channel doping concentrations.

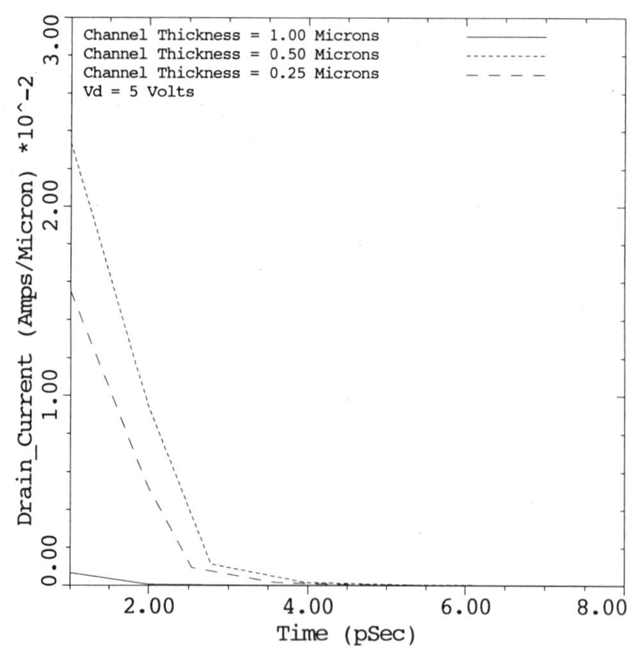

FIGURE 7.19. Turn-off transient response of 0.4-μm MESFETs having different channel layer thicknesses.

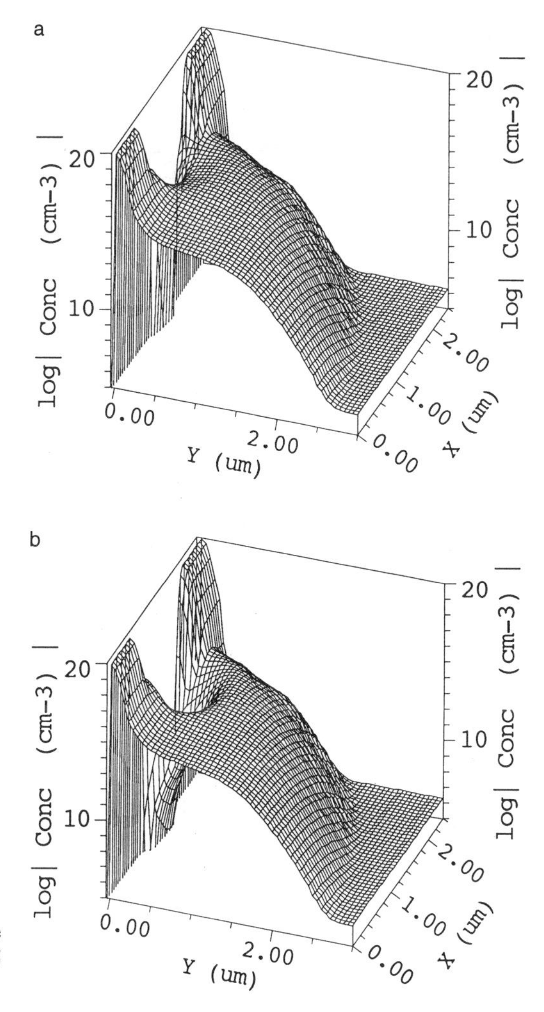

FIGURE 7.20. Electric field contours in the 0.4-μm MESFET at (a) $t = 1$ ps and (b) $t = 5$ ps during the turn-off transients.

From this, the small-signal drain current, conductance, and capacitance at a particular drain voltage can be simulated. The cutoff frequency f_T, which is a figure of merit for ac performance, is calculated. Figure 7.22 shows the cutoff frequency versus the drain current simulated from MESFETs with four gate lengths. At low current levels, f_T is not notably affected by gate length. As current increases, however, the improvement of f_T in the short-channel MESFET (i.e., 0.4 μm) over the longer-channel MESFETs is more apparent.

Figure 7.23 shows f_T simulated from MESFETs having different channel layer thicknesses in the channel. A significantly higher f_T (about 10 times higher) can be obtained if the channel layer thickness is decreased from 1 to 0.25 μm, since f_T is inversely proportional to the source-to-gate capacitance C_{gs}, which is directly proportional to the free-carrier charge storage in channel under the gate region. A thinner channel thus gives rise to a smaller C_{gs} and a higher f_T.

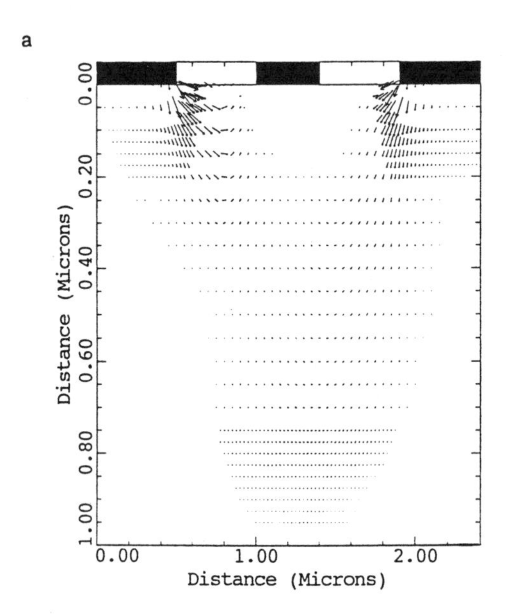

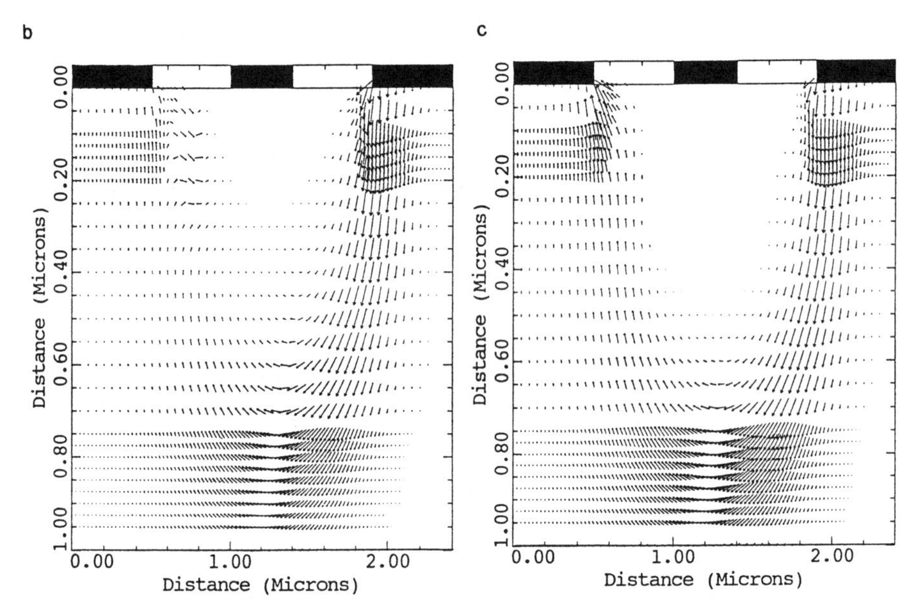

FIGURE 7.21. Electron current vectors in the 0.4-μm MESFET at (a) $t = 1$ ps, (b) $t = 2$ ps, and (c) $t = 4$ ps.

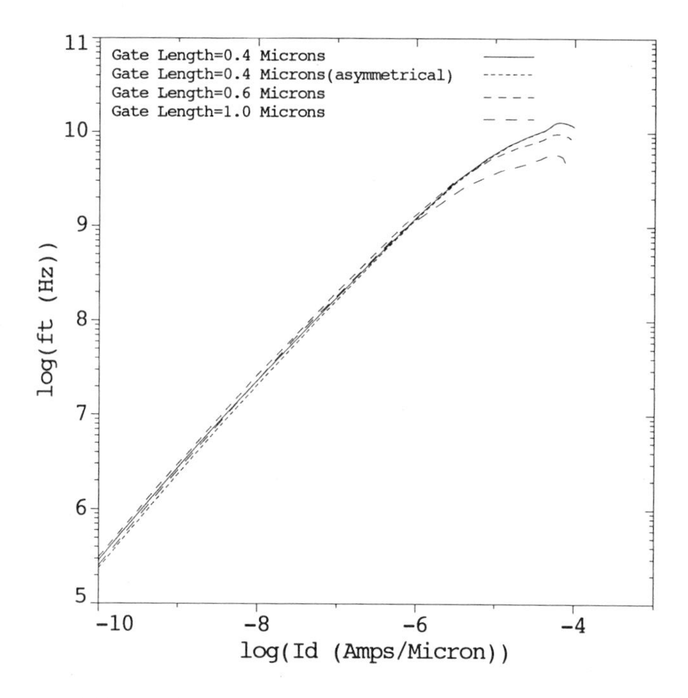

FIGURE 7.22. Cutoff frequency of GaAs MESFETs having different channel lengths.

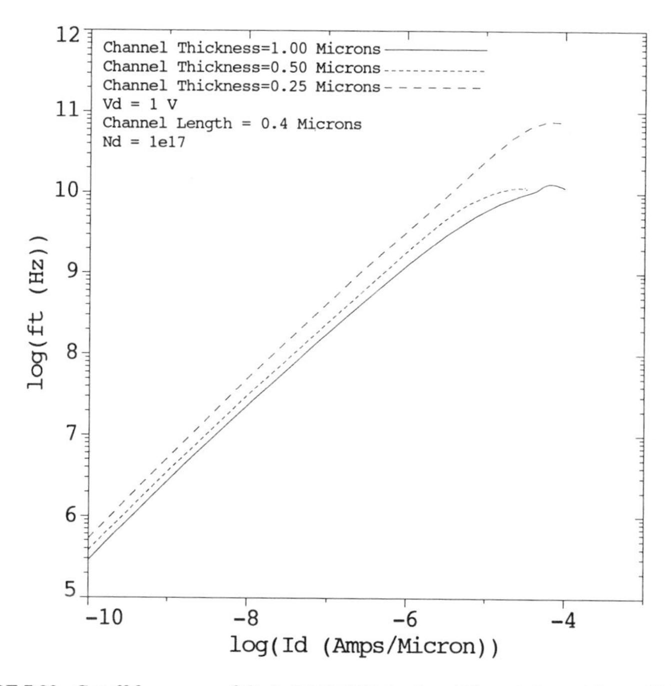

FIGURE 7.23. Cutoff frequency of GaAs MESFETs having different channel layer thicknesses.

7.3.4. Advanced MESFET Structure

So far the performance of GaAs MESFETs using a planar structure have been examined. It is generally believed that the surface potential of a GaAs free surface is between 0.5 and 0.6 V (Shur, 1990). This value is close to the surface potential of the gate Schottky contact at zero bias, due to the fact that the surface Fermi level is pinned at the metal–semiconductor interface due to a high density of surface states. Therefore the depletion region in a GaAs MESFET exists not only under the gate contact but also under the free surface between the source and gate as well as between the gate and drain. As a result, the performance of MESFETs is affected by the geometry of these surfaces (Rocchi, 1985; Lo and Lee, 1994). A recess-gate structure (Fig. 7.24) has been used frequently to minimize the effects of surface states at free surface and to enhance MESFET performance (Yamaguchi and Asai, 1978; Barton *et al.*, 1986).

Figures 7.25a,b present the 2-D electron contours for the recess-gate MESFETs biased at $V_G = 1$ V and $V_D = 1$ V and 5 V. Like the planar MESFET, the edge of the space-charge region can be estimated by the 10^{16} cm^{-3} electron contour. The electric field vectors in the MESFET biased at two V_D are shown in Figs. 7.26a,b, respectively. Very high electric fields are found near the two grooves of the recess gate. Also, as V_D increases, the electric field at the groove near the source becomes less intense and the electric field in the region under the drain is increased.

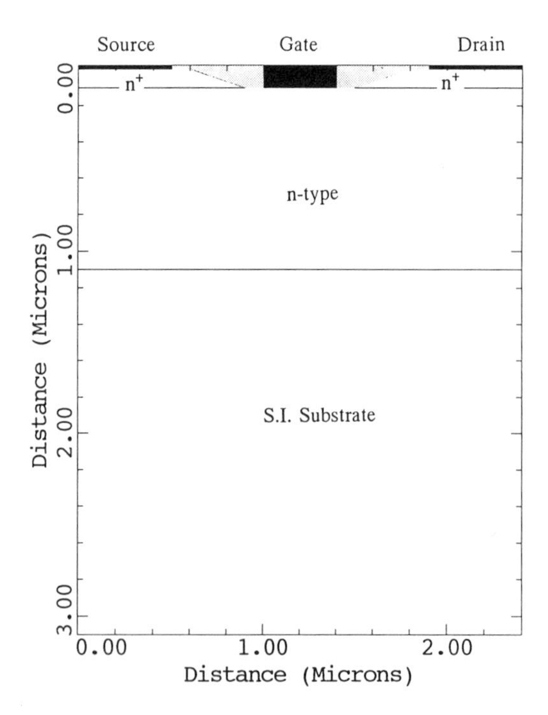

FIGURE 7.24. Schematic of the GaAs MESFET with recess-gate structure.

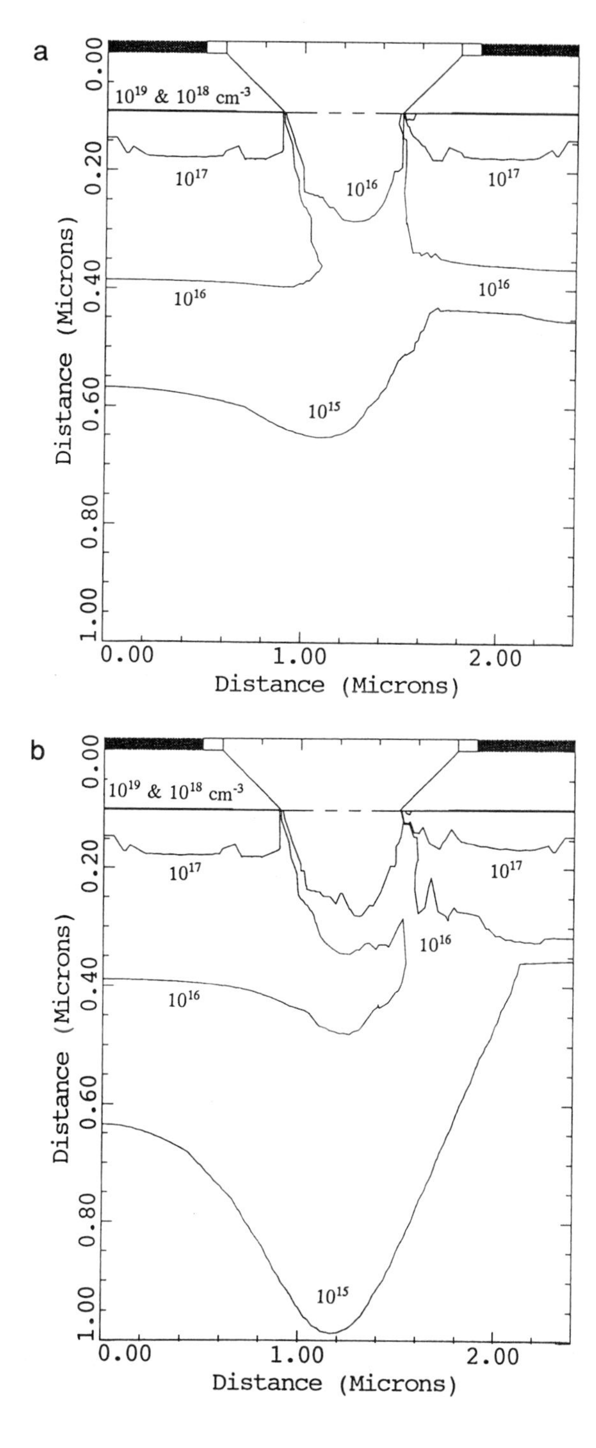

FIGURE 7.25. Two-dimensional electron contours in the recess-gate MESFET biased at $V_G = -1$ V and (a) $V_D = 1$ V and (b) $V_D = 5$ V.

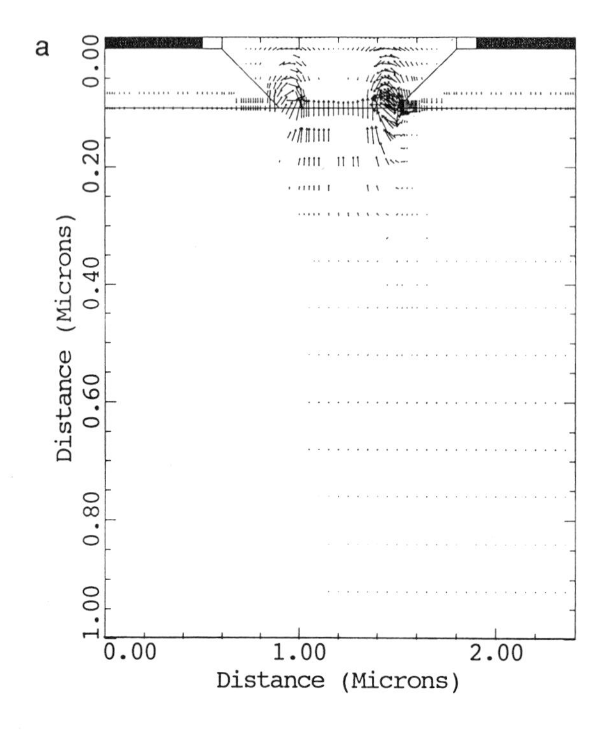

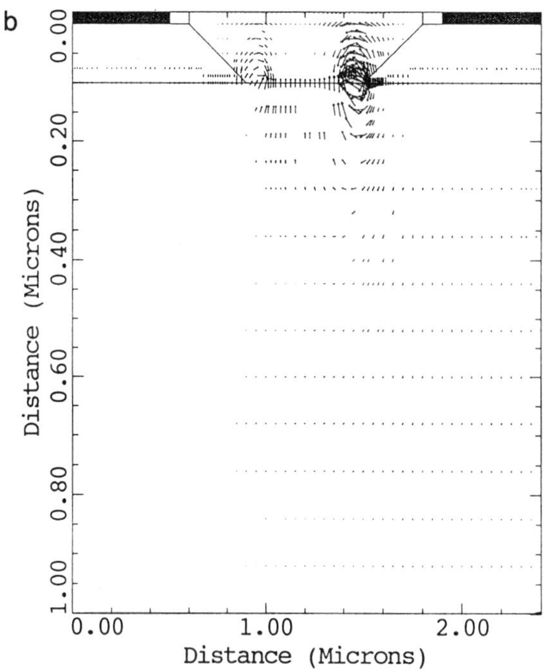

FIGURE 7.26. Electric field vectors in the recess-gate MESFET biased at $V_G = 1$ V and (a) $V_D = 1$ V and (b) $V_D = 5$ V.

The $I–V$ characteristics simulated from the planar and recess-gate MESFETs are given in Fig. 7.27. Both devices have the same makeup: 0.4-μm channel length, 10^{17} cm^{-3} peak channel doping concentration, 1-μm channel layer thickness. It is shown that a smaller recess depth gives rise to a higher drain current, because a larger recess depth results in larger source and drain resistances and consequently a smaller extrinsic transconductance. However, if the recess depth is too small (i.e., approaching planar structure), the effects of the surface states at the free surface become important and can degrade MESFET performance. Comparisons of the turn-off transients and cutoff frequencies of the two devices are also shown in Figs. 7.28 and 7.29, respectively.

Next, the effects of inserting an n^+ buried layer between the channel layer and semi-insulating substrate are considered. Such an approach is used to minimize the effect associated with the voltage applied to the semi-insulating substrate (Choi and Das, 1994). Three buried layer thicknesses are considered, and the $I–V$ characteristics are given in Fig. 7.30. The results suggest that incorporating a buried layer can increase the drain current of MESFET, but only a very thin buried layer (i.e., 0.1 μm) is necessary. Using a thicker buried layer does not provide further enhancement.

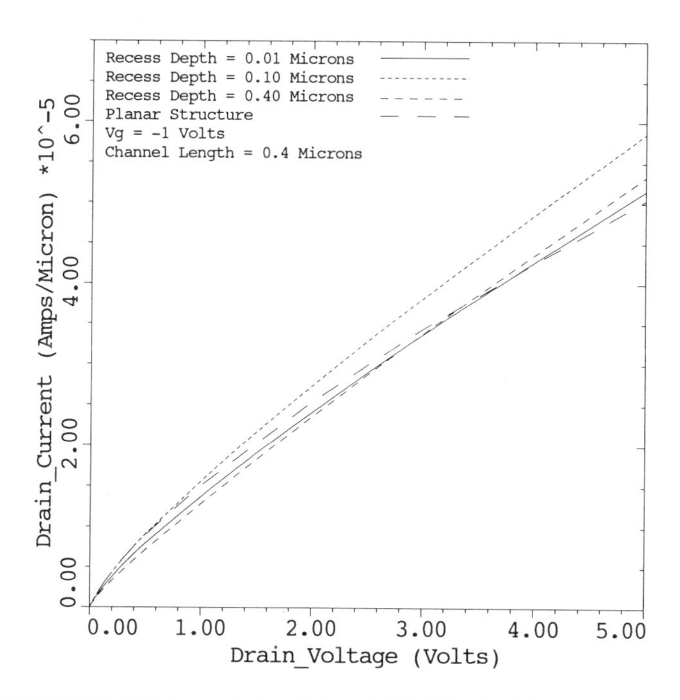

FIGURE 7.27. $I–V$ characteristics of 0.4-μm MESFETs with different recess depths.

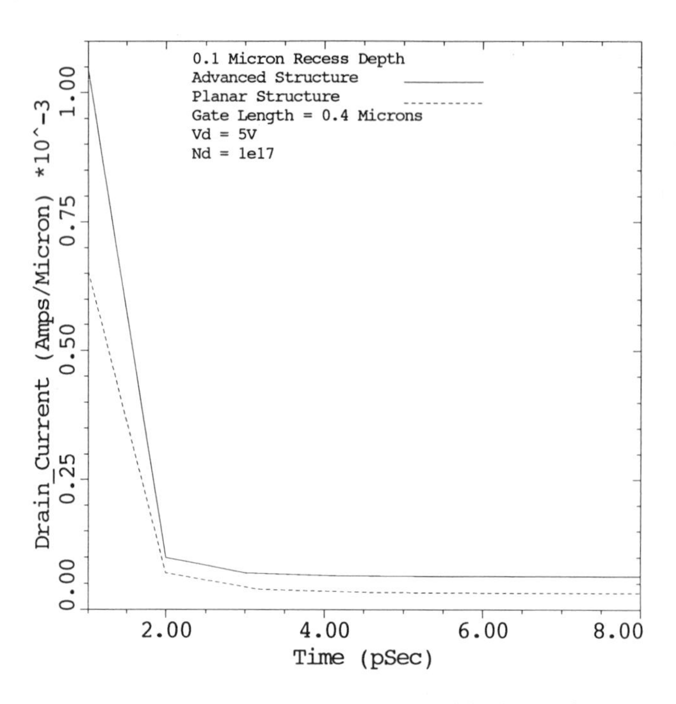

FIGURE 7.28. Comparison of turn-off transient responses of the planar and recess-gate MESFETs.

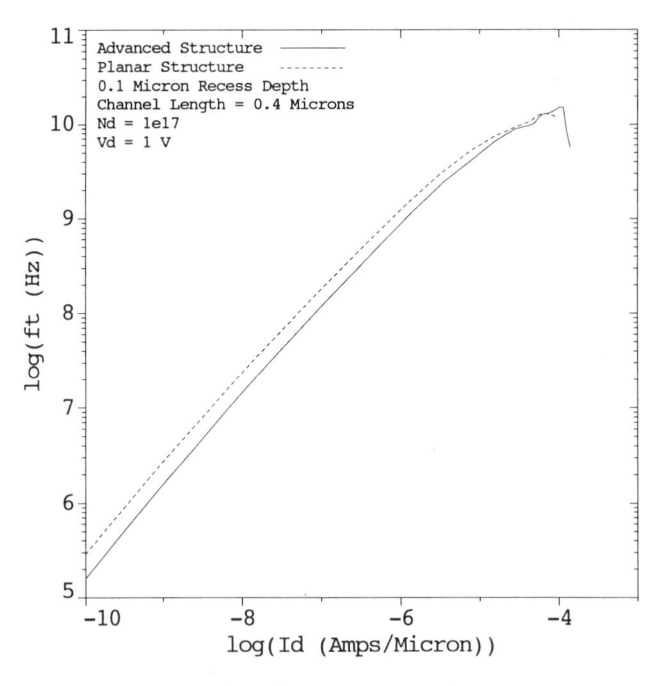

FIGURE 7.29. Comparison of cutoff frequencies of the planar and recess-gate MESFETs.

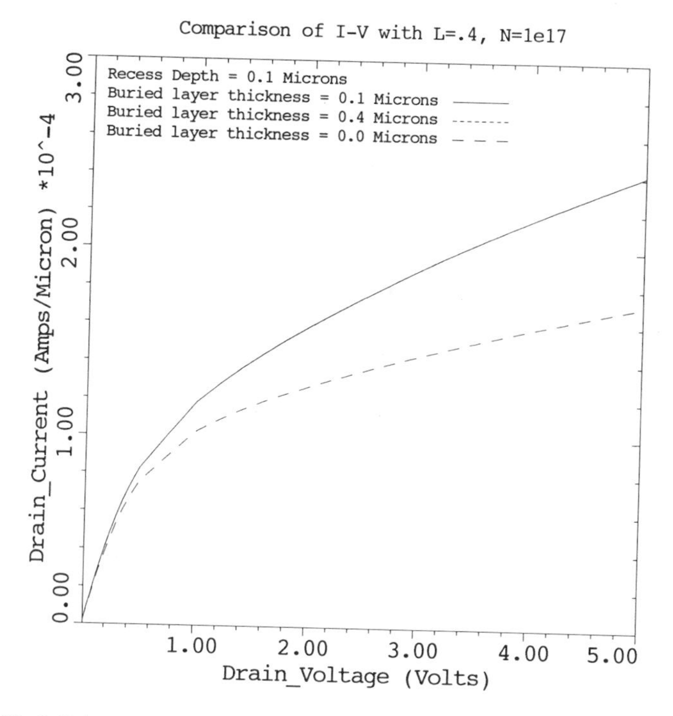

FIGURE 7.30. *I–V* characteristics of recess-gate MESFETs having different buried-layer thicknesses.

7.4. HETEROJUNCTION FETs

Heterojunction FETs or high-electron-mobility transistors (HEMTs) are compound semiconductor transistors. The heterojunctions are formed between semiconductors of different compositions and band gaps (e.g., GaAs/AlGaAs and InGaAs/InP). This is in contrast to conventional GaAs MESFET which utilize junctions between homojunction materials. These heterojunction devices offer advantages in microwave, millimeter-wave, and high-speed digital integrated circuits.

In heterojunction FET, the epitaxial layer structure is designed so that free electrons in the channel are physically separated from the ionized donors, enhancing electron mobility by reducing iodizing impurity scattering. In addition, AlGaAs/InGaAs pseudomorphic heterojunction FETs have several advantages over conventional AlGaAs/GaAs heterojunction FETs because of the superior electron mobility of InGaAs and better electron confinement at the AlGaAs/InGaAs interface due to a larger conduction-band discontinuity (Drummond *et al.*, 1982). The device current capability can be even further improved by inserting in the center of the InGaAs channel layer an additional planar doping (Chao *et al.*, 1989). The adoption of δ doping and planar doping results in improved device performance, such as higher power gain and breakdown voltage (Tan *et al.*, 1991).

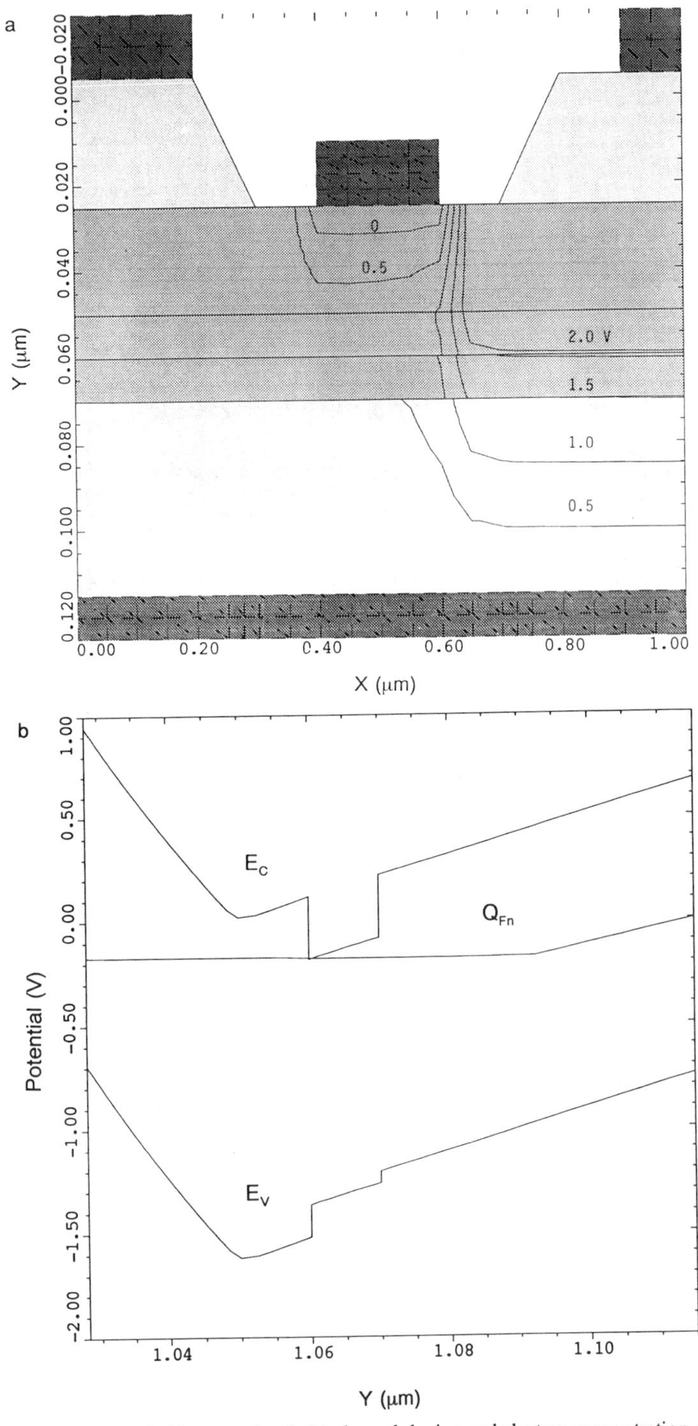

FIGURE 7.31. (a) Potential, (b) energy band, (c) channel doping and electron concentration, and (d) current flow of the heterojunction FET.

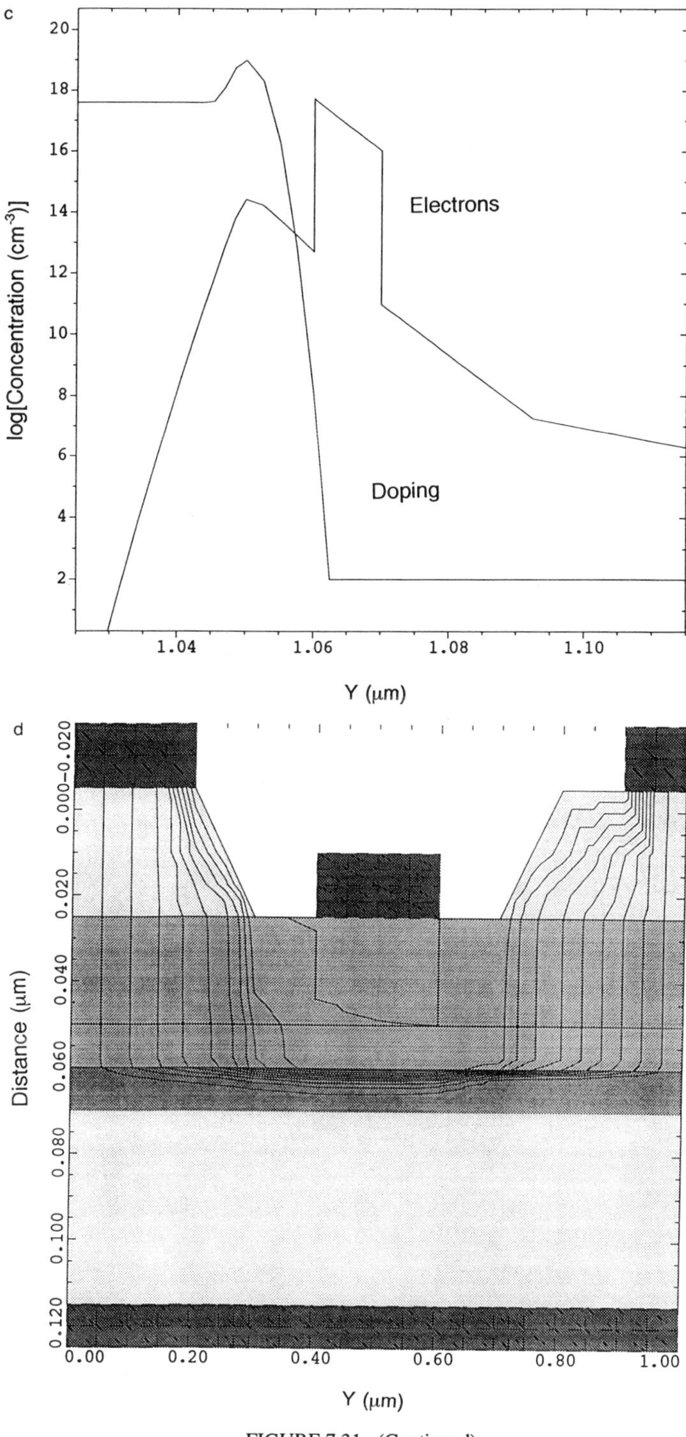

FIGURE 7.31. (Continued)

The physical characteristics of the pseudomorphic heterojunction FET using MEDICI is examined. The heterojunction module is used in conjunction with MEDICI to provide the heterojunction properties. The parameters available for describing the properties of the heterojunction are the usual parameters available for describing the properties of the materials that meet at the heterojunction. Some of these include the energy-band-gap parameters, electron affinity, density of states for the conduction band, density of states for the valence band, and various parameters for describing recombination, mobility, etc. The form of the continuity equations remains unchanged for heterojunctions except that the electric field terms E_n and E_p in the transport equations account for gradients at the conduction- and valence-band edges:

$$E_n = -\nabla\left[\psi + \frac{kT}{q}\ln(n_{ie})\right] = -\nabla\left[\frac{-E_C}{q} + \frac{kT}{q}\ln(N_C)\right] \tag{7.4.1}$$

$$E_p = -\nabla\left[\psi - \frac{kT}{q}\ln(n_{ie})\right] = -\nabla\left[\frac{-E_V}{q} + \frac{kT}{q}\ln(N_V)\right] \tag{7.4.2}$$

The device structure in the MEDICI simulation is largely planar with constant doping in most of the regions. The AlGaAs spacer and InGaAs channel are left undoped. Heterojunctions are used to create a narrow undoped electron well that forms the channel for current flow. Electrons from surrounding doped regions of the device become trapped in the well, resulting in a high concentration of electrons in the channel. The source and drain contact regions are heavily doped (1×10^{20} cm^{-3}) to reduce series resistance. The AlGaAs region under the gate serves as the source of channel electrons and is doped n type to 4×10^{17} cm^{-3}. The thickness of the AlGaAs spacer layer and InGaAs channel is 50 Å. The gate length is 0.2 μm. The band gaps for $Al_xGa_{1-x}As$ and $In_yGa_{1-y}As$ are $1.42 + 1.247x$ and $1.42 - 1.5y - 0.4y^2$ eV, respectively, where x is the aluminum mole fraction and y is the indium model fraction. The gate is made from a Schottky contact, and the work function is set to 5.17 V. Physical models used in the device simulation include SRH recombination, Auger recombination, and concentration-dependent mobility.

Figure 7.31a shows the two-dimensional contours of the heterojunction FET biased at the gate–source voltage $V_{GS} = 0$ V and drain–source voltage $V_{DS} = 2$ V. The highest potential is close to the drain region because the source and gate contacts are tied to ground and the drain contact is connected to 2 V. This contour plot indicates that the electrons must flow from source to drain. To examine the two-dimensional current flow trajectory, the energy-band diagram from the gate to the bottom contact is depicted in Fig. 7.31b. The conduction-band discontinuity between the AlGaAs and InGaAs layers confines electrons at the AlGaAs–InGaAs interface. Further, a high free-electron concentration exists over the two-dimensional electron gas (2-DEG), as evidenced by the concentration plot in Fig. 7.31c. Electrons traveling in this region do not encounter ionized donor atoms because the GaAs is undoped. Current flow lines from the drain to the source displayed in Fig. 7.31d confirm that the drain current I_{DS} moves away from the drain contact and traverses laterally in the InGaAs well to the source contact.

The terminal current–voltage characteristics at different boundary conditions of the pseudomorphic heterojunction FET will be displayed. The drain–source current versus the drain–source voltage at V_{GS} = 0, –0.05, and –0.1 V is shown in Fig. 7.32a. The heterojunction FET exhibits similar characteristics as the MESFET. At low drain–source voltages, the device is in the linear region, and the increase of I_{DS} with respect to V_{DS} is due to the channel length modulation effect. The drain current increases at higher gate–source voltage V_{GS} because of a higher 2-DEG sheet charge concentration at the InGaAs interface. In the heterojunction FET, the confinement of carriers in the 2-DEG increases the transconductance as well. The transconductance and cutoff frequency versus the gate–source voltage at V_{DS} = 0.8 V is depicted in Fig. 7.32b,c. Both the transconductance and cutoff frequency increase with the gate–source voltage at low V_{GS} and decrease with the gate–source voltage at high V_{GS}, which is different from that of the GaAs MESFET. The decrease of g_m and f_T at high V_{GS} can be explained by the potential contour and current vector plots in Fig. 7.33a,b. In these plots the device is biased at V_{DS} = 0.8 and V_{GS} = 2.0 V. In addition to the drain current flowing along the InGaAs layer, the gate begins to inject electron current. This causes a series resistance ohmic drop near the source side of the channel and results in a partial depletion of the channel. With increased gate–source voltage, the gate–source current I_{GS} increases (Fig. 7.33c), and hence more current is subtracted from the intrinsic drain current. Therefore, when the electric field at the source side exceeds the electric field at the drain side of the channel, the transconductance of the heterojunction FET degrades. In addition, since the cutoff frequency is proportional to transconductance, f_T decreases when g_m becomes smaller.

High-field impact ionization of the heterojunction FET is now presented. In the MEDICI simulation, field-dependent impact ionization is initiated. The simulated drain current versus the drain–source voltage up to V_{DS} = 6 V is displayed in Fig. 7.34. The drain current increases significantly at high V_{DS} due to avalanche multiplication. The potential contour plot in Fig. 7.35a shows that the electric field is stronger near the drain edge. Impact ionization resulting from a high electric field generates electron and hole pairs in the drain edge. This triggers the minority (hole) current injection into the substrate (Fig. 7.35b). To demonstrate that the multiplication occurs primarily at the drain edge, two-dimensional contours of impact ionization rates at V_{DS} = 6 V are shown in Fig. 7.35c. Its three-dimensional plot is given in Fig. 7.35d. Impact ionization is stronger near the drain edge where electric field is higher.

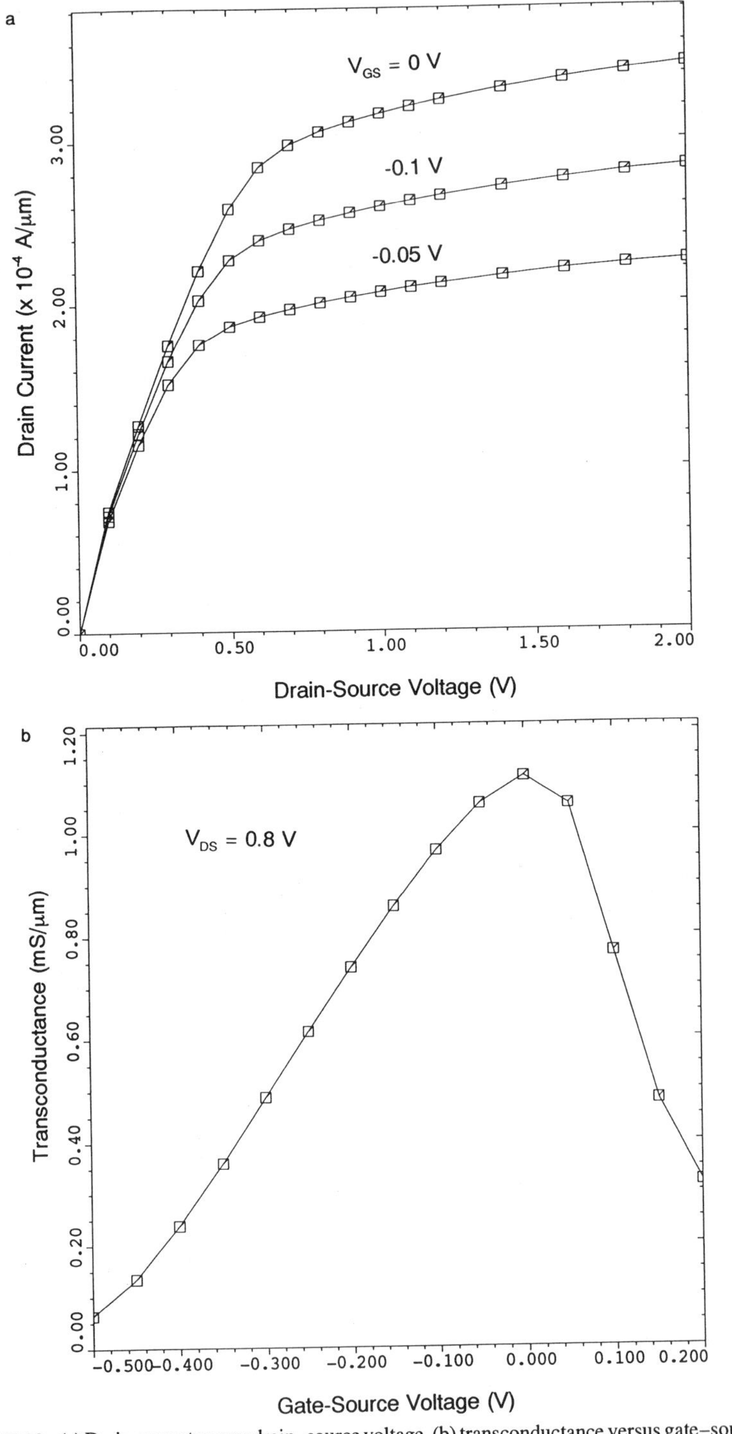

FIGURE 7.32. (a) Drain current versus drain–source voltage, (b) transconductance versus gate–source voltage, and (c) cutoff frequency versus gate–source voltage.

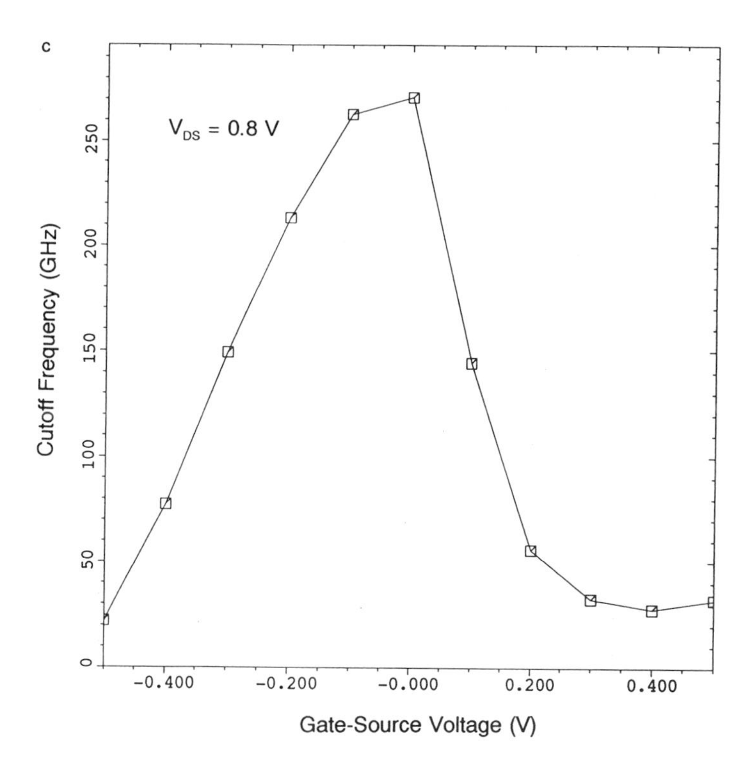

FIGURE 7.32. (Continued)

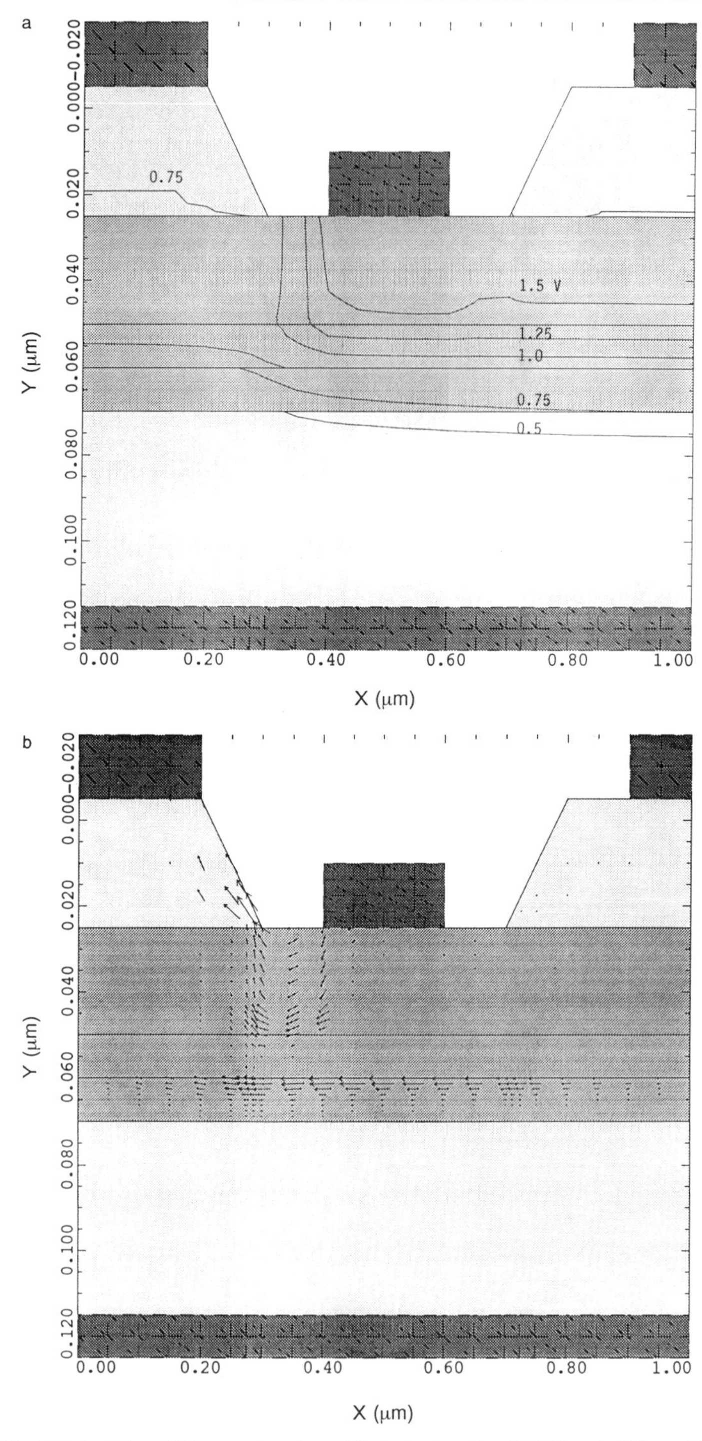

FIGURE 7.33. (a) Potential and (b) current vectors of the heterojunction FET biased at $V_{GS} = 2$ V and $V_{DS} = 0.8$ V, and (c) gate current versus gate–source voltage at $V_{DS} = 0.8$ V.

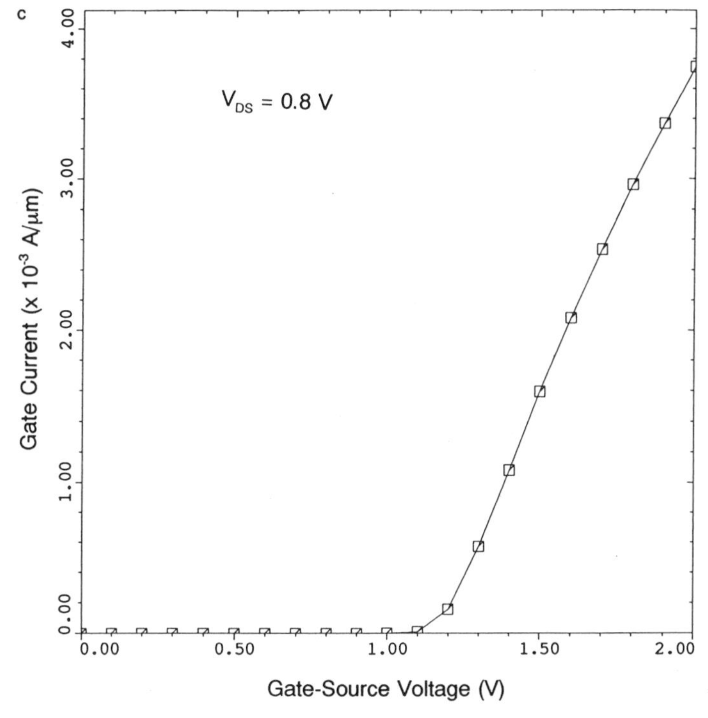

FIGURE 7.33. (Continued)

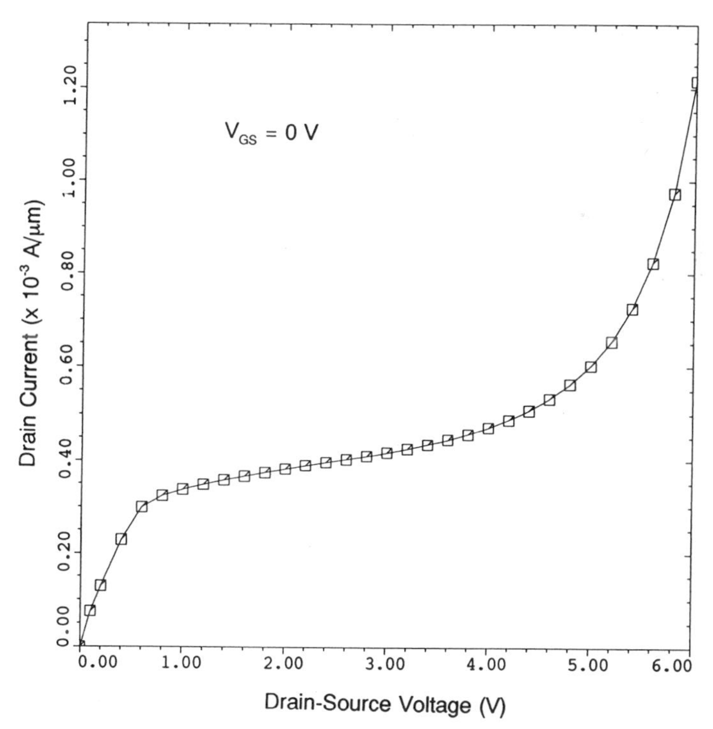

FIGURE 7.34. Drain current versus drain–source voltage.

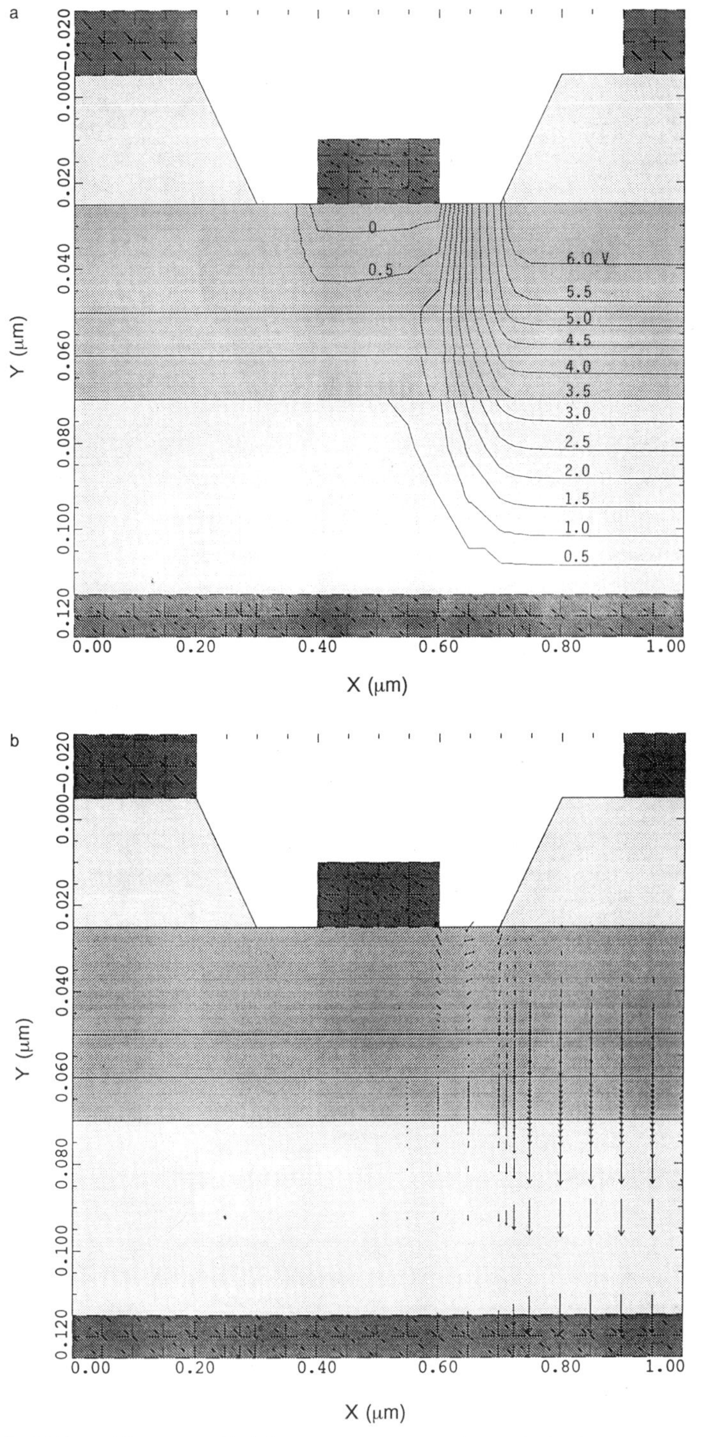

FIGURE 7.35. (a) Potential contour, (b) hole current vectors, (c) impact ionization rate contours, and (d) three-dimensional ionization rates distribution at $V_{DS} = 6$ V.

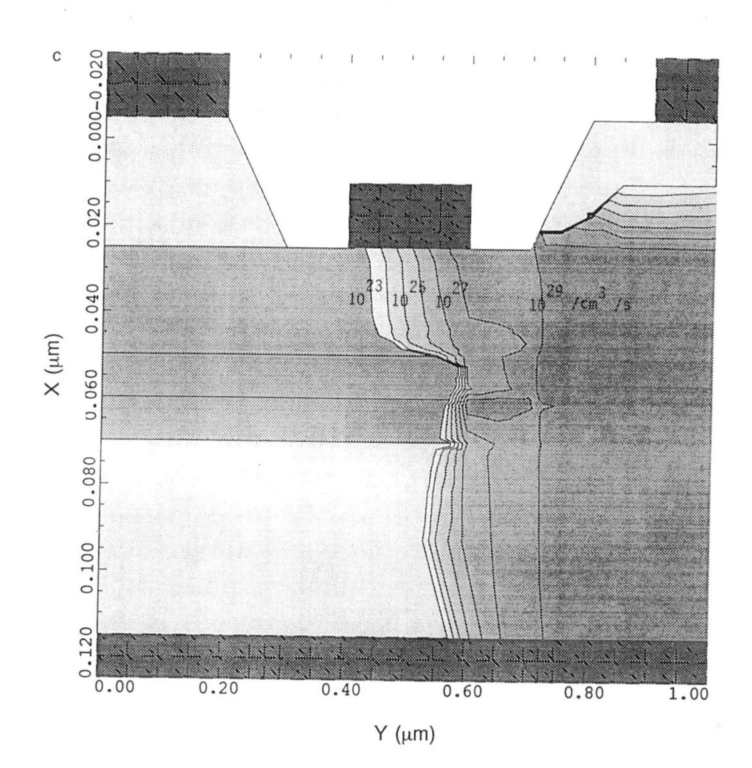

FIGURE 7.35. (Continued)

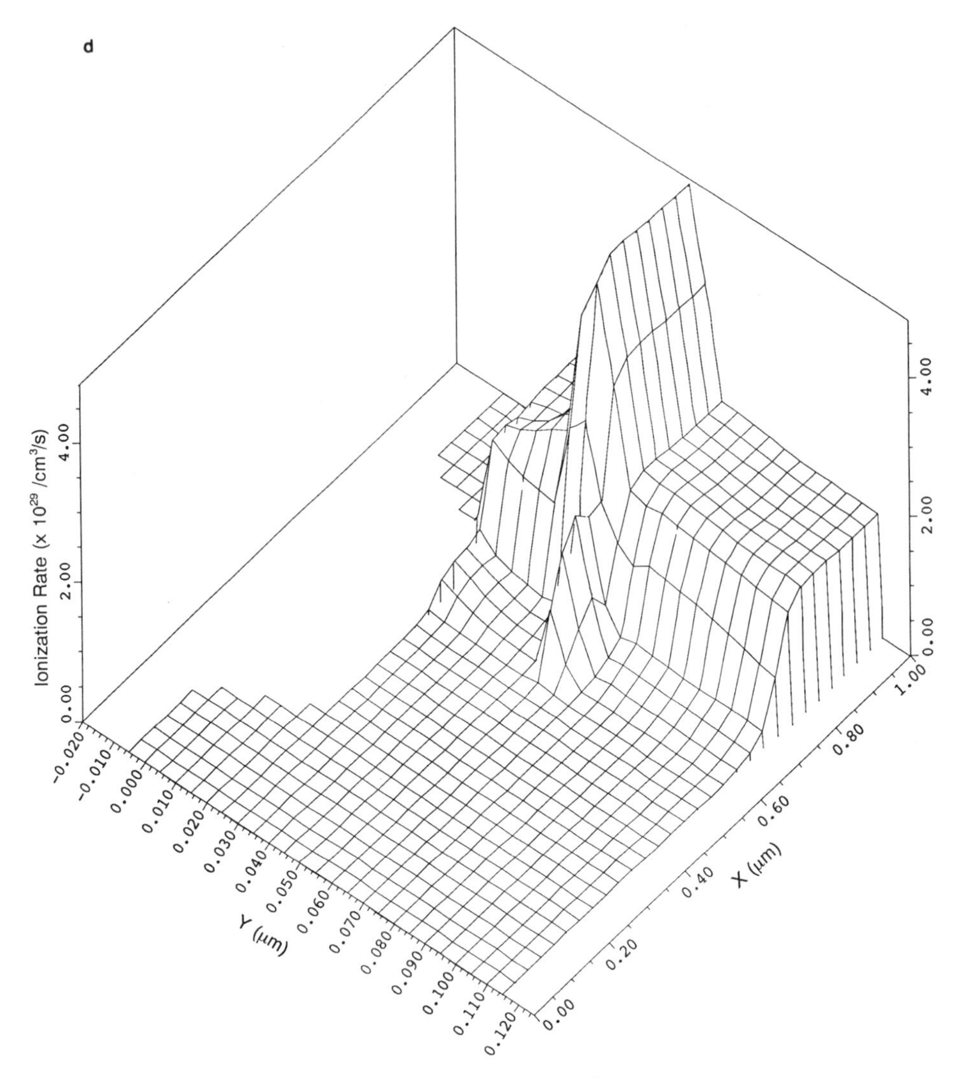

FIGURE 7.35. (Continued)

REFERENCES

Bardeen, J. (1947), *Phys. Rev.* **71**, 717.

Barton, T. M., C. M. Snowden, and J. R. Richardson (1986), in *Simulation of Semiconductor Devices and Processes*, vol. 2, eds. K. Board and D. R. K. Owen (Pineridge Press, Swansea, UK) p. 528.

Bethe, H. A. (1942), MIT Radiation Lab. Report, 43-12.

Cappy, A. (1980), *IEEE Trans. Electron Devices* **ED-27**, 2158.

Chang, C.-S. and D.-Y. Day (1989), *IEEE Trans. Electron Devices* **ED-36**, 269.

Chang, C.-S. and H. R. Fetterman (1986), *Solid-State Electron.* **29**, 1295.

Chao, P.-C., M. S. Shur, R. C. Tiberio, K. H. G. Duhm, P. M. Smith, J. M. Ballingall, P. Ho, and A. A. Jarra (1989), *IEEE Trans. Electron Devices* **ED-36**, 461.

Choi, S. and M. B. Das (1994), *IEEE Trans. Electron Devices* **ED-41**, 1725.
Cowley, A. M. and S. M. Sze (1965), *J. Appl. Phys.* **36**, 3212.
Crowell, C. R. and V. L. Beguwala (1971), *Solid-State Electron.* **14**, 1149.
Crowell, C. R. and S. M. Sze (1966), *Solid-State Electron.* **8**, 979.
Drummond, T. J., W. Kopp, H. Morkoç, and M. Keever (1982), *Appl. Phys. Lett.* **41**, 277–279.
Higgins, A. and D. N. Pattanayak (1982), *IEEE Trans. Electron Devices* **ED-29**, 179.
Himsworth, B. (1972), *Solid-State Electron.* **15**, 1353.
Liao, S. Y. (1985), *Microwave Devices and Circuits*, 2nd ed. (Prentice-Hall, Englewood Cliffs, NJ).
Lo, S. H. and C. P. Lee (1994), *IEEE Trans. Electron Devices* **ED-41**, 1504.
Mulle, R. S. and T. I. Kamins (1986), *Device Electronics for Integrated Circuits*, 2nd ed. (Wiley, New York).
Rhoderick, E. H. and R. H. Williams (1978), *Metal-Semiconductor Contacts* (Oxford Press, London).
Rocchi, M. (1985), *Physica* **129B**, 119.
Schottky, W. and E. Spenke (1939), *Wiss. Veroff. Siemens-Werken* **18**, 225.
Shih, K. M., D. P. Klemer, and J. J. Liou (1992), *Solid-State Electron.* 1639.
Shur, M. (1990), *Physics of Semiconductor Devices* (Prentice-Hall, Englewood Cliffs, NJ).
Sze, S. M. (1983), *Physics of Semiconductor Devices*, 2nd ed. (Wiley, New York).
Tan, K. L., D. C. Streit, R. M. Dia, S. K. Wang, A. C. Han, P. D. Chow, T. Q. Trinh, P. H. Liu, J. R. Velebir, and H. C. Yen (1991), *IEEE Electron Device Lett.* **EDL-12**, 213–214.
Wagner, C. (1931), *Phys. Z.* **32**, 641.
Yamaguchi, K. and S. Asai (1978), *IEEE Trans. Electron Devices* **ED-25**, 362.

EXAMPLE 7.1

```
COMMENT  Rectangular mesh
   MESH
 X.MESH  WIDTH=24  H1=1
 Y.MESH  DEPTH=4.0  H1=0.5  H2=1
 Y.MESH  Y.MAX=20  H1=1
 Y.MESH  Y.MAX=50  H1=2
COMMENT  Eliminate some unwanted substrate nodes
 ELIMIN  COLUMNS  Y.MIN=21
COMMENT  Specify the regions
 REGION  NUM=1  GaAs
COMMENT  Electrodes: #1=gate #2=substrate #3=source #4=drain
 ELECTR  NUM=1  X.MIN=9  TOP
 ELECTR  NUM=2  BOTTOM
 ELECTR  NUM=3  X.MAX=7  TOP
 ELECTR  NUM=4  X.MIN=17  TOP
PROFILE  N.TYPE  N.PEAK=1E4  UNIF  OUTF=SAINT_DOPE
PROFILE  N.TYPE  N.PEAK=1E16  Y.JUNC=20  XY.RAT=0.75
PROFILE  N.TYPE  N.PEAK=1E20  Y.JUNC=4  X.MIN=0.0  WIDTH=7
      +  XY.RAT=0.75
PROFILE  N.TYPE  N.PEAK=1E20  Y.JUNC=4  X.MIN=17  WIDTH=7
      +  XY.RAT=0.75
COMMENT  Regrid on doping
 REGRID  DOPING  LOG  RATIO=1  SMOOTH=1  IN.FILE=SAINT_DOPE
 REGRID  DOPING  LOG  RATIO=1  SMOOTH=1  IN.FILE=SAINT_DOPE
COMMENT  Specify contact parameters
CONTACT  NUM=1  SURF.REC
COMMENT  Physical models: Concentration-dependent mobility, field-dependent
```

```
       +  mobility, and surface mobility models
  MODELS  CONMOB FLDMOB SUFMOB2
 COMMENT  Symbolic factorization. Solve regrid on potential
    SYMB  CARRIERS=0
  METHOD  ICCF DAMPED
   SOLVE
  REGRID  POTEN IGNORE=2 RATIO=0.2 MAX=1 SMOOTH=1
       +  DOPF=pisaintds OUTF=pisaintms
 COMMENT  Solve using the refined grid
    SYMB  CARRIERS=0
   SOLVE  OUTF=pisaints
 COMMENT  Calculate the drain characteristics
 COMMENT  Solve Poisson equation only
    SYMB  CARRIERS=0
  METHOD  ICCG DAMPED
   SOLVE  V1=0.0
 COMMENT  Use Newton's method to solve for electrons
    SYMB  NEWTON CARRIERS=2
  METHOD  ITLIMIT=60
 COMMENT  Setup a log file for I–V data
     LOG  IVFILE=MDEX1DI
 COMMENT  Ramp the drain
   SOLVE  V4=0 V2=0 V1=0 V3=0 ELEC=4 VSTEP=0.2 NSTEP=3
   SOLVE  V4=0.6 ELEC=1 VSTEP=0.2 NSTEP=4
 COMMENT  Plot Ids vs Vds
 PLOT.1D  Y.AXIS=I4 X.AXIS=V4 POINTS TOP=1.6E–4 COLOR=2
 COMMENT  Potential contour at the last bias point
 PLOT.2D  BOUND JUNC DEPL FILL SCALE
 CONTOUR  POTENTIA MIN=–1 MAX=4 DEL=.25 COLOR=6
```

EXAMPLE 7.2

```
 COMMENT  HEMT Simulation
    MESH  IN.FILE=HEMT.MSH ASCII.IN
       $  REGION=1 GaAs Body
       $  REGION=2 AlGaAs InGaAs channel
       $  REGION=3 AlGaAs n-AlGaAs (under gate)
       $  REGION=4 = Electrode #1 Electrode Gate
       $  REGION=5 GaAs Source n+
       $  REGION=6 GaAs Drain n+ Source Contact
       $  REGION=7 = Electrode #2 Drain Contact
       $  REGION=8 = Electrode #3 Substrate
       $  REGION=9 = Electrode #4 AlGaAs Spacer
       $  REGION=10 AlGaAs AlGaAs Spacer
 COMMENT  Doping profiles
 PROFILE  REGION=1 N.TYPE CONC=1E2 UNIF
 PROFILE  REGION=2 P.TYPE CONC=1E2 UNIF
 PROFILE  REGION=3 N.TYPE CONC=4E17 UNIF
```

```
     PROFILE  REGION=5 N.TYPE CONC=2E20 UNIF
     PROFILE  REGION=6 N.TYPE CONC=2E20 UNIF
     PROFILE  REGION=10 N.TYPE CONC=1E2 UNIF
     PROFILE  N.TYPE CONC=1E19 Y.MIN=0.05 CHAR=0.002
        FILL  ^NP.COL SET.COL C.GAAS=2 C.ALGAAS=3
     COMMENT  Assign band gaps for the materials
      ASSIGN  NAME=AL N.VAL=0.17
      ASSIGN  NAME=IN N.VAL=0.17
      ASSIGN  NAME=ALG N.VAL=1.424+@AL*1.247
      ASSIGN  NAME=ING N.VAL=@ALG-(1.247*@AL+1.5*@IN-0.4*@IN*@IN)
      ASSIGN  NAME=ALM N.VAL=8000-22000*@AL+10000*@AL*@AL
    MATERIAL  REGION=2 EG300=@ING AFFIN=4.37
    MATERIAL  REGION=(3,10) GG300=@ALG
    MOBILITY  REGION=2 UN,MAX=13800
    MOBILITY  REGION=(3,10) MUN,MAX=@ALM
     CONTACT  NUM=1 SCHOTTKY WORK=5.17
        SAVE  MESH OUT.F=MES.MSH
       MODEL  CONSRN AUGER ANALYTIC
        SYMB  NEWTON CARR=0
       SOLVE  INI
        SYMB  NEWTON CARR=2
       SOLVE  ELEC=1 VSTEP=0.5 NSTEP=4
       SOLVE  ELEC=3 VSTEP=0.2 NSTEP=3
       SOLVE  V1=2.0 V3=0.8
     PLOT.2D  FILL BOUND
+Title="Current Flow"
        FILL  REGION=2 COLOR=5 ^NP.COL
     CONTOUR  FLOW
     PLOT.2D  FILL BOUN
     CONTOUR  POTENTIAL
```

8

Heterojunction Bipolar Transistors

The concept of the heterojunction bipolar transistor (HBT) was introduced by William Shockley in 1948 (U.S. patent: 2569347). The detailed theory of the device was developed by Kromer in 1957 (Kromer, 1957). The great potential advantages of such a heterostructure design over the conventional homostructure design have long been recognized (Kromer, 1957, 1982), but it was not until the early 1970s, that technology evolved to build practically useful transistors of this kind. The situation began to change with the emergence of liquid-phase epitaxy (LPE) as a technology for III–V compound semiconductor heterostructures. Since the mid-1970s, two additional very promising technologies have appeared: metal–organic chemical vapor deposition (MOCVD) (Dupuis *et al.*, 1979) and molecular beam epitaxy (MBE) (Cho and Arthur, 1975). Impressive results on MOCVD and MBE-grown heterojunction bipolar transistors have been attained (Konnzai *et al.*, 1979; Beneking and Su, 1980).

AlGaAs–GaAs heterojunction bipolar transistors are promising devices for high-speed and high-frequency operation (Asbeck *et al.*, 1987). In comparison with homojunc-

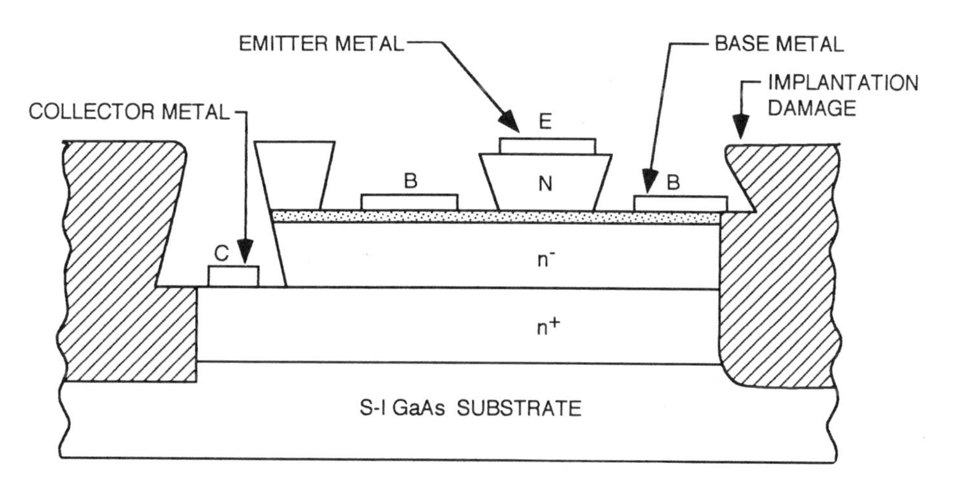

FIGURE 8.1. HBT cross section (Asbeck *et al.*, 1987).

tion Si bipolar transistors, they benefit from higher cutoff frequency, reduced base resistance, lower emitter–base and collector–base junction capacitances, and higher Early voltage. In comparison with GaAs field-effect transistors, heterojunction bipolar transistors have higher transconductance, higher current and power density, better threshold voltage matching, and lower $1/f$ noise.

Figure 8.1 shows a representative HBT device cross section. Emitter layers consist of Al_xGa_{1-x} As with AlAs mole fraction x chosen to be about 0.3. The base layer is typically made 0.05 to 0.1 μm thick, with values of doping from 5×10^{18} to 10^{20} cm^{-3}. High base doping is used to reduce the base resistance and to improve the switching speed. The collector layer is made about 0.5 μm thick, with values of doping from 2×10^{16} to 5×10^{17} cm^{-3}. High collector doping reduces collector resistance and prevents base pushout effect with a decreased collector breakdown voltage.

8.1. HETEROJUNCTION PHYSICS

The central design principle of heterostructure devices utilizes energy-gap variations in addition to electric fields as forces acting on electrons and holes, to control the distribution and flow of charge particles (Kromer, 1982). By a careful combination of energy-gap variations and electric fields, it becomes possible, within wide limits, to control the forces acting on electrons and holes, separately and independently of each other, a design freedom not achievable in homostructures. The resulting greater design freedom permits a reoptimization of doping levels and geometries of the HBT, leading to higher-speed devices.

In HBTs a wide-energy-band-gap emitter is used. The basic theory behind a wide-gap emitter is as follows. Consider the energy-band structure of an *N–p–n* bipolar transistor (capital *N* represents the wide-band-gap emitter) (Fig. 8.2). The emitter–base band edge is sufficiently graded to eliminate any band-edge discontinuities in the conduction band. The dc currents flowing in such a transistor include I_n (electrons injected from the emitter into the base), I_p (holes injected from the base into the emitter), and I_s (due to electron–

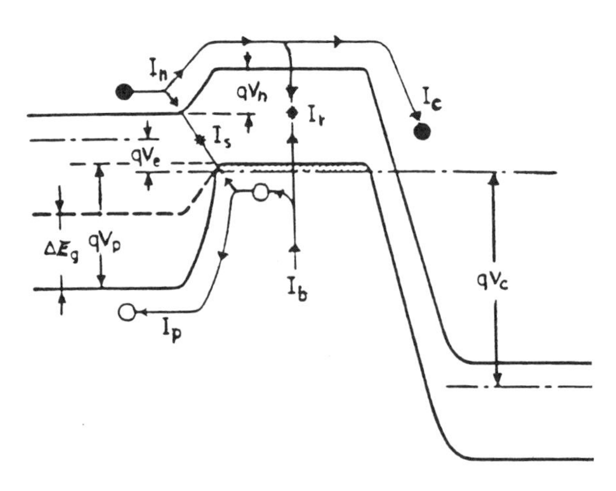

FIGURE 8.2. Energy-band diagram of a heterojunction bipolar transistor (Kromer, 1982).

hole recombination within the emitter–base space-charge layer). In addition, a small part I_r of I_n is lost due to bulk recombination. Neglecting recombination effects, the maximum current gain β_{max} is $\approx J_n/J_p$. The electron and hole injection current densities are of the form

$$J_n = N_e v_{nb} e^{-qv_n/kT} \tag{8.1.1}$$

$$J_p = p_e v_{pe} e^{-qv_p/kT} \tag{8.1.2}$$

where v_{nb} and v_{pe} are the mean speeds, N_e and p_e are emitter and base doping levels, and qV_n and qV_p are the heights of the potential energy barriers for electrons and holes between emitter and base. Since the energy gap of the emitter is larger than that of the base, we obtain

$$\beta_{max} = \frac{J_n}{J_p} = \frac{N_e v_{nb} e^{\Delta E_g/kT}}{p_b v_{pe}} \tag{8.1.3}$$

As a result of a large energy-gap difference between emitter and base, very high values of I_n/I_p can be achieved regardless of the doping ratio.

It is clear that the emitter–base heterojunction plays an important role in current transport in the HBT. In AlGaAs/GaAs HBTs, there are three basic types of emitter–base heterojunctions: the abrupt heterojunction, the graded heterojunction, and heterojunctions with a setback layer. Modern epitaxial technologies produce abrupt heterojunctions in which band-edge discontinuities are present. The presence of spikes necessitates the inclusion of thermionic emission and tunneling mechanisms for the carrier transport from emitter to base. The conduction-band spike can be removed by using a compositional grading over the heterojunction and results in monotonically varying band edges (Grinberg *et al.*, 1984a). If the graded junction is completely smooth, conventional drift and diffusion mechanisms are sufficient to describe carrier transport across the heterojunction. The insertion of a thin spacer or setback layer of intrinsic GaAs between the emitter and base can reduce the barrier potentials on both sides of the heterojunction (Hafizi *et al.*, 1990), thus reducing the importance of thermionic and tunneling mechanisms on the free-carrier transport across the heterojunction. An additional advantage of the setback layer is that it can prevent impurity out-diffusion from the heavily doped base to the emitter.

With the ability to control semiconductor band-gap variation, a gradual change in band gap across the base from E_{g0} near the emitter to $E_{g0} - \Delta E_g$ near the collector can now be designed. This energy gradient constitutes a quasi-electric field that assists electrons across the base by drift. Electrons can be driven by these high quasi-electric fields to velocities in excess of the value predicted by steady-state velocity-field curves. This results in a considerable improvement in base transit time and cutoff frequency.

A double-heterojunction bipolar transistor has a wide-band-gap emitter and a wide-band-gap collector. Valence-band discontinuities exist in both the emitter–base and collector–base heterojunctions. Double heterojunctions have the beneficial effect of eliminating the injection of holes from the base into the collector when the base–collector junction becomes forward biased. Increasing the band gap of the collector greatly reduces the charge storage in the collector and speeds up device turn-off time from saturation. With double-heterojunction devices, symmetrical operation in upward and downward directions can be established, which leads to circuit design flexibility. Additional advantages of the wide-band-gap collector are the increase in breakdown voltage and the reduction in leakage current.

8.2. DC CHARACTERISTICS

8.2.1. Collector and Base Currents

The minority carrier transport in the base of a bipolar transistor is adequately described by the drift–diffusion equation, provided the base width is sufficiently large. For shorter bases, the drift–diffusion equation is invalid and a more refined Boltzmann transport equation is required. For heterojunction bipolar transistors, the minority transport should be carefully distinguished between the cases of a graded and an abrupt emitter–base heterojunction. The graded HBT is essentially similar to the homojunction transistor, except that forces acting on the electrons and holes must include energy-gap variations in addition to electric fields. For the abrupt HBT, the transport mechanism across the emitter–base heterojunction is thermionic emission (Grinberg *et al.*, 1984b). Exact analytical formulas of the collector current are derived for the current–voltage characteristics of a double-heterojunction HBT, valid for arbitrary levels of injection and base doping, including the degenerate case (Grinberg and Luryi, 1993).

The base current in a HBT consists of recombination currents in the quasi-neutral base and emitter, at the emitter–base heterojunction, and at the emitter surface. The emitter recombination current is usually very low due to valence-band discontinuity, which prevents hole injection into the emitter. The surface recombination current of the HBT is due to high surface states at the emitter surface and the extrinsic base. The surface recombination can be a major component to the overall base current, especially for small-geometry devices where the device area-to-perimeter ratio is small (Hayama and Honjo, 1990). This surface recombination current increases exponentially with the base–emitter voltage and has an ideality factor which is closer to 1 than 2, as evidenced by experimental data (Liu and Harris, 1992).

Because of the valence-band discontinuity, which suppresses hole injection into the emitter, the current gain of the heterojunction bipolar transistor is increased significantly compared to that of the silicon bipolar transistor. The base doping of the HBT is thus increased to maintain a reasonable current gain. The increase of base doping decreases the base resistance. This effect results in an increase of cutoff frequency and improved switching speed of the HBT circuit.

8.2.2. Offset Voltage

AlGaAs–GaAs heterojunction bipolar transistors exhibit significant collector–emitter offset voltage V_{CE}, as shown by experimental data (Won *et al.*, 1989) in Fig. 8.3.

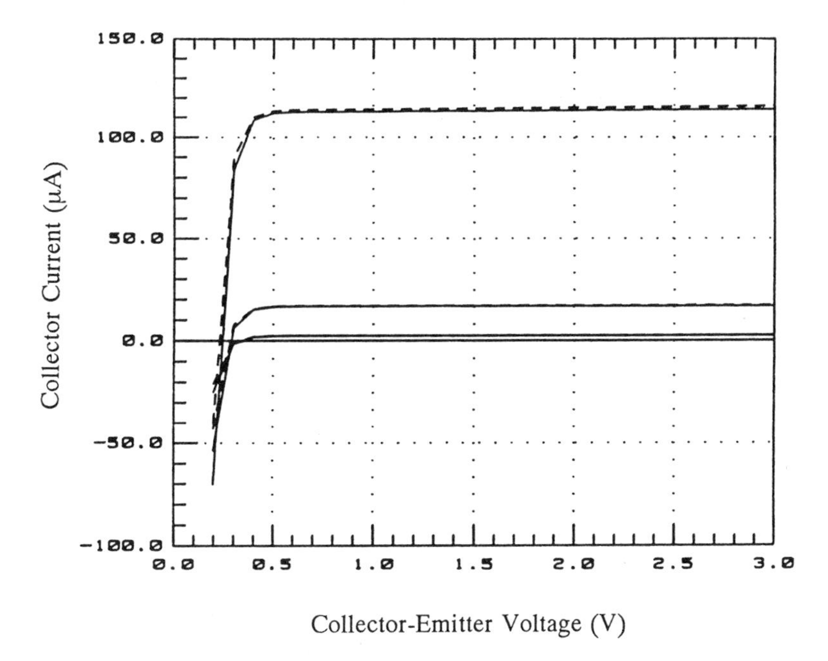

FIGURE 8.3. Collector current versus collector–emitter voltage.

This offset voltage reduces logic swing in digital circuits and increased power dissipation in analog amplifiers. The V_{CE} is attributed to the difference between the turn-on voltages of the emitter–base and collector–base heterojunctions and the difference in size of the collector and emitter areas. In the offset region, both emitter–base and collector–base junctions are forward biased, and the overall collector current is negative. Typically, V_{CE} is about 200 mV for a single-heterojunction bipolar transistor. This voltage can be significantly reduced if a double-heterojunction bipolar transistor is used.

8.2.3. Velocity Overshoot and Ballistic Transport

For GaAs materials, the velocity-field relationship is quite complicated. As the field in the collector space-charge region increases, the electrons in the lower valley can be field-excited to the normally unoccupied upper valley (Sze, 1981). This results in a differential negative resistance and electron velocity overshoot in GaAs material. For AlGaAs–GaAs HBTs, electron velocity overshoot occurs at the collector–base space-charge region. Velocity overshoot contributes to reduced base–collector transit time, reduced capacitance, and increased cutoff frequency.

When the base thickness of the bipolar transistor is made very thin, electrons injected into the base encounter only a few collisions before reaching the collector. High-energy electrons traverse across the thin base ballistically. Conventional drift and diffusion mechanisms used to describe carrier transport become invalid. To accurately describe the movement of electrons in the thin-base bipolar transistor, Monte Carlo methods must be used (Bandyopadhyay *et al.*, 1987). Monte Carlo simulation is a numerical technique that can be used to solve the Boltzmann transport equation. This technique simulates the electron motion in k-space when subjected to the applied electric field and specified scattering mechanisms. The scattering events are selected according to probabilities describing the microscopic processes. Subsequently, the Monte Carlo simulation generates a sequence of random numbers with a given probability distribution used to simulate the random nature of electron motions. If the number of simulated electrons is large enough, the average or ensemble results provide a good approximation of the average behavior of the electrons within a thin-base HBT. Since Monte Carlo simulations are extremely time consuming, most device simulators are developed based on drift–diffusion equations. Some device simulators, however, use hydrodynamic equations to account for nonlocal carrier transport effects.

8.2.4. Self-heating Effect

Heterojunction bipolar transistors using GaAs semi-insulating substrates exhibit self-heating effects due to low thermal conductivity of the GaAs semi-insulating substrate (Maycock, 1967). The collector current of the heterojunction bipolar transistor shows a negative resistance slope at high power dissipation levels (Fig. 8.4). Unlike the silicon bipolar transistor, which has a positive temperature coefficient for current gain, the heterojunction bipolar transistor exhibits a decrease in current gain with increasing collector–base voltage due to self-heating. The negative temperature coefficient of the

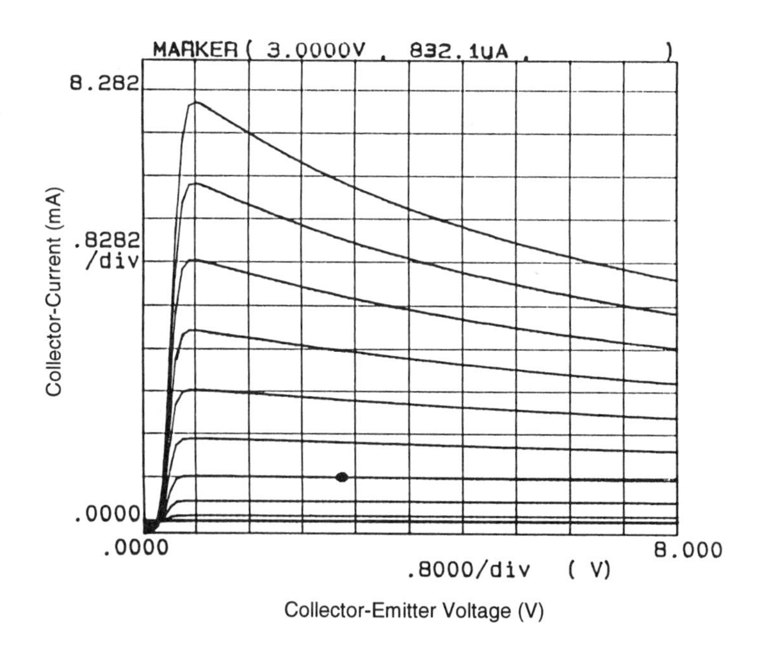

FIGURE 8..4. Collector current versus collector–emitter voltage showing negative resistance.

current gain results from band-gap narrowing in the base of the HBT. For example, self-heating increases base and collector currents. The increase of base current is larger than that of the collector current. Thus, for a constant base current in the I_C–V_{CE} plot, the collector current decreases with increasing collector–emitter voltage. The negative-resistance effect becomes significant when the power dissipation is large. That is why self-heating is very important for power heterojunction bipolar transistor circuit design.

The base current reversal due to impact ionization at the collector–base junction discussed in Chapter 3 for Si bipolar transistors has also been observed experimentally for GaAs heterojunction bipolar transistors (Zanoni *et al.*, 1992). Unlike the Si bipolar transistor, the base current reversal of the GaAs HBT is sensitive to device self-heating. Figure 8.5 shows the base current of the heterojunction bipolar transistor versus the collector–base voltage. The base current in the HBT increases with increasing collector–base voltage and decreases with V_{CB} at high collector–base bias. The initial increase of the base current is attributed to the junction temperature rise due to self-heating. The precipitous drop of the base current at higher V_{CB} is due to the exponential dependence of impact ionization coefficient on the electric field.

Self-heating effects on a current mirror are evaluated. The current mirror in Fig. 8.6 consists of two discrete heterojunction bipolar transistors Q_1 and Q_2 and two external resistors R_1 and R_2. Since the collector–emitter voltage of Q_1 equals the base–emitter voltage V_{BE1}, the bipolar transistor Q_1 usually does not experience self-heating. However, the collector–emitter voltage of the bipolar transistor Q_2 is $V_{CC} - I_{C2}R_2$ and Q_2 can have significant self-heating if the collector current I_{C2} and the collector–emitter voltage of Q_2 are high. Self-heating of Q_2 can thus result in a totally different output current much higher than the designed current at the isothermal condition. For a fixed resistor R_1, for instance, I_{C1} of the reference transistor Q_1 is constant. Current I_{C2} of Q_2 follows I_{C1} when $R_1 = R_2$. If R_2 is decreased, I_{C2} increases due to the increase of collector–emitter voltage at Q_2. The output current I_{C2} will then be much larger than the reference current.

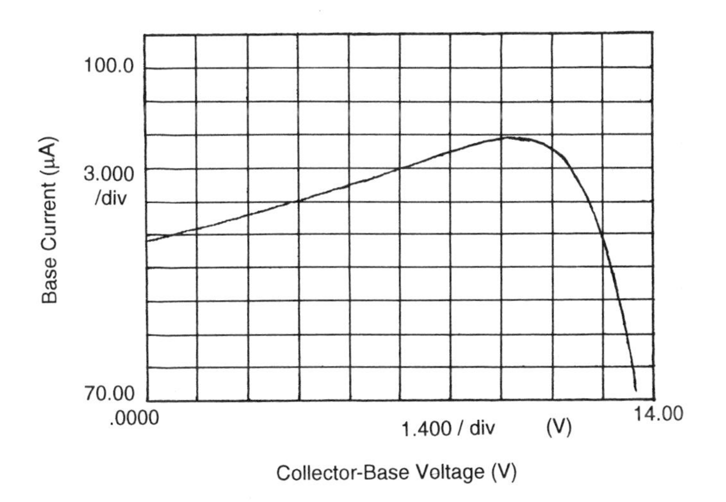

FIGURE 8.5. Base current versus collector–base voltage.

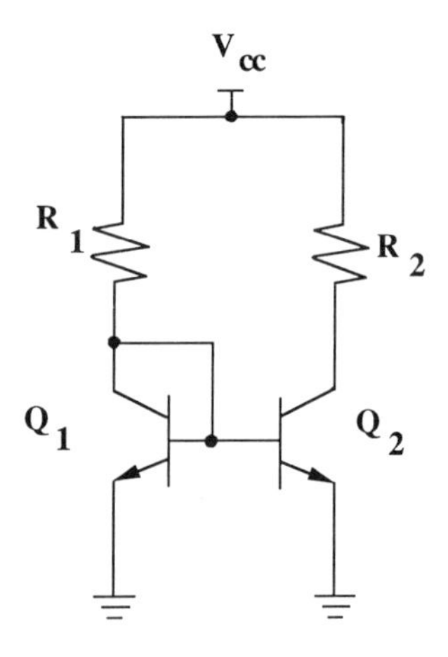

FIGURE 8.6. Schematic of a current mirror.

Figure 8.7 shows the sensitivity of the collector currents for reference transistor Q_1 and mirror transistor Q_2 as a function of $1/R_2$ at $V_{CC} = 15$ V. Resistor R_1 is selected to be 2 kΩ in order to make the base–emitter voltage V_{BE} equal 1.4 V. The mirror current designed to operate at the isothermal condition should follow the reference current (solid line in Fig. 8.7). The predicted mirror current with self-heating, however, increases with

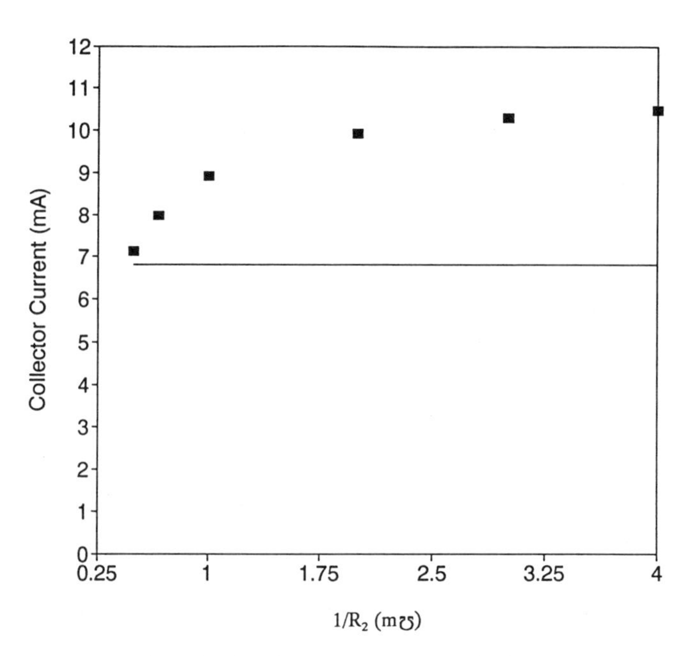

FIGURE 8.7. Collector current versus $1/R_2$.

decreasing R_2 due to an increase of collector–emitter voltage for Q_2. The solid squares of Fig. 8.7 represent R_2 simulated at 2k, 1.5k, 1k, 500, 333, and 250 Ω, respectively. The corresponding relative error $(I_{C2} - I_{C1})/I_{C1}$ is 4.85%, 17.3%, 31%, 46%, 51.3%, and 54.1%, respectively.

In device simulation, it is typically assumed that drift and diffusion transport is always isothermal, and that the carriers are always in a quasi-static state such that a direct relation between the carrier energy and the local electric field exists. The first assumption eliminates the carrier flux component driven by the temperature gradient, while the second allows the use of local field-dependent relationships for carrier transport parameters such as carrier mobility. Since only potential and carrier concentrations are included in the solutions, it is implicitly assumed that dissipated power density inside devices is so low that no significant lattice temperature increase occurs.

These assumptions, however, begin to fail for heterojunction semiconductor devices on the GaAs semi-insulating substrate where self-heating is significant. Moreover, hot carrier nonlocal transport, such as velocity overshoot, begins to emerge due to the large field gradient and the finite energy relaxation rate. Consequently, the drift–diffusion model becomes inadequate, and a full thermodynamic solution and lattice temperature are needed for more accurate device performance prediction.

"Energy balance" models have become popular as a method of accounting for nonlocal transport effects, and nonisothermal models have been developed to account for

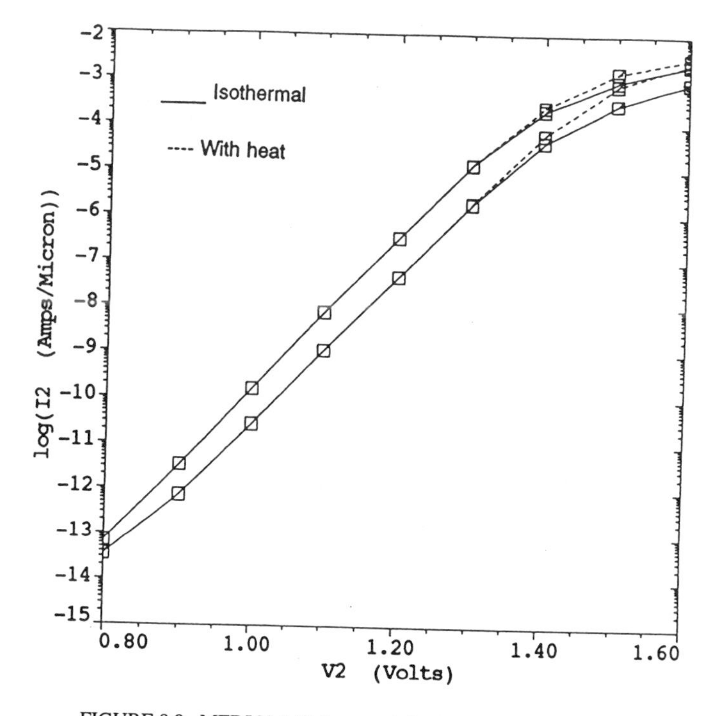

FIGURE 8.8. MEDICI I–V characteristics including self-heating.

lattice heating. The complete thermodynamic system consists of six partial differential equations and uses the nonisothermal energy-balance model (Apanovich *et al.*, 1995). These equations are

Poisson's equation:

$$\nabla(\varepsilon\nabla\psi) = -q(p - n + N_D^+ - N_A^-) \tag{8.2.1}$$

particle continuity equations for electron and holes:

$$J_n = qU \tag{8.2.2}$$

$$J_p = -qU \tag{8.2.3}$$

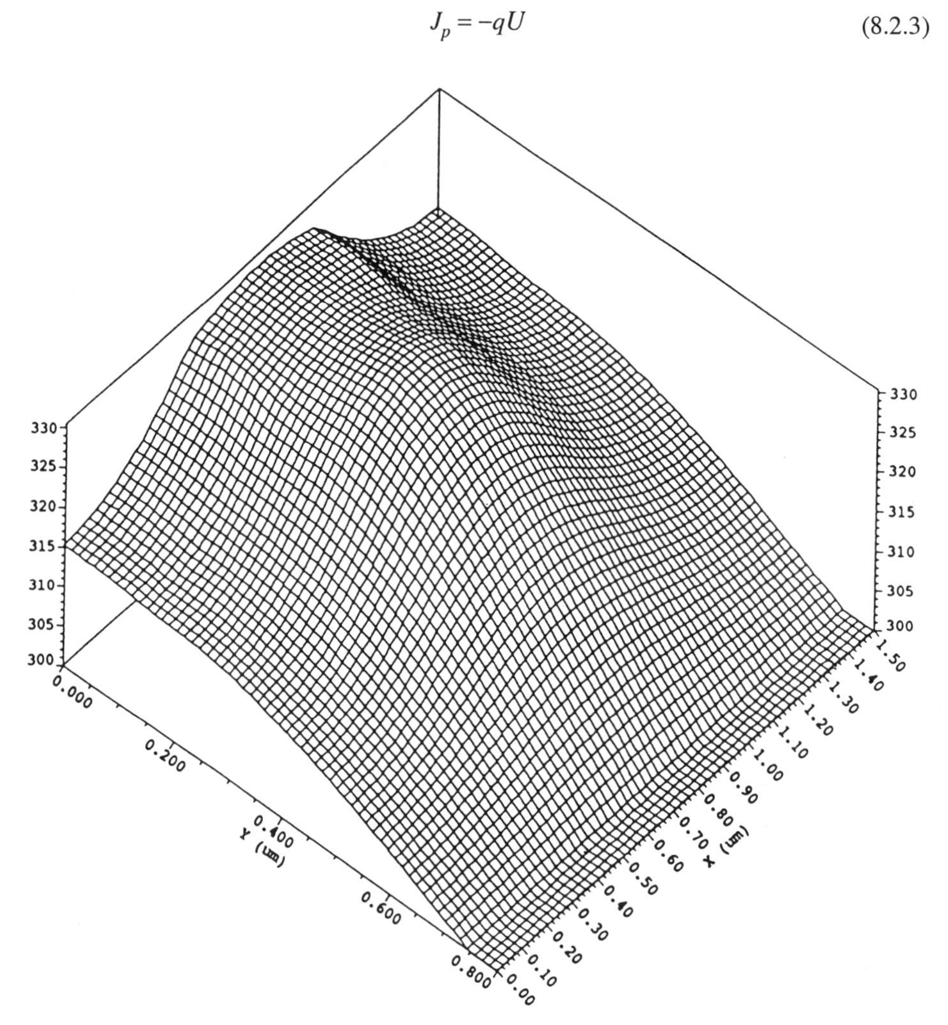

FIGURE 8.9. Nonuniform temperature distribution in a HBT.

particles energy balance equations for electrons and holes:

$$\nabla S_n = J_n \nabla (E_C/q) - W_n \tag{8.2.4}$$

$$\nabla S_p = J_p \nabla (E_V/q) - W_p \tag{8.2.5}$$

and a lattice energy-balance equation:

$$\nabla (\kappa \nabla T_L) = -H \tag{8.2.6}$$

S_n and S_p are the energy flux densities associated with electrons and holes, W_n and W_p are the energy density loss rates for electrons and holes, and H is the heat source for the lattice. For HBTs the collector and base currents predicted by using the nonisothermal and energy-balance models are higher than those predicted by drift and diffusion models alone.

Figure 8.8 shows the *I–V* characteristics of MEDICI HBT simulation including self-heating effect. Self-heating increases collector and base current at high V_{BE}. The device nonuniform temperature distribution is shown in Fig. 8.9.

8.3. RF BEHAVIOR

8.3.1. Junction Capacitance

Heterojunction capacitances play an important role in determining device switching speed and high-frequency response. Because of heavy doping in the quasi-neutral base and emitter regions, minority carrier charge storage in the emitter and base is small. The junction capacitance can be much larger than the diffusion capacitance of the heterojunction bipolar transistor. The emitter–base heterojunction capacitance applicable for the reverse-bias and low-forward-bias regions for different heterojunction systems was derived in Chapter 2. Emitter–base heterojunction capacitance for high forward bias has also been examined (Liou *et al.*, 1988). In this section, the effect of impact ionization on the collector–base junction capacitance of the HBT is presented.

The typical collector concentrations of heterojunction bipolar transistors are about 10^{17} cm^{-3}. As a result of high collector doping, the peak electric field at the collector–base metallurgical junction increases. If the field is sufficiently high, impact ionization could be triggered. Since impact ionization generates many free electrons and holes in the collector–base space-charge region, modulation of the mobile charge with respect to the base–collector voltage is significant in the avalanche regime. This phenomenon is consistent with a forward-biased *p–n* junction where excitation produces a change in volume charge density. The modulation of free-carrier charge creates a free-carrier capacitance.

Figure 8.10 shows the collector–base junction capacitance versus collector–base voltage at V_{BE} = 1.4 V. The collector–base junction capacitance decreases with collector–base voltage and then increases with collector–base voltage when the device is in close

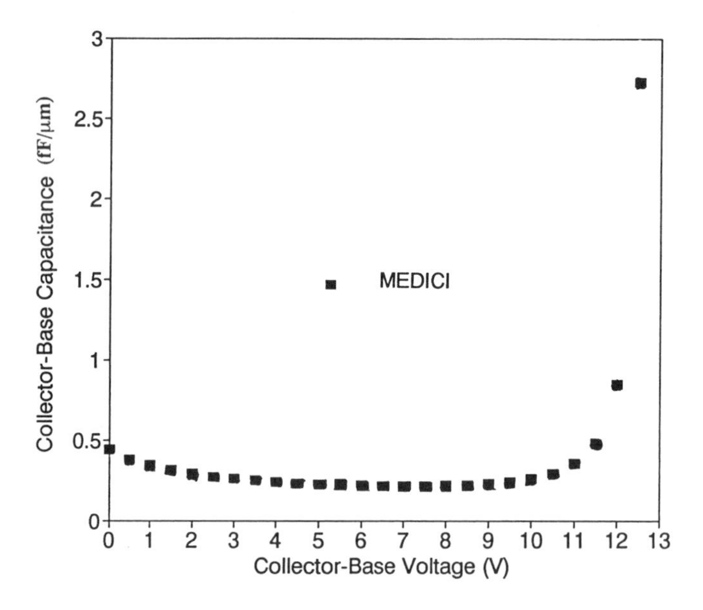

FIGURE 8.10. Collector–base junction capacitance versus collector–base voltage.

proximity to avalanche breakdown. To provide greater insight into electron and hole concentrations in the collector graphs obtained from MEDICI for the HBT at $V_{BE} = 1.4$ V and $V_{CB} = 6$ and 12 V are shown in Fig. 8.11. Both electron and hole concentrations and collector–base depletion region thickness at $V_{CB} = 12$ V are much larger than those at $V_{CB} = 6$ V. The difference is clearly due to avalanche multiplication at a large reverse-biased junction. Impact-ionization-induced holes drift toward the base terminal, which reduces the base current at high V_{CB} and leads to a negative base current shown in Fig. 8.12.

8.3.2. Transconductance

The transconductance is important in achieving small input-voltage-swing circuit operation, low output impedance for fast charging of load capacitances, and high voltage gain. For GaAs HBTs, high-current transconductance is enhanced by the high base doping and high electron velocity, which minimize high injection and base pushout effects to which bipolar transistors are more susceptible. However, the benefits of operating at high transconductance will be limited by the effects of metal migration, emitter ohmic contact, and HBT reliability.

8.3.3. Output Conductance

Output conductance describes the base-width modulation effect on the collector current. Because of high base doping, GaAs HBTs have negligible base-width modulation effect. They thus have much lower output conductance and higher Early

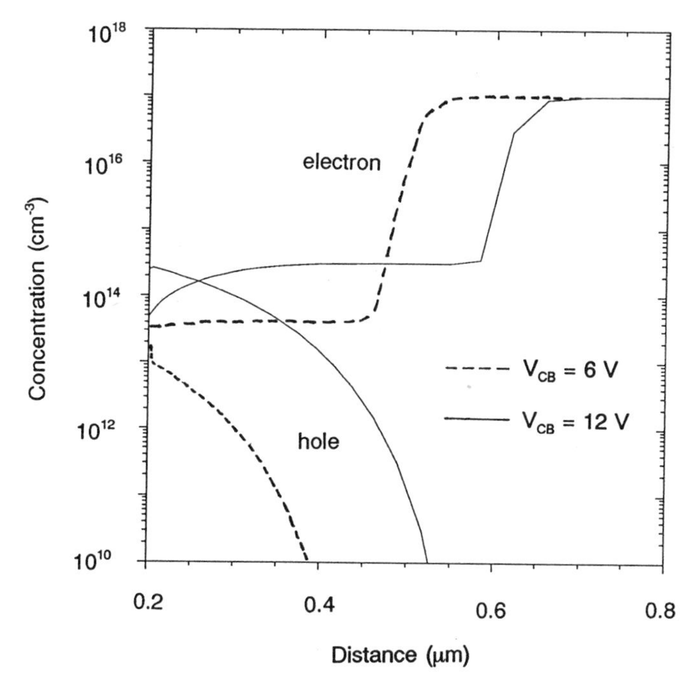

FIGURE 8.11. Electron and hole concentrations in the collector.

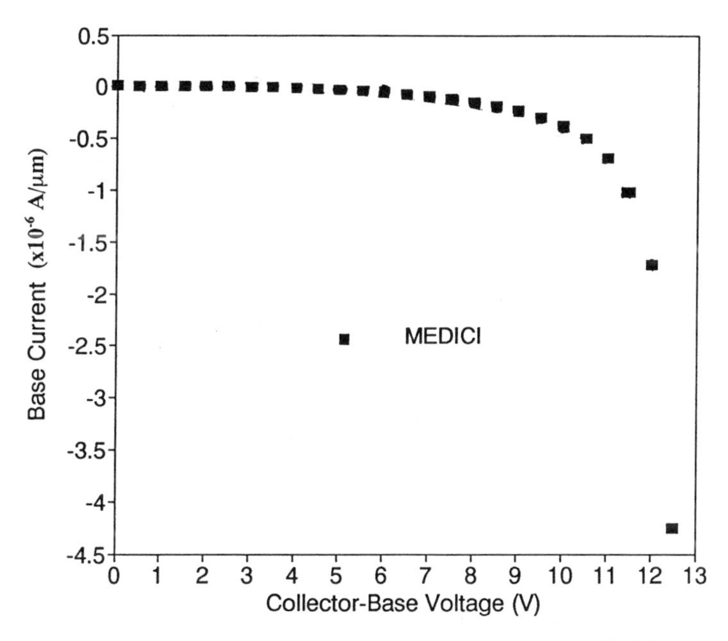

FIGURE 8.12. Base current versus collector–base voltage (MEDICI).

voltage (about 10 times higher) compared to silicon bipolar transistors. The output conductance of the HBT, however, suffers significant self-heating effect. For a given base current, self-heating decreases the collector current and causes a negative differential output resistance.

Figure 8.13 shows the output conductance versus collector–emitter voltage, including effect of self-heating from MEDICI. At very low V_{CE}, the output conductance decreases with V_{CE}. This is due to a transition from saturation to forward-active mode of operation. The output conductance increases with V_{CE} at high collector–emitter voltage due to self-heating effect.

8.3.4. *Cutoff Frequency*

The cutoff frequency of the HBT is a figure of merit. This parameter is important for analyzing the bandwidth of small-signal amplifiers and the power gain of power amplifiers. The cutoff frequency is given by the well-known approximation in Eq. (3.3.3).

The thermal effect also affects the cutoff frequency f_T of the heterojunction bipolar transistor. The cutoff frequency versus the collector current density (Liou *et al.*, 1993) is shown in Fig. 8.14. The increase of cutoff frequency at low collector current density is due to the decrease of emitter charging time ($\tau_E \propto 1/J_C$). When the collector current density is sufficiently high, the cutoff frequency decreases with J_C due to an increase of junction temperature resulting from self-heating effect. Self-heating decreases the satu-

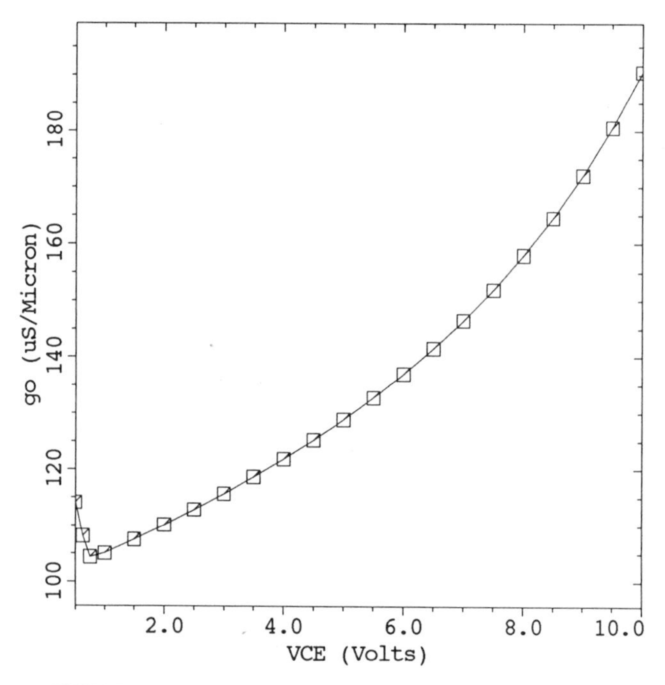

FIGURE 8.13. Output conductance versus collector–emitter voltage.

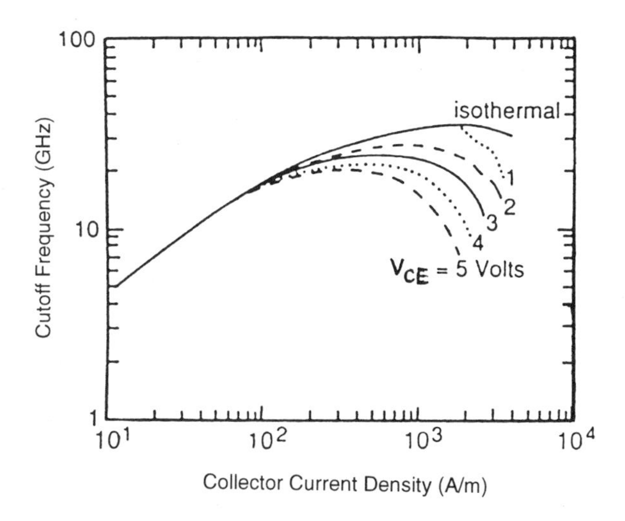

FIGURE 8.14. Cutoff frequency versus collector current density at different V_{CE} (Liou *et al.*, 1993).

ration velocity in the collector–base depletion layer and electron mobility in the base, which increases the collector–base depletion layer transit time and base transit time, respectively.

8.3.5. Flicker Noise

The HBT, like the Si BJT, has noise composed of base resistance thermal noise, flicker ($1/f$) noise, and shot noise from the base and collector currents. While shot noise depends on the HBT biasing, thermal noise and $1/f$ noise depend on device design and processing.

At low frequencies, noise processes occur that are predominantly $1/f$ in nature. Under low-forward-bias conditions, recombination in the emitter–base space-charge layer controls the conduction. The generation–recombination mechanism contributes $1/f$ noise (van der Ziel *et al.*, 1986). Also, the HBT's vertical current flow through well-shielded junctions leads to small trapping effects and low $1/f$ noise. Although the measured $1/f$ noise corner frequency, on the order of tens to several hundred kilohertz, is higher than Si bipolar transistors, it is a factor of 10 to 100 times lower than GaAs FETs, which suffer significant surface states and surface trapping effects. A direct effect is observed in the phase noise performance of the HBT in an oscillator circuit. Oscillator phase noise is dependent on the overall quality factor of the resonator and the flicker noise of the active devices.

Noise is an important parameter in designing analog circuits such as low-noise amplifiers and oscillators. When the noise of an oscillator is added to the carrier and fed back through the resonator, the resulting signal is both amplitude- and phase-modulated. The limiting amplifier removes the amplitude-modulated component but does not affect the phase modulation. Phase noise dominates the near-carrier spectrum of the oscillator and often determines one endpoint of the dynamic range in a system. The maximum undistorted signal depends on system linearity, while the minimum is set by noise. Presently, the phase noise spectrum of bipolar oscillators is dominated by $1/f$ noise of the device (Kim, 1988).

Phase noise corresponds to instability in the phase or frequency of a signal. It is measured as the ratio of power in the noise to that in the carrier. Noise power is measured at a specified offset frequency and a specified bandwidth. Measurement of phase noise requires good instrumentation setup and tedious experiment. Phase noise may be extracted from a small-signal equivalent circuit. The analysis of oscillator phase noise shows that the phase noise of a bipolar oscillator decreases with collector current until saturation occurs. The bipolar transistor biased at high collector current before saturation provides the lowest phase noise. Phase noise decreases with cutoff frequency and increases with forward transit time and packaging inductance of the bipolar transistor. The InGaAs heterojunction bipolar transistor can provide better phase noise characteristics since the HBT has smaller transit time and larger cutoff frequency.

8.4. TRANSIENT CHARACTERISTICS

The transient characteristics of the GaAs HBT could be different from the silicon BJT. For example, the heating effects related to the temperature differences of the input and latching transistors of the HBT voltage comparator were identified as the major source of comparator inaccuracy in analog-to-digital converters (Wang *et al.*, 1987). Also, self-heating causes nonuniform temperature distribution in the steady-state and transient operations. The emitter-up HBT has shorter heating and cooling processes than those of the collector-up HBT during switch-on and switch-off transients (Zhou *et al.*, 1995).

8.4.1. Emitter-Coupled Logic Using HBTs

In circuit applications, the highest-speed performance is achieved by using nonsaturating transistor operation. For digital and A/D applications, emitter-coupled logic and its variant, current-mode logic, provide the highest-speed performance. This is opposed to the slower saturating I^2L logic. In the ECL-type of logic, differential pair transistors steer current from a current source through load resistors based on the logic state of the inputs. In CML, the emitter-follower output drive transistor of ECL is eliminated; the differential pair transistors serve as both the logic switch and the output driver. ECL and CML typically have high power dissipation due to the use of a constant-current source and vertical logic structure. However, power dissipation can be traded for speed. Because of the HBT's high intrinsic speed, low power-delay products can be achieved. Furthermore, the HBT's high-performance switching and drive capabilities make the simpler CML approach attractive for many circuit applications.

A novel logic approach, diode–HBT logic, that is implemented with AlGaAs–GaAs HBTs and Schottky diodes to provide high-density and low-power digital circuit operation was proposed (Wang *et al.*, 1992). This logic family has been realized with the same technology used to produce ECL–MCL circuits. A gate delay of 160 ps was measured with 1.1 mW of power per gate. The new logic approach allows monolithic integration of high-speed ECL–MCL circuits with high-density diode–HBT circuits.

8.4.2. HBT Turn-off Transient

The basic principle of switch-off transient discussed in Chapter 3 is still valid for heterojunction bipolar transistors. When the AlGaAs–GaAs single-heterojunction bipolar transistor is switched from saturation to cutoff, the minority charge storage will be determined primarily by the hole injection stored in the lightly doped collector, since the base doping of the HBT is much larger than the collector doping. However, the collector–base junction of the HBT could be a heterostructure. For instance, the SiGe HBT is a double-heterojunction bipolar transistor. The collector–base heterojunction of

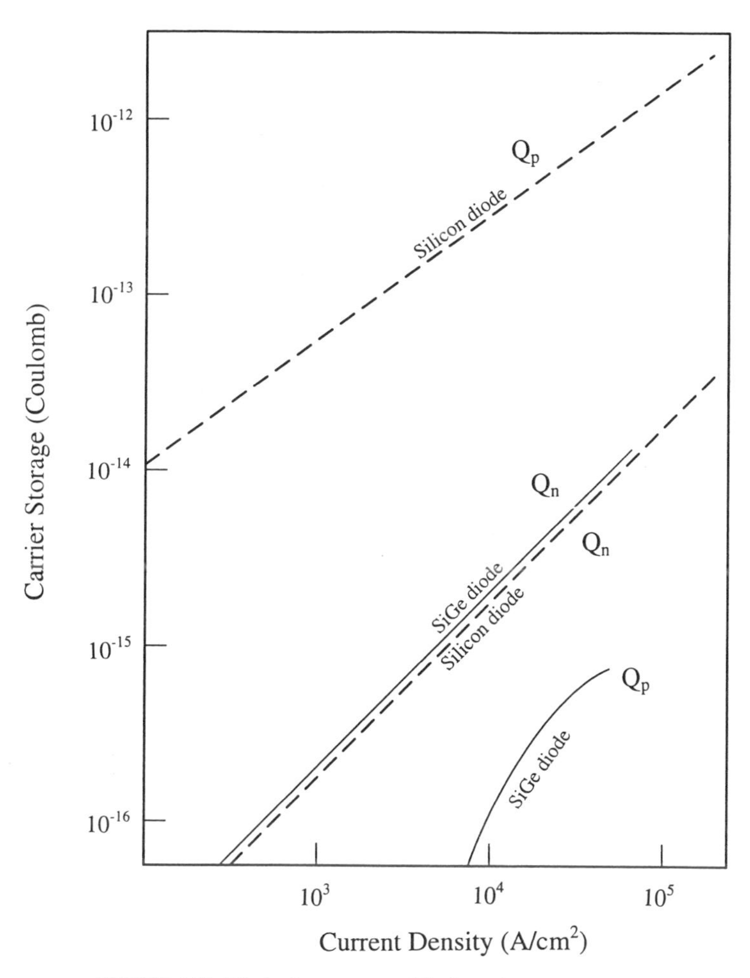

FIGURE 8.15. Diode charge storage (Ugajin and Amemiya, 1991).

the SiGe HBT hardly shows minority carrier accumulation in forward conduction, so high-speed diodes for LSI or VLSI circuits can be realized using this junction. The SiGe-base HBTs also show little carrier accumulation even when operated in the saturation region, so they are capable of operating at high speed in saturated mode switching applications. This is because the hole injection is prevented by the large potential barrier at the heterojunction interface and holes are hardly accumulated in the *n*-type collector region. Figure 8.15 shows the simulation result of carrier storage in the SiGe diode and the silicon homojunction diode as a function of forward current density (Ugahin and Amemiya, 1991). The SiGe band-gap decrement is 0.3 eV. Q_n shows the absolute value of electron storage in the *p*-type region, and Q_p shows hole storage in the *n*-type region. It is clear from this figure that the SiGe diode has a total hole storage two orders less than that of the silicon diode. The detail of SiGe HBT is discussed in Sec. 8.6.

8.5. InP HBTs

InGaAs-base heterojunction bipolar transistors lattice-matched to InP substrates are emerging as an alternative technology to GaAs-base HBTs. InGaAs heterojunction devices offer numerous advantages over GaAs devices for high-speed, low-power analog, digital, and optoelectronic applications due to the following superior properties (Jalali and Pearton, 1995; Asbeck *et al.*, 1987):

- A higher Γ–L valley separation, which gives pronounced velocity overshoot and shorter space-charge layer transit time
- A lower surface recombination velocity, which reduces $1/f$ noise and surface recombination current density
- Smaller band gap, which reduces the turn-on voltage and minimize power dissipation
- Higher carrier mobilities, which translate into superior minority carrier transport and bulk resistance
- Excellent specific contact resistance for nonalloyed ohmic contacts to *n*- and *p*-type InGaAs
- Compatibility with 1.3–1.55-μm lightwave communication systems

InP-base HBTs can be fabricated with several different combinations of epitaxial InP, GaInAs, and AlInAs lattice-matched to InP. For the wide-band-gap emitter, both AlInAs and InP have been used with comparable device performance. To achieve the lowest turn-on voltage while maintaining a large valence-band discontinuity at the base–emitter junction, it is necessary to compositionally grade the base–emitter junction between the GaInAs base and the AlInAs or InP emitter. A compositionally graded base–emitter junction also significantly improves the reliability of the HBT with respect to current gain and turn-on voltage stability. For applications requiring higher base–collector breakdown voltages and lower output conductances, a wide-band-gap InP collector is the material of choice.

Because of the unique material properties of InP and the ternary compounds lattice-matched to it, InP-based HBT technology has a number of advantages over the more widely used AlGaAs–GaAs HBT technology. From a device standpoint, the low-band-gap energy of GaInAs combined with a compositionally graded base–emitter junction results in device turn-on voltages lower than those of silicon bipolar transistors. This translates to a potential for lower IC power dissipation than in GaAs technologies, with the added advantage that the In–P based HBTs can be "dropped in" to integrated circuit applications requiring silicon bipolar technology voltage levels. AlInAs–GaInAs HBTs have stable dc and RF characteristics over a wide temperature range from cryogenic to over 125°C.

8.6. SiGe HBTs

AlGaAs heterojunction bipolar transistors are being actively investigated for microwave–millimeter and high-speed digital applications and are receiving most of the attention in HBT research. Other III–V compound materials with smaller surface recombination velocities, such as InP and InGaAs, are being used to reduce base surface recombination current and to eliminate emitter size effects (Sze, 1990). Material systems in column V and growth techniques compatible with silicon technology have been developed (Iyer *et al.*, 1987). The progress in the MBE and CVD growth of strained SiGe on Si has allowed band-gap engineering technology to be transferred to the silicon-based system. SiGe alloys have smaller fundamental band gaps compared to silicon because of a larger lattice constant and altered lattice constituents. The smaller band gap in the HBT base increases minority carrier injection. This results in an increase in collector current and current gain even though the base doping is high. High base doping yields a smaller base resistance, lower noise figure, and reduced gate propagation delay.

The addition of Ge reduces the band gap of Si leading to an alloy that can be used to form the narrow-band-gap SiGe base of the HBT. The lattice constant of the alloy differs considerably from that of Si if the thickness of epitaxial layers is kept below a critical thickness. Then the mismatch between the alloy and the Si substrate is compensated for elastically and no misfit dislocations form. The resulting SiGe pseudomorphic layers are considerably strained. The overall band-gap shrinkage of strained SiGe is increased over that of unstained material with the same Ge content. The band-gap shrinkage in the *p*-type SiGe base of the HBT increases electron injection into the base. The band-gap difference ΔE_G between the emitter and the base consists of the valance-band discontinuity ΔE_V and the conduction-band discontinuity ΔE_C (Fig. 8.16). In the SiGe HBT, most of the band-gap reduction results from a shift in the valence-band edge, and the conduction-band discontinuity is usually a small fraction of the total band-gap difference (King *et al.*, 1989).

The germanium profiles with the same total germanium content and graded base width include uniform, trapezoid, and triangle profiles (Fig. 8.17). The grading of the Ge across the quasi-neutral base induces a drift field in the base that accelerates the electrons injected from the emitter to the collector, thereby decreasing the base transit time compared to a Si BJT. Figure 8.18 shows the ratio of SiGe to Si for the four parameters of current gain, Early voltage, gain–Early voltage product, and inverse base transit time. The *x* axis represents the total Ge grading across the base. At $\Delta E_{g,Ge}$(grade) = 0, a pure

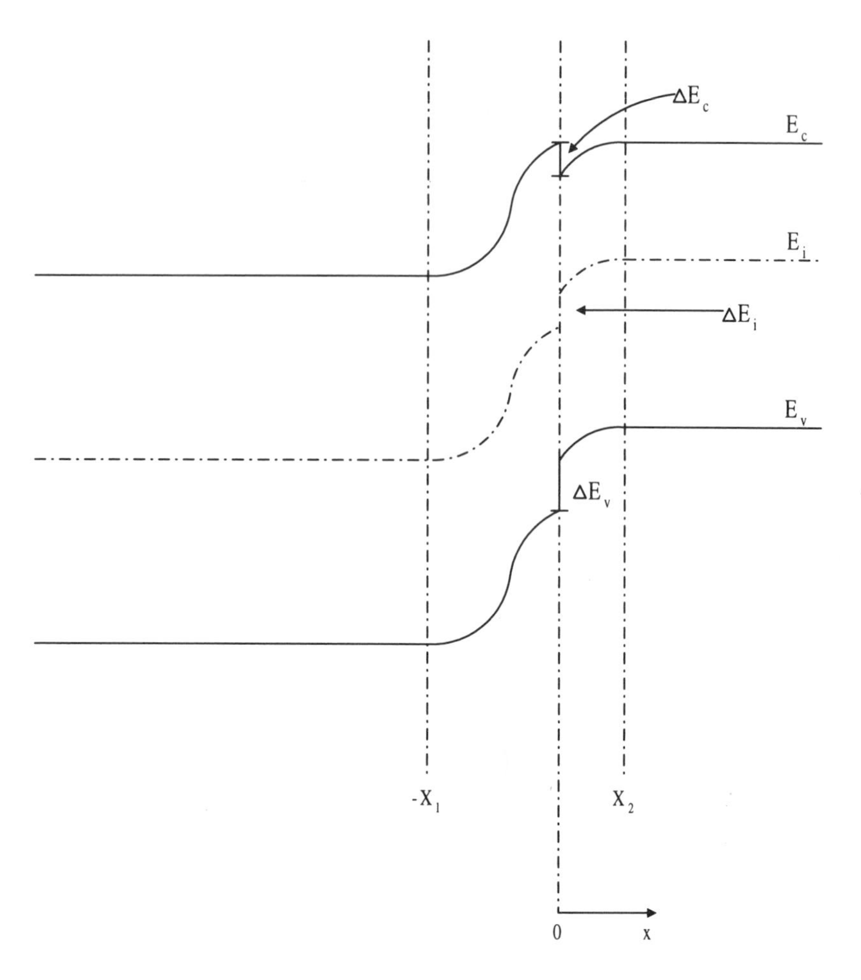

FIGURE 8.16. Energy-band diagram of a SiGe heterojunction.

Ge box profile of 8.4% Ge is obtained, while at $\Delta E_{g,Ge}$(grade) = 125 meV, a purely triangular profile of 0 to 18.6% Ge is obtained. The intermediate Ge trapezoid being located between these two extremes. The triangular profile has the largest Early voltage, gain–Early voltage product, and smallest base transit time. The box profile, however, has the highest current gain, since the enhancement of current gain depends exponentially on the boundary value in band-gap reduction at the emitter–base junction.

Historically, the use of bipolar transistors at liquid nitrogen temperature (LNT) was not viable because the bipolar transistor suffered severe current gain degradation with cooling (Kauffman and Bergh, 1968). By increasing the base doping or decreasing the emitter doping, the current gain at low temperature is improved (Dumke, 1981; Woo and Plummer, 1987). The recent advance of high-quality low-temperature epitaxial processes offers great potential for achieving the very thin, heavily doped, and abrupt base profiles required for continued vertical scaling of bipolar transistors. Because of the smaller band

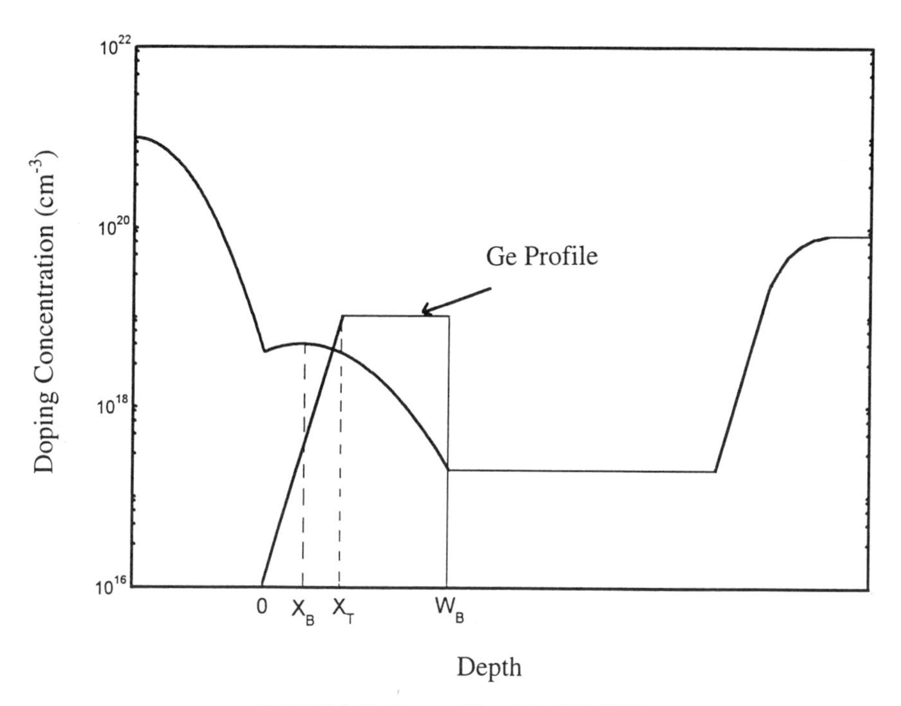

FIGURE 8.17. Base profiles of the SiGe HBT.

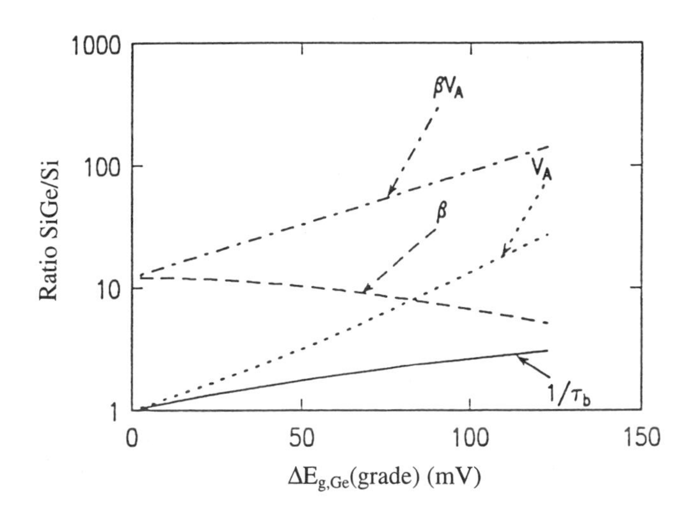

FIGURE 8.18. Early voltage and Early voltage-gain product.

gap and graded germanium content in the base region, the current gain and cutoff frequency of the SiGe HBT are superior to those of Si bipolar transistors and are comparable to those of its GaAs counterpart (Cressler *et al.*, 1993). Figure 8.19 shows the transistor current gain as a function of reciprocal temperature for Si and SiGe bipolar transistors (Harame *et al.*, 1995). Recent technology produces a current gain at LNT that is essentially unchanged from its room-temperature (RT) value. The high current gain at low temperature is a result of both the heavily doped base and the presence of SiGe, which offsets the band-gap narrowing in the emitter. These results suggest that silicon–germanium bipolar technology is suitable for very high speed applications in cryogenic computer systems.

For an *n–p–n* double-heterojunction HBT, the valence-band discontinuity at the collector–base heterojunction has little effect on the device characteristics until base pushout occurs. When the collector current is larger than the onset current for base pushout, holes tend to move into the epitaxial collector, when the electric field at the collector–base junction changes from negative to positive. The valence-band discontinuity ΔE_V at the base–collector heterojunction, however, prevents holes from moving into the epilayer. Thus, electrons in the epicollector are not compensated with holes, and an electric field is established that creates a potential barrier (Cottrell and Yu, 1990). The valence-band barrier prevents holes from flowing into the collector. Charge neutrality cannot be maintained, and a barrier to electron transport is induced in the conduction band. The collector–base heterojunction barrier effect severely degrades the current gain and transconductance of SiGe-base transistors operating at low temperatures, as evidenced by the experimental data in Fig. 8.20. The induced barrier is thermally activated. Thus, even though the small barrier may be virtually undetectable at RT, its effects are greatly increased at low temperatures.

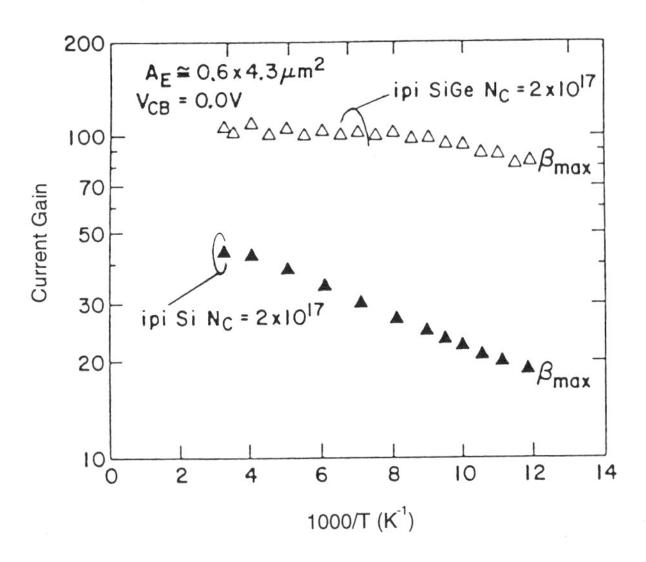

FIGURE 8.19. Current gain versus inverse temperature (Cressler *et al.*, 1993).

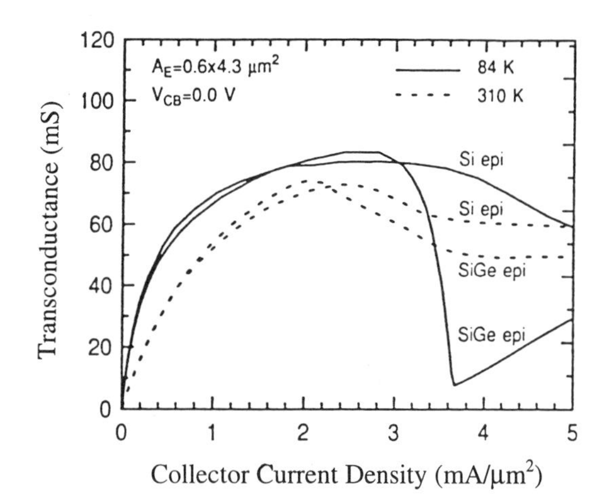

FIGURE 8.20. Transconductance versus collector current (Cressler *et al.*, 1993).

The heterojunction capacitance as a function of bias is shown in Fig. 8.21. In this figure the solid line represents the analytical result, and the dashed line represents the depletion approximation. At low bias, the junction capacitance predicted by the analytical approach is larger than the depletion approximation (Yuan, 1992). This is because the analytical result accounts for holes and electrons in the depletion region, which increase the junction capacitance. The heterojunction capacitance predicted by the depletion

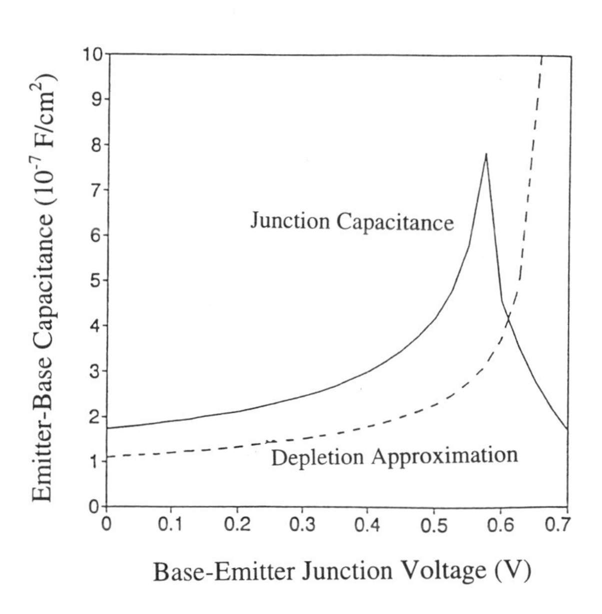

FIGURE 8.21. Junction capacitance versus applied voltage.

approximation increases monotonically and approaches infinity as the base–emitter voltage approaches the built-in voltage. Note that the built-in voltage of a SiGe heterojunction is smaller than that of the Si junction because the Ge has a smaller intrinsic concentration. The depletion capacitance causes the total free-carrier charge stored in the emitter–base space-charge region to approach infinity even though the space-charge region vanishes when the base–emitter voltage approaches the built-in voltage. The analytical result, however, correctly represents the device physics of the heterojunction at high biases.

The cutoff frequency of the SiGe HBT as a function of collector current is shown in Fig. 8.22. The cutoff frequency falls off at current densities well below the onset current for base pushout. The early degradation is triggered by the collector space-charge region widening and inverse base-width modulation effects (Ugajin and Fossum, 1994; Ugajin *et al.*, 1994). The collector space-charge widening results from an increase of collector current, which increases mobile carriers in the space-charge region. Mobile carriers modulate the space-charge and increase the space-charge layer width. The inverse base-width modulation is caused by a widening of the quasi-neutral base width that results when the collector–base junction space-charge region on the base side shrinks as the collector current increases.

The average measured ECL gate delay as a function of temperature is shown in Fig. 8.23 (Cressler *et al.*, 1993). ECL gate delays are nearly constant with cooling between room temperature and liquid-nitrogen temperature, beginning at 28.8 ps at 310 K, reaching a minimum of 25.4 ps at 150 K, and achieving 28.1 ps at 84 K. For the SiGe bipolar transistor, the total base resistance increases slowly with cooling, the collector capacitance decreases slowly with cooling, and the peak cutoff frequency increases slowly until the temperature reaches about 150 K and then gradually decreases. From 310 to 150 K the increasing cutoff frequency and decreasing collector–base junction capacitance effectively offset the increase in base resistance. Below 150 K the decrease in f_T can no longer balance the rising R_B, and the gate delay begins to degrade very slightly.

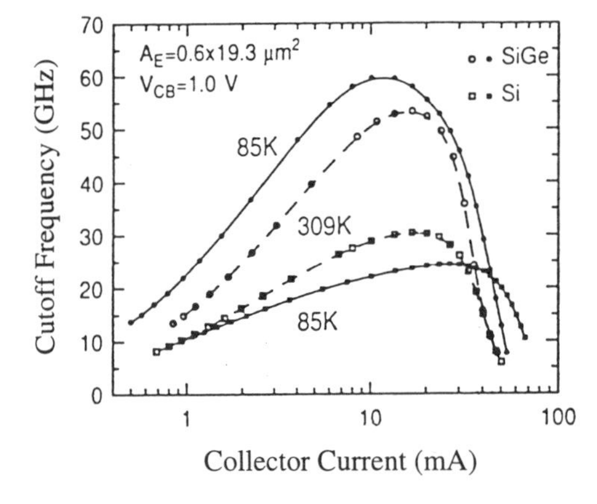

FIGURE 8.22. Cutoff frequency versus collector current (Cressler *et al.*, 1993).

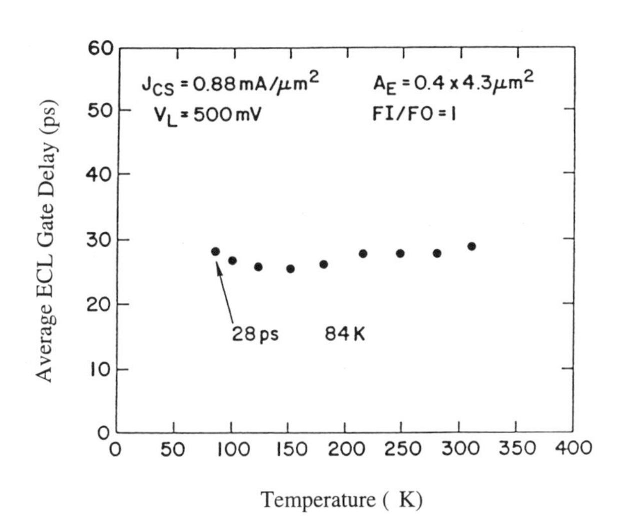

FIGURE 8.23. ECL gate delay versus temperature (Cressler *et al.*, 1993).

8.7. RELIABILITY OF HBTs

Reliability requirements are dictated by the application. Reliability qualification work is aimed at establishing robust process and design rules for achieving high mean time between failure (MTBF). General HBT reliability testing includes stabilization bake, accelerated life testing, and step-stress testing. An initial stabilization bake is performed to stabilize the device parameters prior to accelerated aging and step-stress tests and to screen for infant mortality. Devices are typically subjected to 200 to 240°C for 48 h in a nitrogen ambient atmosphere to stabilize dc parameters.

Accelerated life tests include aging at 200 to 260°C under dc bias with step-stress testing designed to extract the activation energy and MTBF for degradation designed to explore accelerated lifetimes that will be more realistic because high-temperature aging induces a failure mechanism that may never be encountered in normal operation conditions. High-temperature testing is performed to reduce the time needed to achieve median lifetimes.

Reliability testing of GaAs HBTs has focused on accelerated aging of discrete devices under normal operating dc bias. The packaged devices are periodically removed and tested for dc current gain, base–emitter turn-on voltage, ohmic contact resistance, and junction leakage.

Figures 8.24 to 8.27 show the experimental data of base–emitter leakage current, current gain, collector–emitter offset voltage, and base–collector leakage versus stress time (Hafizi *et al.*, 1993). These key parameters enable test engineers to assess the stability and integrity of the base–emitter junction, the base layer, and the base–collector junction. The base–emitter leakage current in Fig. 8.24 steadily and uniformly decreases with increasing stress time in the test population. The current gain in Fig. 8.25 has an initial increase of approximately 10%, which is attributed to reduced surface recombination

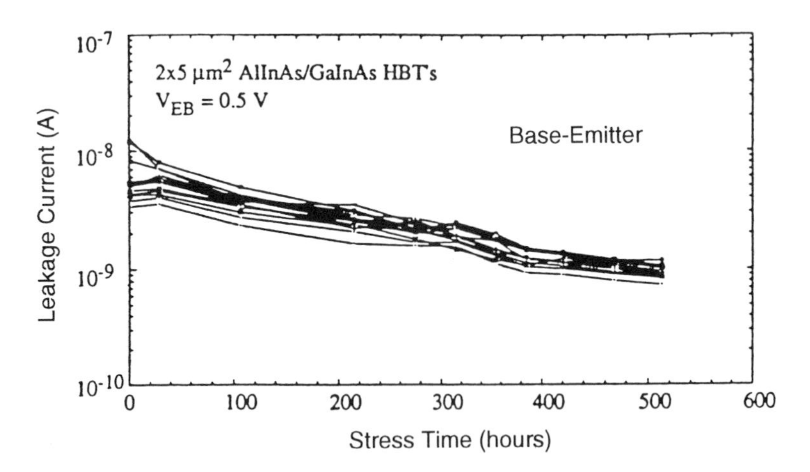

FIGURE 8.24. Base–emitter leakage current versus stress time (Hafizi *et al.*, 1993).

velocity as a result of the exposure to high temperature. As stress increases, this process reverses and the current gain gradually decays. The offset voltage increases with stress time, as shown in Fig. 8.26. The offset voltage is initially unchanged. It is then followed by a period of stress time and displays an identical trend to that of the base–collector leakage increase in Fig. 8.27. It is clear that the leakage current steadily decreases in approximately the first 300 h of stress. This is then followed by a rapid increase in the leakage current of the base–collector junction. The increase in the leakage current occurs very uniformly over a short interval of time, as opposed to the case where the degradation spreads over the entire length of the stress period. This feature is highly desirable. The confinement and the uniformity of device degradation results in a failure distribution for the life test population, which has a very low dispersion. Another apparent feature is the

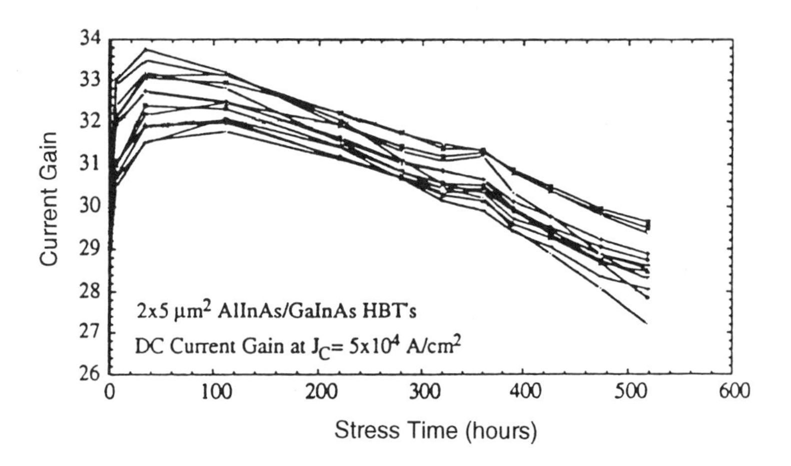

FIGURE 8.25. Current gain versus stress time (Hafizi *et al.*, 1993).

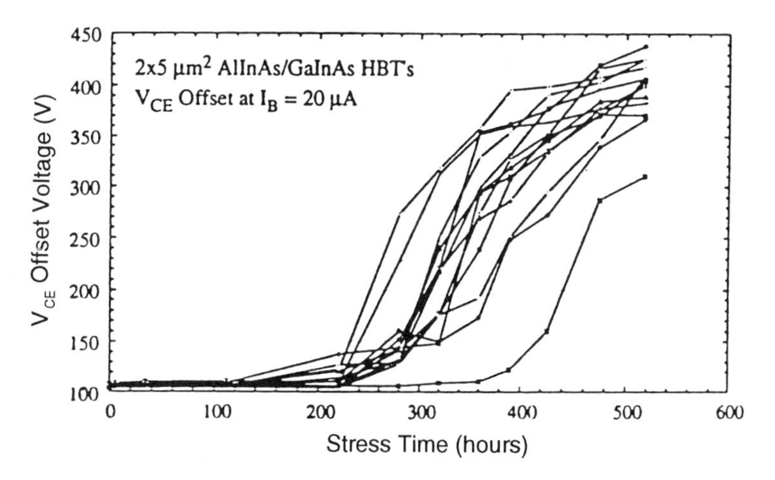

FIGURE 8.26. Collector–emitter offset voltage versus stress time (Hafizi *et al.*, 1993).

lack of early device failures or rapid degradation in the early hours of stress, known as "infant mortality."

8.7.1. Multiemitter Fingers

Figure 8.28 shows the schematic of a conventional HBT chip with multiemitter fingers fabricated on the top surface of a GaAs substrate. Each emitter finger is further divided into a number of unit areas. Each unit area is assumed to be a heat source with constant temperature, with the exception of the substrate bottom surface where the temperature is kept at heat sink temperature T_0 and the heat sources on the top. The

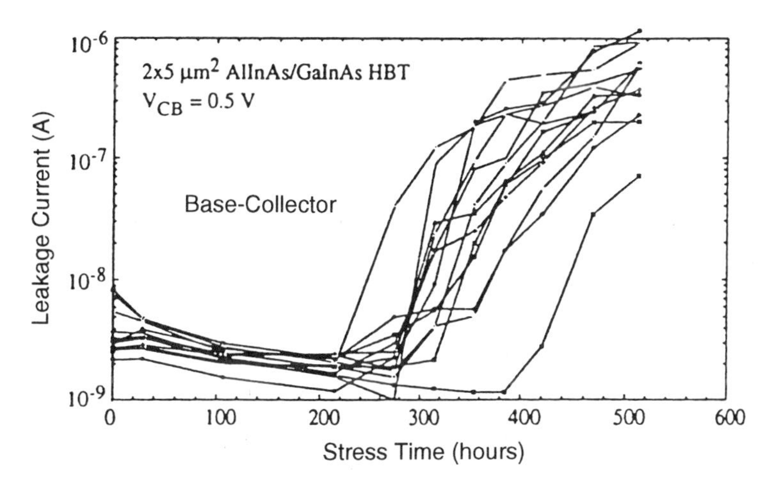

FIGURE 8.27. Base–collector leakage current versus stress time (Hafizi *et al.*, 1993).

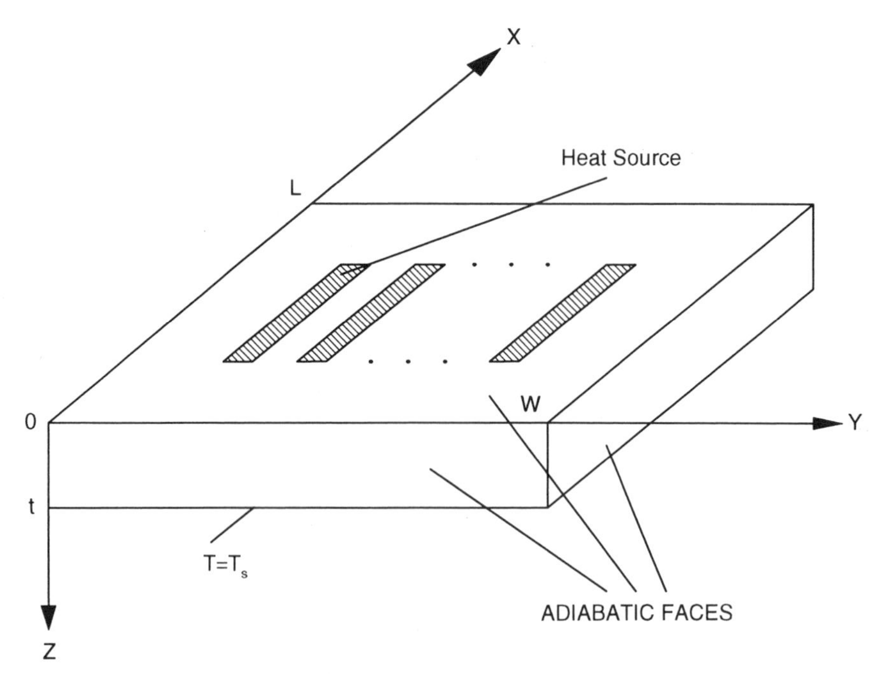

FIGURE 8.28. Multiemitter fingers on the top surface of a GaAs substrate.

remaining surfaces are assumed to be adiabatic. With these boundary conditions, the temperature distribution at the top surface (i.e., $z = 0$) can be obtained by solving the heat transfer equation

$$\nabla\kappa(T)T = 0 \tag{8.7.1}$$

If thermal conductivity is temperature independent ($\kappa(T) = \kappa_0$), the solution is (Liou and Bayraktaroglu, 1994; Gao *et al.*, 1989)

$$\theta(x,y,0) = T_0 + \sum_{i=1}^{N} \left\{ \frac{q_i \Delta_x \Delta_y d}{L_x L_y \kappa_0} + \sum_{\mu,\nu=0}^{\infty} \left[C_{\mu\nu} q_i \cos\left(\frac{\mu\pi x_i}{L_y}\right) \cos\left(\frac{\nu\pi y_i}{L_y}\right) \frac{\tanh(\gamma_{\mu\nu} d)}{\kappa_0 \gamma_{\mu\nu}} \right] \cos\left(\frac{\mu x}{L_x}\right) \sin\left(\frac{\nu y}{L_y}\right) \right\} \tag{8.7.2}$$

where

$$\gamma_{\mu\nu} = \pi\sqrt{(\mu/L_x)^2 + (\nu/L_y)^2}$$

$$C_{\mu\nu} = \frac{16}{\mu\nu\pi^2} \sin(0.5\mu\pi x) \sin(0.5\nu\pi y), \qquad \mu \neq 0, \quad \nu \neq 0$$

$$= \frac{4y}{\mu\pi L_y} \sin\left(\frac{0.5\nu\pi x}{L_x}\right), \qquad \mu \neq 0, \ \nu = 0$$

$$= \frac{4x}{\mu\pi L_x} \sin\left(\frac{\nu\pi y}{L_y}\right), \qquad \mu = 0, \ \nu \neq 0$$

$q_i\Delta x\Delta y$ is the heat generated at the ith unit area with the center located at (x_i,y_i), $\Delta x\Delta y$ is the unit area, N is the number of unit areas on the surface, κ_0 is the thermal conductivity at the heat sink temperature, L_x, L_y, and d are the x, y, and x dimensions respectively. Kirchhoff's transformation can be used to account for the temperature-dependent thermal conductivity. Assuming the thermal conductivity is proportional to $(T/T_0)^{-1.22}$, the corrected temperature is

$$T = \left(\frac{1}{T_0^{0.22}} - \frac{0.22(\theta - T_0)}{T_0^{1.22}}\right)^{-4.5454} \tag{8.7.3}$$

where θ is given in (8.7.2).

The effect of chip thickness on HBT thermal behavior is shown in Fig. 8.29. Both the active region and substrate are assumed to have a square shape. L_a/L_c is the ratio of

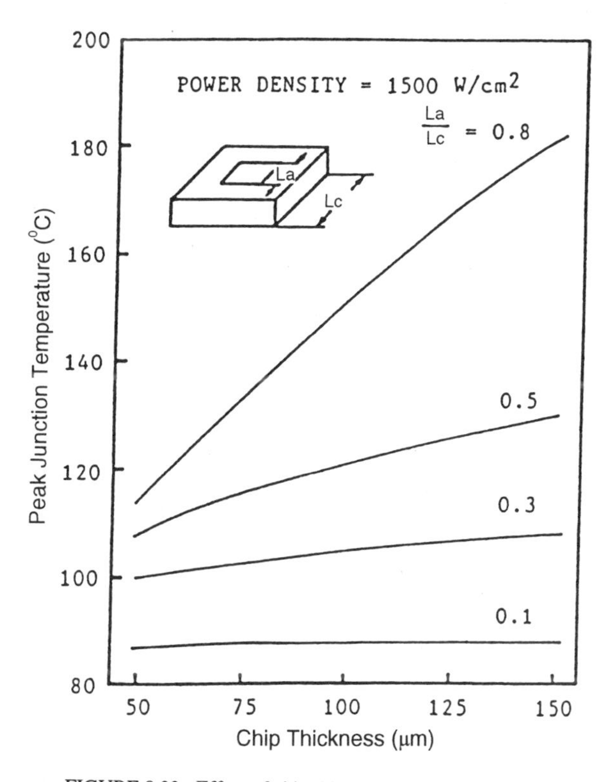

FIGURE 8.29. Effect of chip thickness (Gao *et al.*, 1989).

side lengths between the active region and the substrate. If this ratio is 0.8, which is typical of conventional Si power transistors, the peak junction temperature decreases almost linearly with decrease of chip thickness. However, if the ratio is less than 0.5, thinning the substrate has little impact on peak junction temperature based on the result of the three-dimensional thermal analysis. When the dimension of the active region is much less than the chip size, the heat flow is distributed in a cone shape, and the junction temperature is mainly determined by the lateral size of the device rather than the chip thickness.

A three-dimensional transmission-line matrix simulation was used to evaluate the thermal behavior of power HBTs with multiple emitter fingers (Gui *et al.*, 1992). The power device consists of five cells. Each cell has two emitter fingers grouped closely together by the emitter–base self-alignment process. The simulation result of temperature distribution at a dissipated power density of 2 mW/μm^2 under steady state is shown in Figs. 8.30 and 8.31. Figure 8.30 shows the isometric view of temperature distribution of five emitter fingers, and Fig. 8.31 is the contour plot with interval of 15°C. The isometric projection and the corresponding temperature contours clearly show that two peak values of junction temperature occur at the centers of the two central emitter fingers. The junction temperature drops sharply around the central cell, with an average temperature gradient as high as 9°C/μm in these areas. From a device reliability point of view, the peak junction temperature is the most important parameter for the device thermal design and should be used to characterize the device thermal behavior.

The nonuniform junction temperature distribution of HBTs for different substrate materials is depicted in Fig. 8.32. The peak junction temperatures are 180, 140, and 101°C for the HBTs on GaAs, InP, and Si substrates, respectively. The HBT on a Si substrate obviously operates better than others for the same power dissipation.

In the design of power heterojunction bipolar transistors, the emitter finger length is primarily determined by the emitter metallization resistance and the junction temperature nonuniformity along the emitter length. In general, the length and the length-to-width

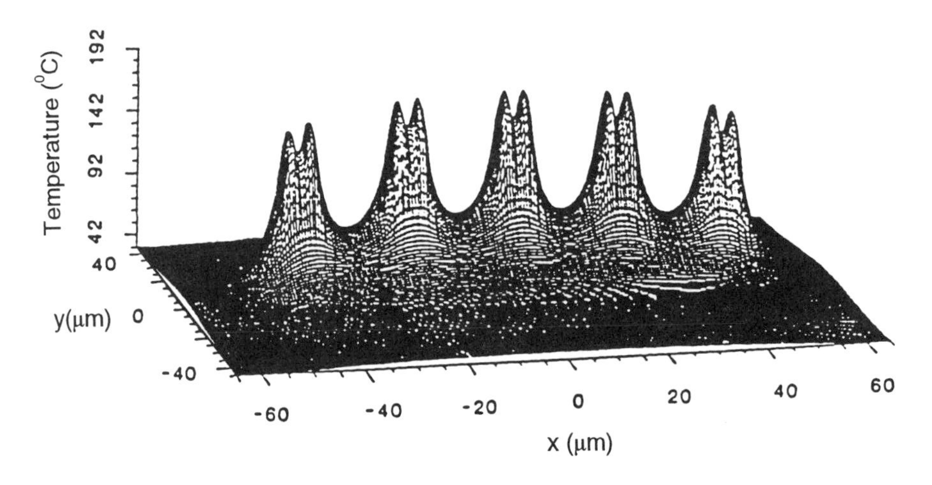

FIGURE 8.30. Isometric view of temperature distribution of five emitter fingers (Gui *et al.*, 1992).

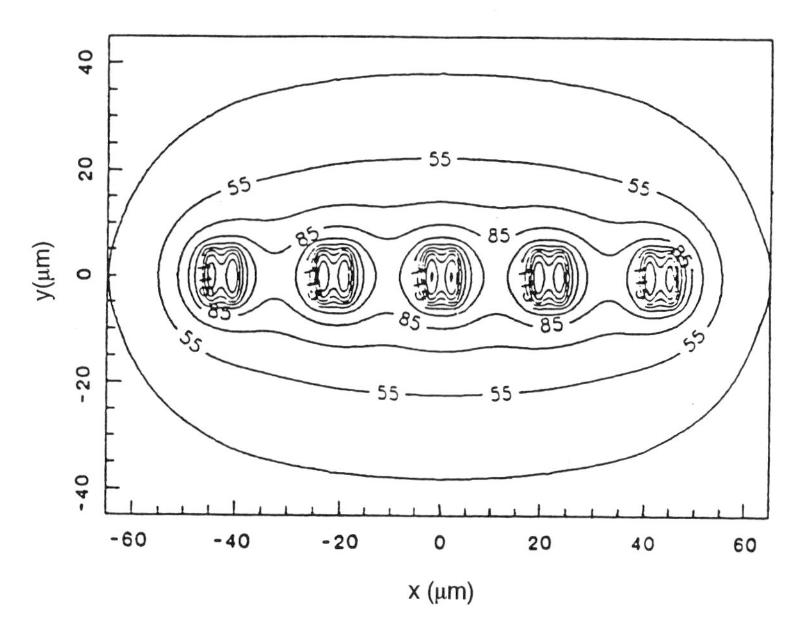

FIGURE 8.31. Contour plot of temperature distribution (Gui *et al.*, 1992).

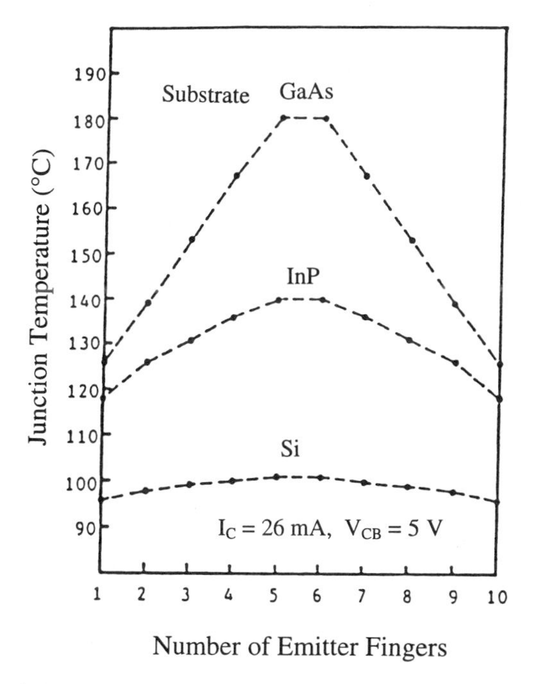

FIGURE 8.32. Junction temperature distribution for different substrate materials (Gao *et al.*, 1989).

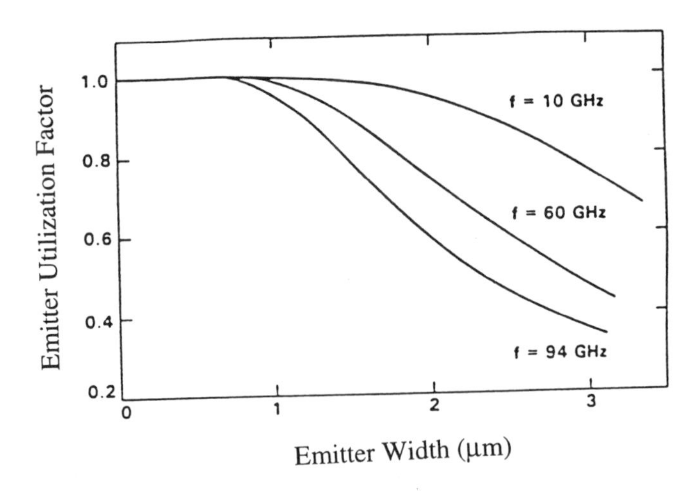

FIGURE 8.33. Emitter utilization factor as a function of frequency (Gao *et al.*, 1992).

ratio of an emitter finger should not exceed 30 μm and have a 20-to-1 ratio in order to minimize the junction temperature and the voltage drop along the length of the finger. From thermal considerations, a power HBT with a longer emitter length results in a higher junction temperature and in turn causes a reduction in the output power density.

The total emitter area is determined by the desired output power or the maximum peak collector current. For silicon bipolar transistors, the peak current is strongly dependent on the emitter periphery due to current crowding at high currents. For HBTs with high base doping concentration, emitter crowding effect is reduced significantly. The calculated emitter utilization factor as a function of operating frequency and emitter size is depicted in Fig. 8.33 (Gao *et al.*, 1992). Clearly, an emitter width of 2 μm is sufficient for an operating frequency of 10 GHz. In this case, the emitter utilization factor should be greater than 95%. This allows the current capability and the output power for HBTs to depend on the emitter area rather than the emitter periphery.

Another very important parameter for multiple-emitter-finger HBT design is the emitter spacing, which is defined as the center distance between two adjacent emitter fingers. A larger spacing results in a higher base resistance and capacitance. Both degrade the f_{max} or the power gain. On the other hand, a larger spacing reduces the peak value of junction temperature at a given power dissipation (Fig. 8.34). When the spacing increases from 4.4 μm to 6, 10, and 14 μm, the peak value of junction temperature decreases by 17.5%, 35.0%, and 42.5%, respectively, for a power density of 2 mW/μm^2.

8.7.2. Emitter Collapse Phenomenon

Power heterojunction bipolar transistors are designed to deliver large amounts of power at high frequencies. Because HBTs are operated at high power densities, the ultimate limits on the performance of HBTs are imposed by thermal considerations. Recently, a thermal phenomenon was observed for the case when a multifinger power

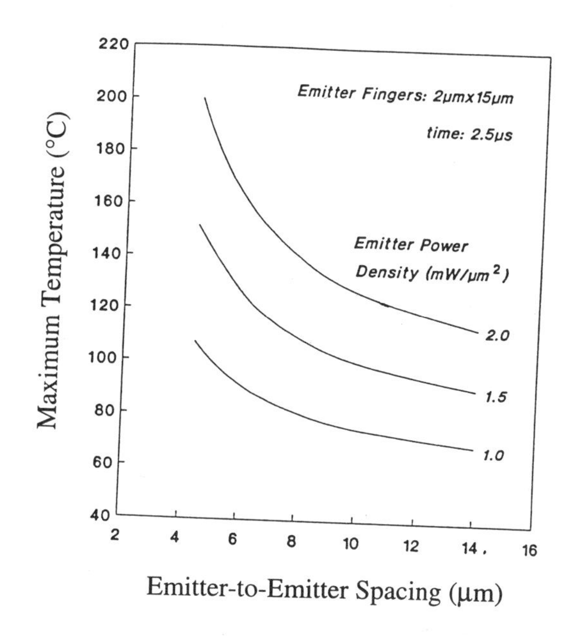

FIGURE 8.34. Peak junction temperature versus emitter spacing (Gao *et al.*, 1992).

HBT was operating at high power densities (Gui *et al.*, 1992). This phenomenon, referred to as "emitter collapse," occurs when one finger of a multifinger HBT suddenly conducts most of the device current. This leads to an abrupt decrease in current gain. The collapse does not cause HBTs to fail immediately, and the device can be biased in and out of the collapse by adjusting the level of power dissipation. This collapse, however, degrades the device performance significantly and should be avoided.

The collapse phenomenon is illustrated by the common–emitter I–V characteristics measured at constant base currents. As the collector–emitter voltage increases, the power dissipation in the HBT increases. The increase of power dissipation elevates the junction temperature above the ambient temperature and causes an increase in base and collector currents. Since the increase in base current is more than that of the collector current, the collector current decreases with collector–emitter voltage for a given base current. The negative-resistance effect is observed in Fig. 8.35 for $V_{CE} < 8$ V. As V_{CE} increases beyond 8 V, the negative differential resistance is replaced by the collapse phenomenon, as marked by an abrupt, dramatic lowering of the collector current. The collapse loci, as a function of V_{CE}, at which collapse occurs, is shown as the dashed curve in Fig. 8.35.

Consider the case of a multifinger HBT. If one finger becomes slightly warmer than the others, this particular finger conducts more current at a given base–emitter bias. This increased collector current, in turn, increases the power dissipation in the junction and raises the junction temperature and collector current even further. The collapse occurs when the junction temperature in one finger becomes much hotter than the rest of the fingers.

Note that the collapse does not cause the HBTs to fail immediately. In fact, the collapse is reversible and the same device can be measured for several experiments. This

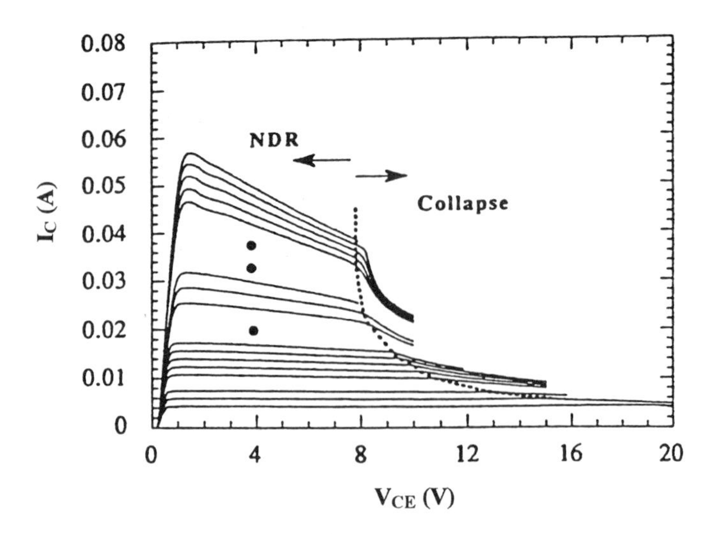

FIGURE 8.35. Collector current versus collector–emitter voltage (Liu *et al.*, 1993).

is markedly different from thermal runaway in silicon BJTs, which causes the devices to fail irreversibly. Both thermal runaway of BJTs and the collapse of HBTs result from an increase in junction temperature. Since the current gain in Si bipolar transistors increases rather than decreases with temperature, the increased current in that finger of the BJT will draw more and more current. This is in contrast to the collapse when I_C decreases after the critical condition occurs.

The collector current collapse (or crush) effect has also been observed by using numerical electrothermal simulation (Liu *et al.*, 1993). The contour plots of the temperature distribution on the top surface of a three-finger HBT under a bias of $V_{CE} = 2$, 3, 5, and 6 V are shown in Fig. 8.36. The first two operation points are before the current crush, and the temperature variation is small across the three emitter fingers. The last two operating points occur after the current crush, and the temperature distribution shows a drastic difference between the center and the other two emitter fingers. As V_{CE} increases, the center finger will eventually conduct the largest base current, and the other two fingers become nearly inactive. The collector current will continue to decrease due to the smaller current gain at higher junction temperatures.

The interdependence between the collapse phenomenon and the avalanche breakdown has been examined and modeled (Liu, 1995). The measured and calculated I–V characteristics for $I_B = 5\ \mu\text{A}$ are shown in Fig. 8.37. The calculated collector current flowing in the hot finger is I_{C1}, and in one of the cold fingers it is I_{C2}. The calculated total current is $I_{C1} + 5I_{C2}$. Significant avalanche multiplication starts at $V_{CE} \approx 22.5$ V for both curves. The collector current reaches a maximum value of 5.3 mA before the collapse occurs and the collector current decreases.

At point A which corresponds to $V_{CE} \approx 22.5$ V and $I_C \approx 0$ A, the collector current level is well below the collapse locus. Therefore, the total device current at point A is

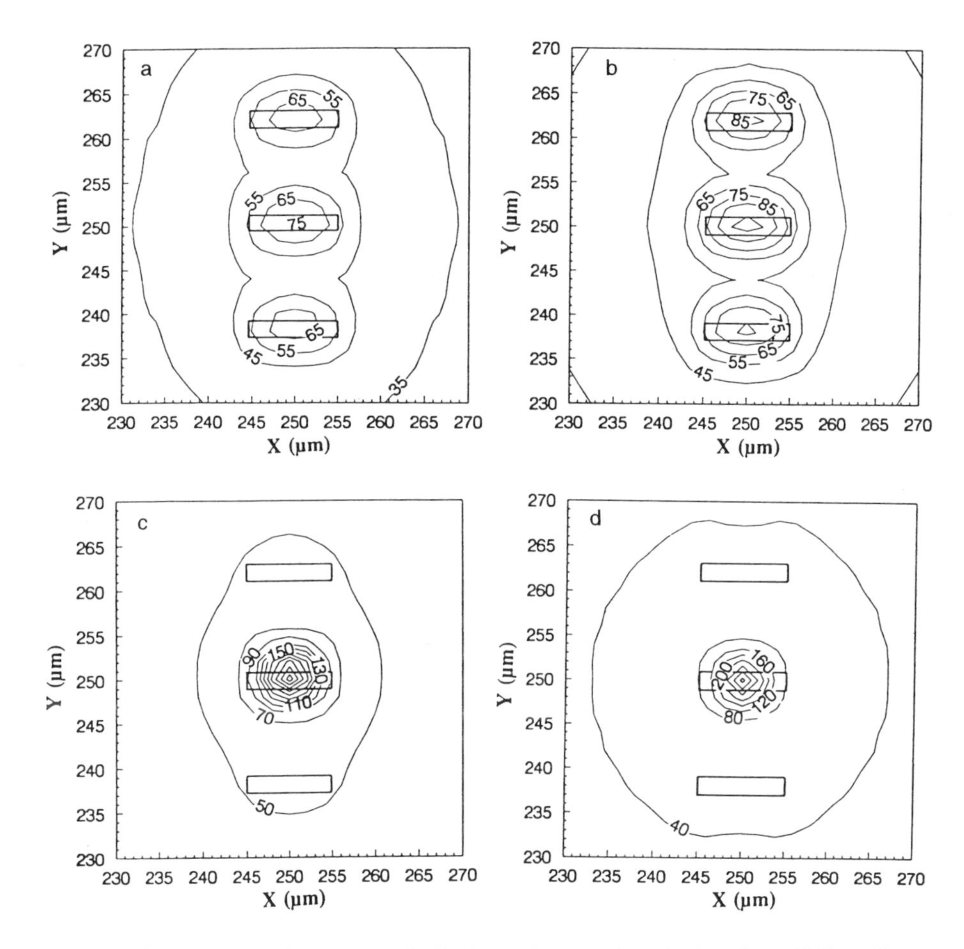

FIGURE 8.36. Contour plot of temperature distribution on the top surface of a three-finger HBT: (a) $V_{CE} = 2$ V, (b) $V_{CE} = 3$ V, (c) $V_{CE} = 5$ V, and (d) $V_{CE} = 6$ V (Liou and Bayraktaroglu, 1994).

uniformly distributed among the six fingers. As V_{CE} increases slightly past point ***A***, impact ionization in the base–collector junction becomes noticeable. The product of avalanche multiplication factor M_n and current transfer ratio α_0 approaches unity and causes the collector current to increase rapidly toward the collapse locus. This well-known trend of increasing I_C, however, stops abruptly upon initiation of the collapse at point ***B***, corresponding to $V_{CE} \approx 24.1$ V. Beyond this point, a further increase in collector–emitter voltage results in a greater proportion of collector current being conducted in the hot finger until all of the available device current is conducted through the hot finger.

Between point ***C***, which marks the onset of the collapse, and point ***D***, where the device burns out, device operation can be separated into two phases according to the features in the junction temperature of the hot finger (T_1). Between points ***B*** and ***C***, T_1 quickly increases toward some drastically higher temperature than that prior to collapse. The rapid increase of T_1 in the hot finger corresponds to the rapid decrease of T_2 in the

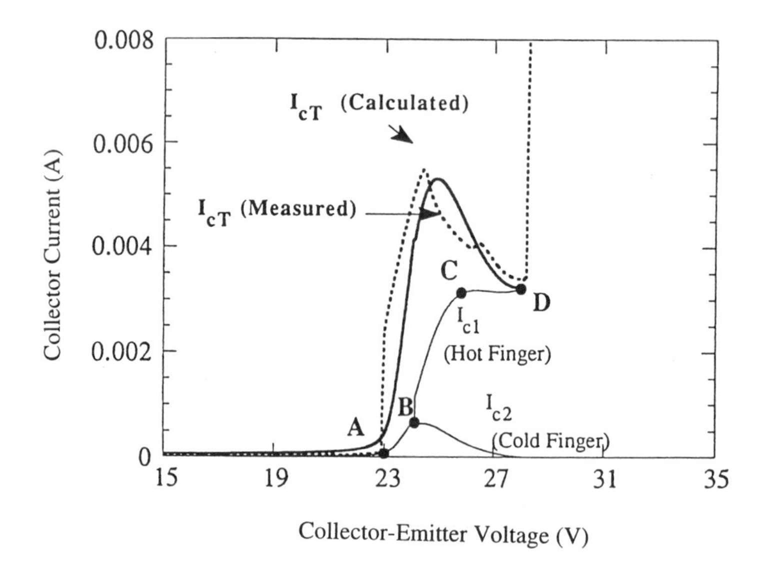

FIGURE 8.37. Measured and calculated *I–V* characteristics showing the interaction between the collapse and avalanche breakdown (Liu *et al.*, 1995).

cold finger. During this phase, the incremental increase of I_{C1} with respect to V_{CE} is high, since a greater amount of current flows into the hot finger. The surge in power density further increases the hot finger junction temperature. In contrast, the cold finger loses more and more of its current with increasing V_{CE}, and the cold finger temperature plummets. Between points ***C*** and ***D***, most of the available current has been directed to the hot finger. The six-finger HBT is effectively a one-finger HBT with the hot finger conducting the entire device current.

8.7.3. *Emitter Ballasting Resistors*

To reduce the likelihood of the emitter collapse phenomenon and to ensure a more uniform current distribution, emitter ballast resistances can be inserted in series with each emitter finger. However, the determination of the value of ballast resistance is not simple. There are a number of conflicting requirements. Too large a resistance will cause degradation of power gain, while too small a resistance will not effectively protect the HBT from thermal instability. The optimal value of the total ballasting resistance is determined by (Gao *et al.*, 1991)

$$R_E = \frac{-\Xi V_{CE} R_T - (R_{EC} + nkT/qI_E)[1 - (T_j - T_c)\zeta]}{1 + (T_j - T_c)(\varsigma - \zeta)} \tag{8.7.4}$$

where R_T is the total thermal resistance from active device area to the point where the case temperature T_c is measured, $T_j = I_C V_{CE} R_T + T_c$, R_{EC} is the emitter contact resistance, $\Xi = \partial V_{BE}/\partial T$, $\varsigma = 1/R_e\, dR_e/dT$, and $\zeta = 1/\kappa(T)\, d\kappa/dT$.

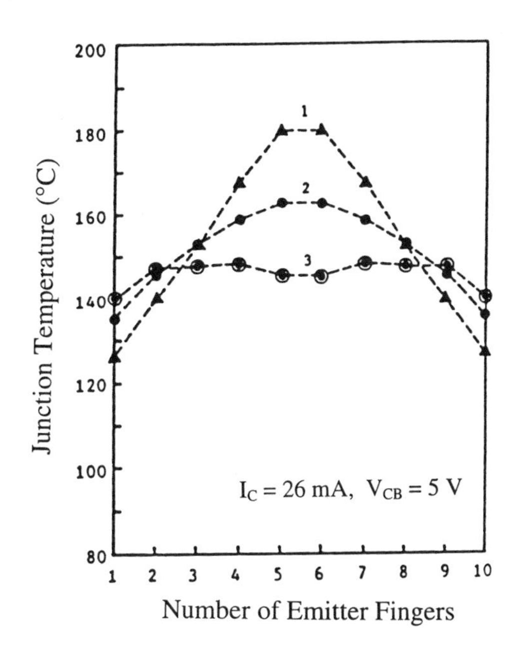

FIGURE 8.38. Junction temperature for different emitter ballasting resistors (Gao *et al.*, 1989).

The effect of emitter ballasting resistors on the junction temperature of the HBT is shown in Fig. 8.38. Curve 1 represents the HBT without ballasting resistors, curve 2 the HBT with equally valued emitter ballasting resistors, and curve 3 the HBT with unequally valued emitter ballasting resistors. As shown, nonuniform junction temperature distribution with multiple emitter fingers over the entire active region can be improved by using ballasting resistors, especially unequally valued emitter resistors. It is anticipated that highly thermostable HBTs with homogeneous junction temperature can be fabricated by this technique.

In the numerical simulation, the effects of emitter finger layout, emitter resistance, and substrate thermal conductivity on the collector current crush are examined (Liou and Bayraktaroglu, 1994). The dependence of the current instability on the spacing between emitter fingers is shown in Fig. 8.39. The maximum temperature on the chip as a function of V_{CE} is also given. The case of zero spacing corresponds to an HBT with only one emitter finger with an area of $10 \times 6\ \mu m^2$. As the spacing between the emitter finger increases, V_{CE} shifts slightly to a smaller value, while I_C shifts slightly to a larger value, at the current crush point. This result shows that the threshold power current instability is only minimally improved by increasing the emitter finger spacing beyond 10 μm and not increasing the number of emitter fingers. The dependence of current crush due to thermal instability on the emitter specific resistance is shown in Fig. 8.40. As the emitter specific resistivity ρ_e increases, threshold power increases. This result verifies the reduction in thermal instability by implementing a ballast resistor at each emitter node. The dependence of current crush on substrate thermal conductivity is shown in Fig. 8.41. A reduction in the thermal resistance due to the increase in the thermal conductivity results in a smaller temperature rise and a larger critical voltage for current

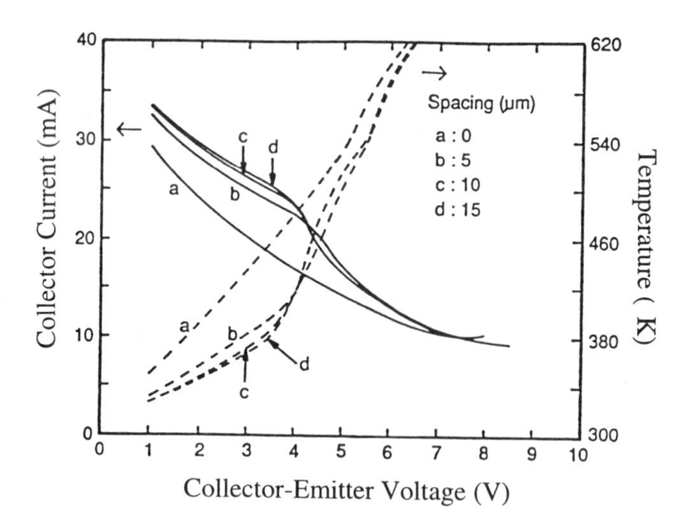

FIGURE 8.39. *I–V* characteristics for different emitter spacing (Liou and Bayraktaroglu, 1994).

crush. Improvement of thermal instability by increasing substrate conductivity is the most efficient way of increasing power capability without encountering high-frequency limitations and consuming extra semiconductor area. This suggests that the power HBT on a Si substrate provides a smaller negative differential resistance and a larger current crush voltage and it is more thermally stable than the HBT on a GaAs substrate.

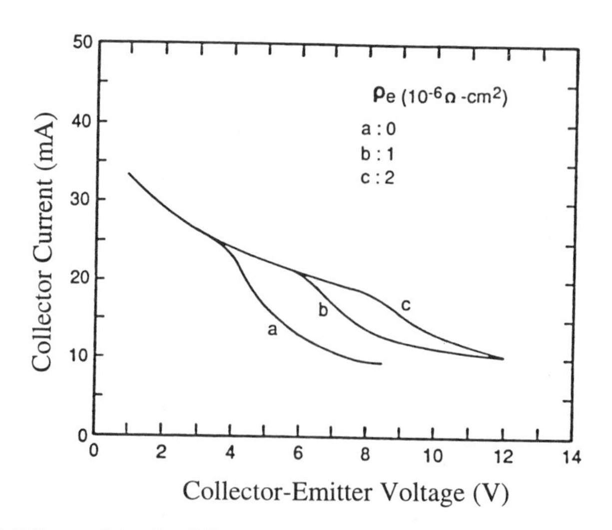

FIGURE 8.40. *I–V* characteristics for different emitter specific resistances (Liou and Bayraktaroglu, 1994).

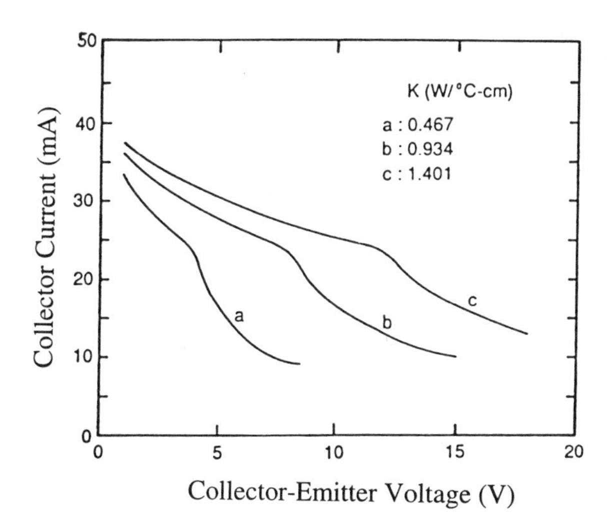

FIGURE 8.41. *I–V* characteristics for different substrate conductivities (Liou and Bayraktaroglu, 1994).

REFERENCES

Apanovich, Y., P. Blakey, R. Cottle, E. Lyumkis, B. Polsky, A. Shur, and A. Tcherniaev (1995), *IEEE Trans. Electron. Devices* **ED-42**, 890.

Asbeck, P. M. *et al.* (1987), *IEEE Microwave Theory Tech.* **MTT-35**, 1462.

Bandyopadhyay, S. *et al.* (1987), *IEEE Trans. Electron. Devices* **ED-34**, 391.

Beneking, H. and L. M. Su (1980), *Electron. Lett.* **16**, 41.

Cho, A. Y. and J. R. Arthur (1975), *Proc. Solid-State Chem.* **10**, 157.

Cottrell, P. E. and Z. Yu (1990), *IEEE Electron. Device Lett.* **EDL-11**, 431.

Cressler, J. D., E. F. Crabbe, J. H. Comfort, J. M. C. Stork, and J. Y.-C. Sun (1993), *IEEE Trans. Electron. Devices* **ED-40**, 525.

Cressler, J. D., J. H. Comfort, E. F. Crabbé, G. L. Patton, J. M. C. Stork, J. Y.-C. Sun, and B. S. Meyerson (1993), *IEEE Trans. Electron. Devices* **ED-40**, 525.

Dumke, W. P. (1981), *IEEE Trans. Electron. Devices* **ED-28**, 494.

Dupuis, R. D., L. A. Moudym, and P. D. Dapkus (1979), *1st Phys. Conf. Gallium-Arsenide and Related Compounds* **45**, 1.

Gao, G.-B., H. Morkoç, and M.-C. Chang (1992), *IEEE Trans. Electron. Devices* **ED-39**, 1987.

Gao, G.-B., M. S. Unlu, H. Morkoç, and D. Blackburn (1991), *IEEE Trans. Electron. Devices* **ED-38**, 185.

Gao, G.-B., M.-Z. Wang, X. Gui, and H. Morkoç (1989), *IEEE Trans. Electron. Devices* **ED-36**, 854.

Grinberg, A. A. and S. Luryi (1993), *IEEE Trans. Electron. Devices* **ED-40**, 859.

Grinberg, A. A., M. S. Shur, R. J. Fisher, and H. Morkoç (1984), *IEEE Trans. Electron. Devices* **ED-31**, 1758.

Gui, X., G.-B. Gao, and H. Morkoç (1992), *IEEE Device Lett.* **EDL-13**, 411.

Hafizi, M. E., C. R. Crowell, L. M. Pawlowicz, and M. E. Kim (1990), *IEEE Trans. Electron. Devices* **ED-37**, 1779.

Hafizi, M., W. W. Stanchina, R. A. Metzger, J. F. Jensen, and F. Williams (1993), *IEEE Trans. Electron. Devices* **ED-40**, 2178.

Harame, D. L., J. H. Comfort, J. D. Cressler, E. F. Crabbé, J. Y.-C. Sun, B. S. Meyerson, and T. Tice (1995), *IEEE Trans. Electron. Devices* **ED-42**, 455.

Hayama, N. and K. Honjo (1990), *IEEE Electron. Device Lett.* **EDL-11**, 388.

Iyer, S. S., G. L. Patton, S. L. Delage, S. Tiwari, and J. M. C. Stork (1987), *IEDM Tech. Dig.* 874.

Jalali, B. and S. J. Pearton, eds. (1995), *InP HBTs: Growth, Processing, and Applications* (Artech House, Norwood, MA).

Kauffman, W. L. and A. A. Bergh (1968), *IEEE Trans. Electron. Devices* **ED-15**, 732.

Kim, M. E., A. K. Oki, J. B. Camou, P. O. Chow, B. L. Nelson, D. M. Smith, J. C. Canyon, C. C. Yang, D. Dixit, and B. R. Allen (1988), *IEEE GaAs IC Symposium* 117.

King, C. A., J. L. Hoyt, and J. F. Gibbons (1989), *IEEE Trans. Electron. Devices* **ED-36**, 2093.

Konnzai, M., K. Katsukawa, and K. Takahashi (1979), *J. Appl. Phys.* **48**, 4389.

Kromer, H. (1957), *Proc. IRE* **45**, 1535.

Kromer, H. (1982), *Proc. IEEE* **70**, 13.

Liou, J. J., F. A. Lindholm, and B. Wu (1988), *J. Appl. Phys.* **63(10)**, 5015.

Liou, L. L. and B. Bayraktaroglu (1994), *IEEE Trans. Electron. Devices* **ED-41**, 629.

Liou, L. L., J. E. Ebel, and C. I. Huang (1993), *IEEE Trans. Electron. Devices* **ED-40**, 35.

Liu, W. (1995), *IEEE Trans. Electron. Devices* **ED-42**, 591.

Liu, W. and J. S. Harris, Jr. (1992), *IEEE Trans. Electron. Devices* **ED-39**, 2726.

Liu, W., S. Nelson, D. Hill, and A. Khatibzadeh (1993), *IEEE Trans. Electron. Devices* **ED-40**, 1917.

Maycock, D. P. (1967), *Solid-State Electron.* **10**, 161.

Sze, S. M. (1981), *Physics of Semiconductor Devices*, 2nd ed. (Wiley-Interscience, New York).

Sze, S. M. (1990), *High-Speed Semiconductor Devices* (Wiley-Interscience, New York).

Ugajin, M. and J. G. Fossum (1994), *IEEE Trans. Electron. Devices* **ED-41**, 1796.

Ugajin, M. and Y. Amemiya (1991), *Solid-State Electron.* **34**, 593.

Ugajin, M., G.-B. Hong, and J. G. Fossum (1994), *IEEE Trans. Electron. Devices* **ED-41**, 266.

van der Ziel, A., X. Zhang, and A. H. Pawlikiewicz (1986), *IEEE Trans. Electron. Devices* **ED-33**, 1371.

Wang, K.-C., P. M. Asbeck, M.-C. F. Chang, D. L. Miller, G. J. Sullivan, J. J. Corcoran, and T. Hornak (1987), *IEEE Trans. Electron. Devices* **ED-34**, 1729.

Wang, K. H., S. M. Beccue, M.-C. F. Chang, R. B. Nubling, A. M. Cappon, C.-T. Tsen, D. M. Chen, P. M. Asbeck, and C. Y. Kwok (1992), *IEEE J. Solid-State Circuits* **SC-27**, 1372.

Won, T., S. Iyer, S. Agarwala, and H. Morkoç (1989), *IEEE Electron. Device Lett.* **EDL-10**, 274.

Woo, J. C. S. and J. D. Plummer (1987), *IEDM Tech. Dig.* 401.

Yuan, J. S. (1992), *Solid-State Electron.* **7**, 921.

Zanoni, E., R. Malik, P. Pavan, J. Nagle, A. Paccagnella, and C. Canali (1992), *IEEE Electron Device Lett.* **EDL-13**, 253.

Zhou, W. Y., Y. B. Liou, and C. Huang (1995), *Solid-State Electron.* **38**, 1118.

9

Photoconductive Diodes

The photoconductive circuit element (PCE), or photoconductive diode, can detect optical signals and convert them to electrical signals. It can be switched on and off rapidly by using an optical source, it can be made fairly compact, and it can conduct large currents in the on state and hold off high voltages in the off state (Auston, 1984). As a result, the PCE is useful in high-speed switching (Frankel *et al.*, 1990), in establishing optical links between circuits (Lee, 1990), and is an attractive alternative or supplement to the high-power mechanical switchgear currently used in the electric utility industry (Triaros *et al.*, 1990). In order to minimize commutation energy losses in PCEs, the inductance in these devices must be reduced. Silicon has been widely used in PCE fabrication because it offers low cost, small circuit inductance, high thermal conductivity, and enhanced power density rating, all of which are critical to switching applications (Rose, 1963).

The PCE consists basically of an intrinsic semiconductor such as silicon, which becomes highly conductive when applying a laser or other types of optical excitation that provides photons with energies exceeding the energy band gap of the semiconductor. In practice, the intrinsic semiconductor is approximated by either a high-resistivity p layer (π layer) or a high-resistivity n layer (i layer). To provide better ohmic contacts, the heavily doped layers are formed underneath the metal contacts. The conductivity in the high-resistivity p layer between the two contacts, referred to as the channel (with a length L), is controlled by the light source. When the light is off (off state), the near intrinsic channel cannot conduct current, thus holding off the voltage applied at the terminals. When the light is on (on state), many excess electrons and holes are generated in the conducting channel, which increase the channel conductivity and make the device conductive.

The performance of a PCE is affected by the following factors: device structure (geometry and heavily doped contact region), device makeup (type of material, resistivity of semiconductor, impurity doping profile, and junction layer thickness), electron–hole recombination mechanism, operating temperature, and light wavelength. In general, the influence of the device structure on the PCE performance is significant if the intrinsic region thickness is comparable to or shorter than the free-carrier diffusion–drift length, as in a high-voltage PCE using a high-purity silicon wafer. This is because the photogenerated free carriers in such a device will be unlikely to recombine in the intrinsic region before they drift or diffuse to the contacts due to (1) a thick depletion layer in the intrinsic region resulting from the large reverse voltage; and (2) a long free-carrier lifetime in the high-purity silicon wafer. On the other hand, if the applied voltage is low or the silicon wafer has a low purity, the effects of the device structure on the PCE performance become less important.

There are three types of PCEs: horizontal PCE, vertical PCE, and edge PCE (Figs. 9.1a–c). In the horizontal PCE, the current flows primarily horizontally between the two contacts placed on the surface, whereas in the vertical and edge PCEs the current flows vertically between two contacts placed on the top and bottom of the wafer. Note, except for the edge PCE the light is illuminated at the surface of the wafer. As a result, the horizontal and vertical PCE surfaces cannot be covered entirely by the metal but should have a window pattern to allow optimal light exposure. While the edge PCE does not require the window pattern on the metal contact and thus is easier to fabricate, its light exposure area is nonetheless confined by the wafer thickness (about 100–500 μm). Consequently, the on-state performance of the edge PCE is severely limited unless a very concentrated light source is used. In this chapter, only horizontal and vertical PCEs are treated. The dc and transient behavior of the silicon PCEs are simulated, and their physical insight and relevance to the switching applications are addressed.

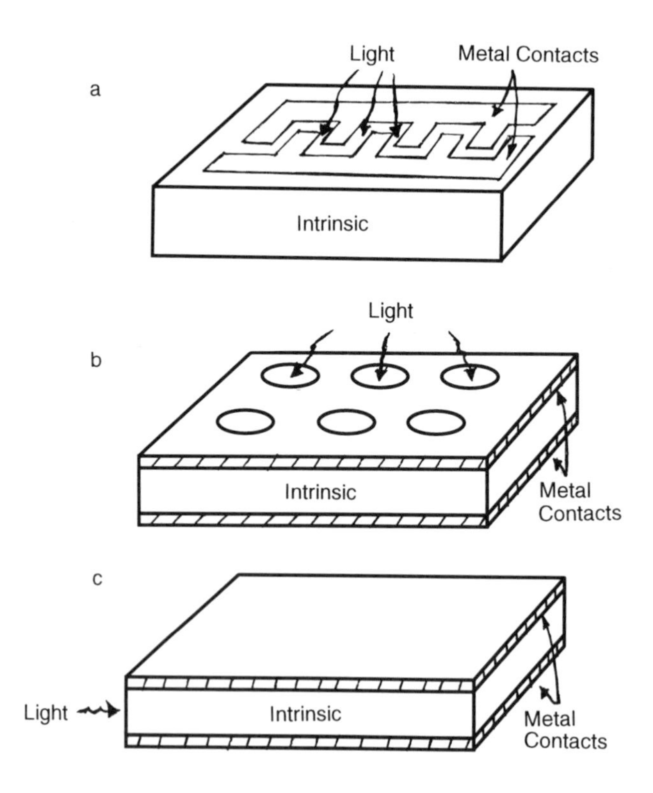

FIGURE 9.1. Schematic structure of (a) horizontal PCE, (b) vertical PCE, and (c) edge PCE.

9.1. HORIZONTAL PCE

Figure 9.2 shows the horizontal PCE structure and its applied voltage and light source. Depending on the type of heavily doped regions and the voltage polarity, three different horizontal PCEs are possible (Figs. 9.3a–c): forward-biased p^+–i–n^+, n^+–i–n^+, and reverse-biased p^+–i–n^+.

9.1.1. Analytical Model

In this section, we provide the mathematical analysis of an n^+–i–n^+ PCE. The energy-band diagram along with the coordinate of the n^+–i–n^+ PCE is shown in Fig. 9.4. In the figure, the region between 0 and L is the intrinsic channel region, the region between L and $L + x_n$ is the space-charge region in the n^+ region, and the region between $L + x_n$ and $L + x_n + W$ is the quasi-neutral region. A positive voltage is applied to the left-hand-side metal contact. As a result, the left-hand-side junction (n^+–i) is reverse biased, and the right-hand-side junction (i–n^+) is forward biased. The reverse-biased n^+–i junction collects electrons that approach it (electrons drift toward the left). In contrast, the forward-biased i–n^+ junction presents a retarding electric field to holes approaching it (holes drift toward the right), which causes the holes to accumulate to the right side of the i–n^+ region. Also, this junction injects electrons into the intrinsic region. Neither junction injects holes (Thompson and Lindholm, 1990).

Assuming electrons and holes in the intrinsic region are Boltzmann gases, the electron and hole continuity equations yield (Thompson and Lindholm, 1990)

$$\frac{d^2n}{dx} - \frac{n}{L_a^2} = -\frac{g}{D_a} \tag{9.1.1}$$

where g is the photogeneration rate, L_a is the ambipolar diffusion length, and D_a is the ambipolar diffusion coefficient. Equation (9.1.1) has a solution of the form

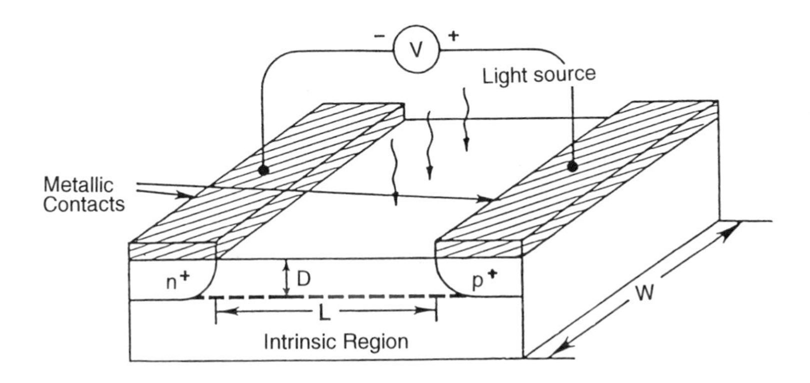

FIGURE 9.2. Simplified horizontal PCE structure.

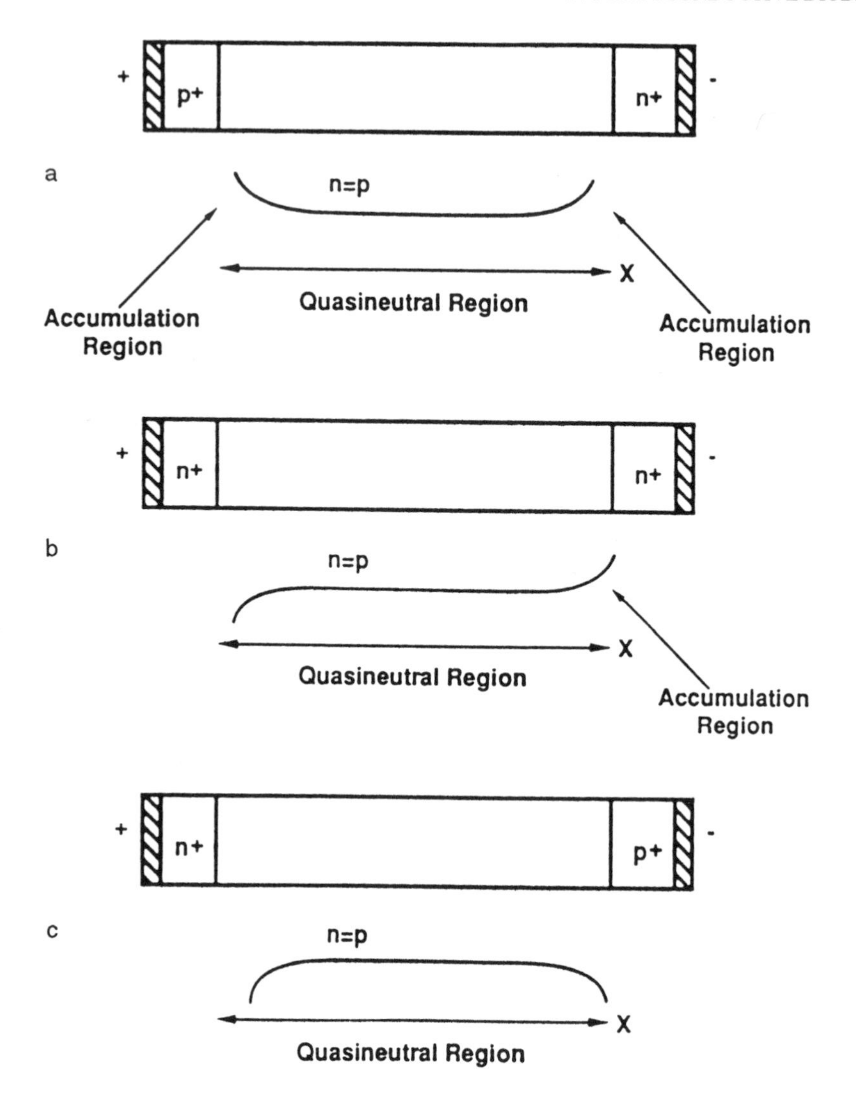

FIGURE 9.3. Three possible PCE systems: (a) forward-biased p^+–i–n^+ PCE, (b) n^+–i–n^+ PCE, and (c) reverse-biased p^+–i–n^+ PCE (Thompson and Lindholm, 1990).

$$p(x) = n(x)$$

$$= \frac{\{[n(0) - g\tau_L]\sinh(L - x/L_a) + [n(L) - g\tau_L]\sinh(x/L_a)\}}{\sinh(L/L_a)} + g\tau_L \tag{9.1.2}$$

where τ_L is the photogenerated carrier lifetime. Here

$$p(0) = n(0) = n_i e^{\Delta E_1/2kT} \tag{9.1.3}$$

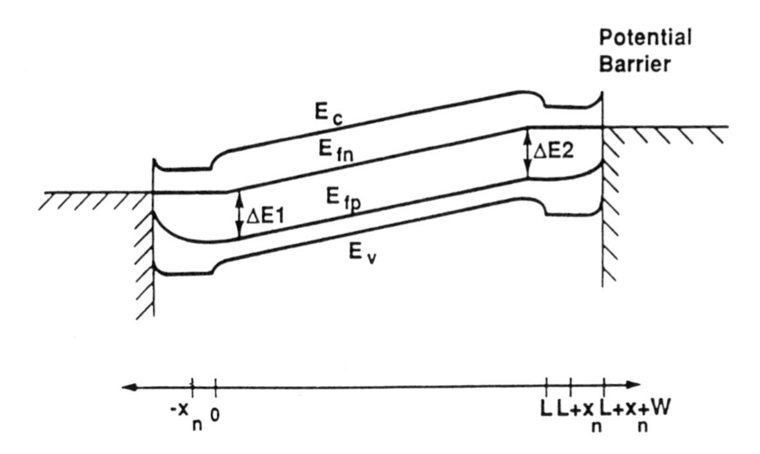

FIGURE 9.4. Energy-band diagram and coordinate system for the n^+–i–n^+ PCE (Thompson and Lindholm, 1990).

$$p(L) = n(L) = n_i e^{\Delta E_2/2kT} \tag{9.1.4}$$

ΔE_1 and ΔE_2 are the separation of the quasi-Fermi energies at $x = 0$ and $x = L$ (Fig. 9.4), respectively. Note that the applied voltage is

$$V_A = \frac{-\Delta E_1}{2q} + \frac{\Delta E_2}{2q} + V_i \tag{9.1.5}$$

Here V_i is the voltage drop in the intrinsic region.

Solving the hole continuity equation in the n^+ quasi-neutral region from $L + x_n$ to $L + x_n + W$ and applying the boundary conditions:

$$J_p(L + x_n + W) = qs_p p(L + x_n + W) \tag{9.1.6}$$

$$p(L + x_n) = n_i e^{\Delta E_2/kT} \tag{9.1.7}$$

we derive the hole current density J_p at $L + x_n$ as

$$J_p(L + x_n) = J_{p0} e^{\Delta E_2/kT} \tag{9.1.8}$$

where

$$J_{p0} = \frac{qD_p}{L_p}\left\{\frac{n_i[D_p/s_pL_p]\sinh(W/L_p) + \cosh(W/L_p)}{[D_p/s_pL_p]\cosh(W/L_p) + \sinh(W/L_p)}\right\} \tag{9.1.9}$$

In these equations, s_p is the surface recombination velocity at the contact, D_p is the hole diffusion coefficient, and L_p is the hole diffusion length. At $x = L$ and $x = 0$,

$$J_p(L) = \frac{J_T}{b+1} - \frac{2kT\mu_n\mu_p}{\mu_n + \mu_p}\frac{dn(x=L)}{dx} \tag{9.1.10}$$

$$J_p(0) = \frac{J_T}{b+1} - \frac{2kT\mu_n\mu_p}{\mu_n + \mu_p}\frac{dn(x=0)}{dx} \tag{9.1.11}$$

where J_T is the total current ($J_T = J_p + J_n$) and $b = \mu_n/\mu_p$.

The electric field $E(x)$ and electrostatic potential $V(x)$ in the intrinsic region have the following form (Thompson and Lindholm, 1990):

$$E(x) = \frac{J_p}{q\mu_p(b+1)p(x)} - \frac{kT(b-1)}{q(b+1)p(x)}\frac{dp(x)}{dx} \tag{9.1.12}$$

$$V(x) = \frac{J_T}{q\mu_p(b+1)}\int_0^L \frac{dx}{p(x)} + \frac{kT(b-1)}{q(b+1)}\ln\left(\frac{p(0)}{p(L)}\right) \tag{9.1.13}$$

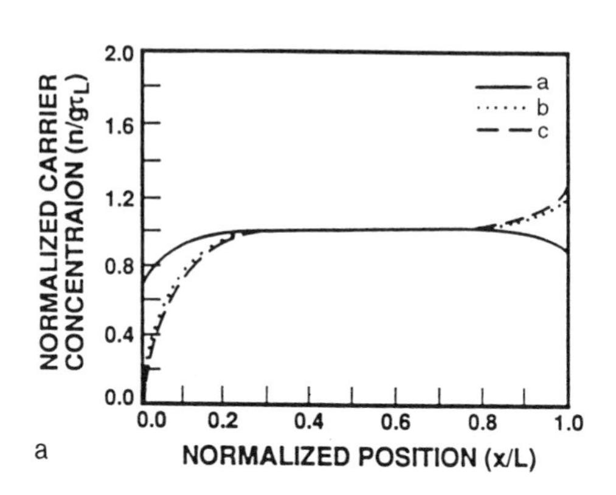

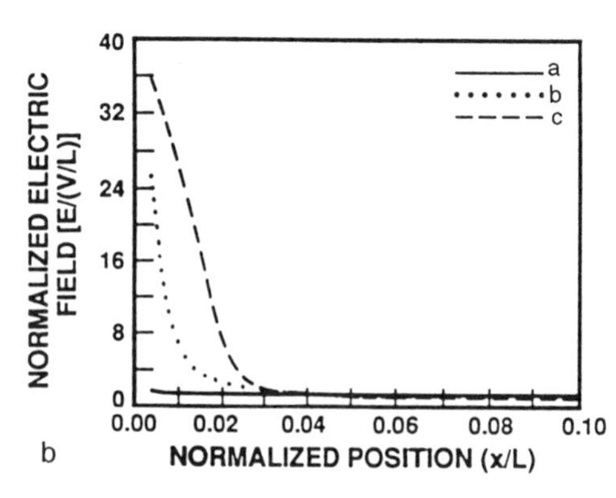

FIGURE 9.5. Normalized (a) free-carrier concentration and (b) electric field in the intrinsic region calculated from the model for $V_A = 0.4$ V (curve *a*), $V_A = 1.0$ V (curve *b*), and $V_A = 1.8$ V (curve *c*) (Thompson and Lindholm, 1990).

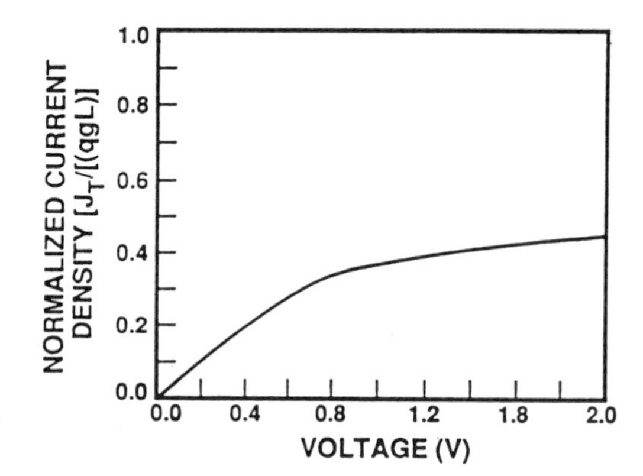

FIGURE 9.6. Normalized total current density versus the applied voltage (Thompson and Lindholm, 1990).

Figures 9.5a,b show the normalized free-carrier concentration and electric field in the intrinsic region for three voltages calculated from the analytic equations. The current–voltage characteristics are given in Fig. 9.6.

9.2. MEDICI SIMULATION

9.2.1. DC Characteristics

In the following simulation, we consider the off- and on-state dc characteristics of n^+–i–n^+ and reverse-biased p^+–i–n^+ PCEs having various channel lengths and bulk resistivities. Note that the forward-biased p^+–i–n^+ case is omitted since it is less suited for power switching because it can carry a large dark current and thus does not have a large resistance in the off state (Shakouri and Liou, 1992). The effects of three intrinsic levels, or bulk resistivities (35 kΩ-cm, 20 kΩ-cm, and 8 kΩ-cm), and three channel lengths (100, 70, and 40 μm) on the n^+–i–n^+ and p^+–i–n^+ PCEs current–voltage characteristics will be studied. The metal at the contact region is aluminum, and a maximum voltage of 60 V will be applied to the PCEs. Since both carriers are important in the intrinsic region, the two-carrier transport option available in MEDICI will be used in simulations.

9.2.1.1. Light-Off State Figure 9.7 shows the *I–V* characteristics of the nine n^+–i–n^+ devices under steady-state conditions. They vary in channel length and bulk resistivity. All nine devices show an analogous diode behavior with different slopes. In these devices, the higher the resistivity of the bulk material the less current is in the channel, which is desirable in the off state. Also the longer the channel length between the two contacts in these devices, the less current is in the channel. The observed characteristics result because the current–voltage relation of the n^+–i–n^+ PCE is inversely proportional to the resistance associated with the channel. As the channel length or bulk resistivity increases, the resistance of the channel increases, which subsequently reduces the current flows through the channel. Three-dimensional plots of the total current density and the electric field magnitude for the 40-μm channel, 8 kΩ-cm n^+–i–n^+ PCE are illustrated in Figs. 9.8 and 9.9, respectively. Note that the current density is higher near

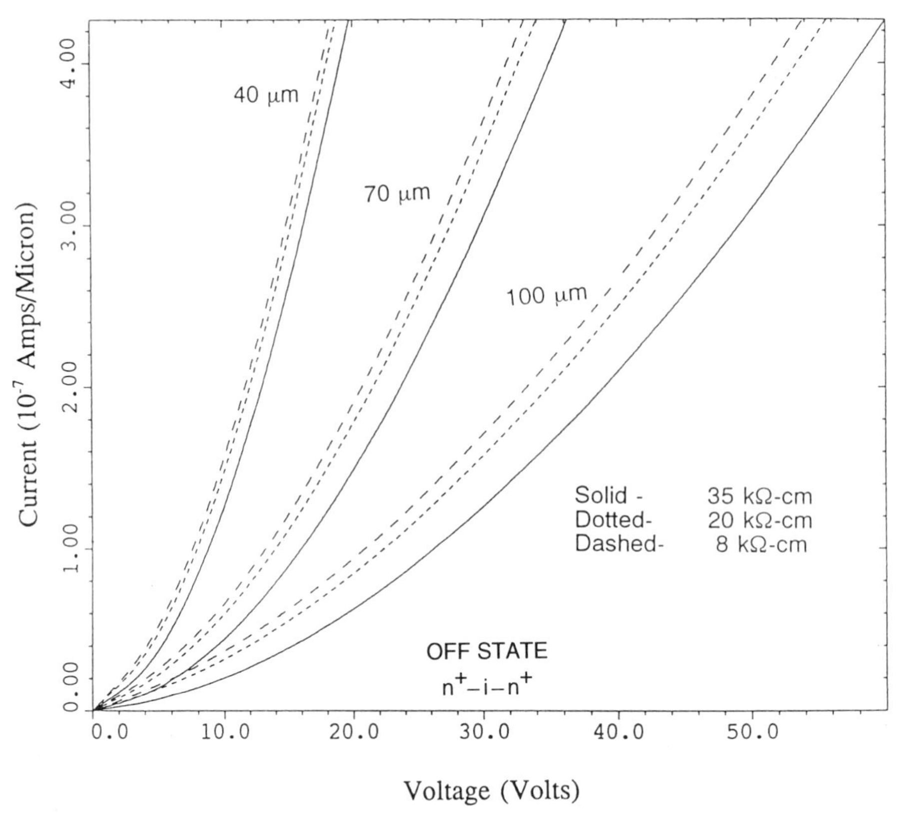

FIGURE 9.7. Current–voltage characteristics of the n^+–i–n^+ PCEs in the off state. Note that current is in units of A/µm, where µm is the length of the PCE's third dimension.

the surface than that in the bulk, indicating the majority of free-carrier flow near the surface between the two contacts. Also, because the positive voltage is applied to the right-hand-side n^+ contact, the maximum electric field exists at the right-hand-side n^+–i junction and decreases toward the left-hand-side i–n^+ junction, as shown in Fig. 9.9.

Figure 9.10 shows the I–V behavior of the reverse-biased p^+–i–n^+ devices where the p-type contact is grounded and the n-type contact is connected to a positive voltage. Unlike its n^+–i–n^+ counterpart, the reverse-biased p^+–i–n^+ PCEs (p^+–i–n^+ RB) show a current saturation behavior when the voltage is large. In addition, the p^+–i–n^+ RB PCE differs from the n^+–i–n^+ PCE in that increasing the channel length or increasing the bulk resistivity will increase the current. To examine the effects of the channel length on the current of the reverse-biased p^+–i–n^+ PCE, we have also simulated the depletion regions (unshaded areas in Fig. 9.11) in the channel of the 40-µm and 100-µm reverse-biased p^+–i–n^+ PCE biased at different voltages. Note that, at 60 V, the entire channel of the 40-µm PCE becomes the depletion region, whereas a large portion of the channel of the 100-µm PCE remains charge neutral. Since very few free carriers exist in the depletion region, the flow of the free carriers in the channel, which constitutes the current, will be

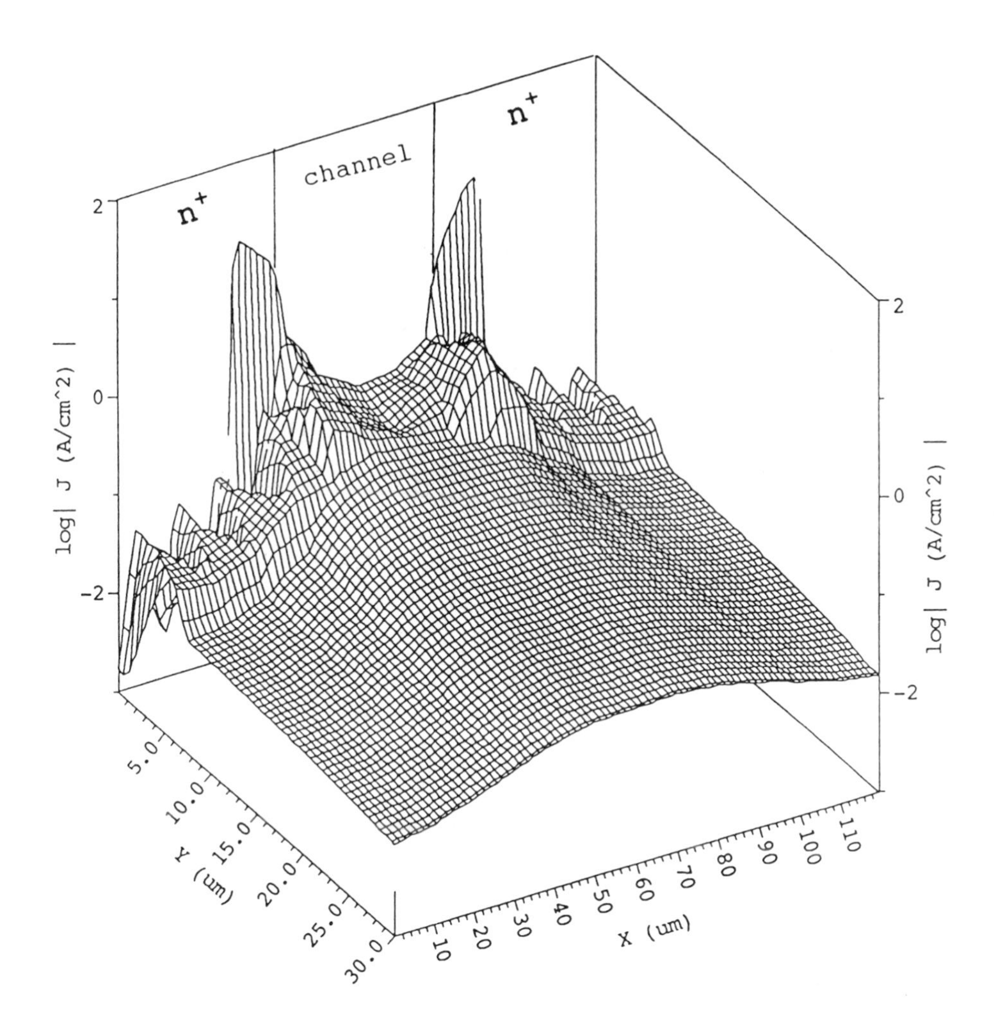

FIGURE 9.8. Three-dimensional current density contour in the 40-μm, 8-kΩ-cm n^+–i–n^+ PCE.

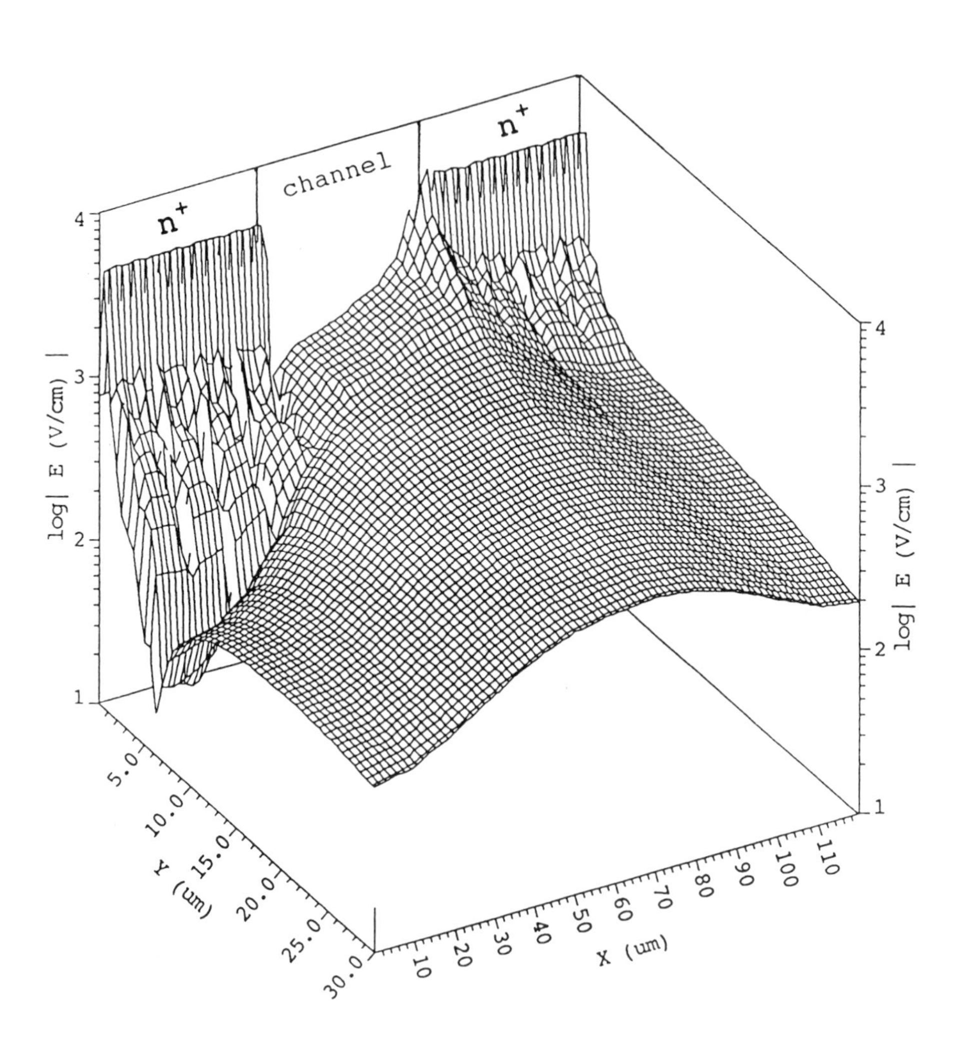

FIGURE 9.9. Three-dimensional electric field contour in the 40-μm, 8-kΩ-cm n^+–i–n^+ PCE.

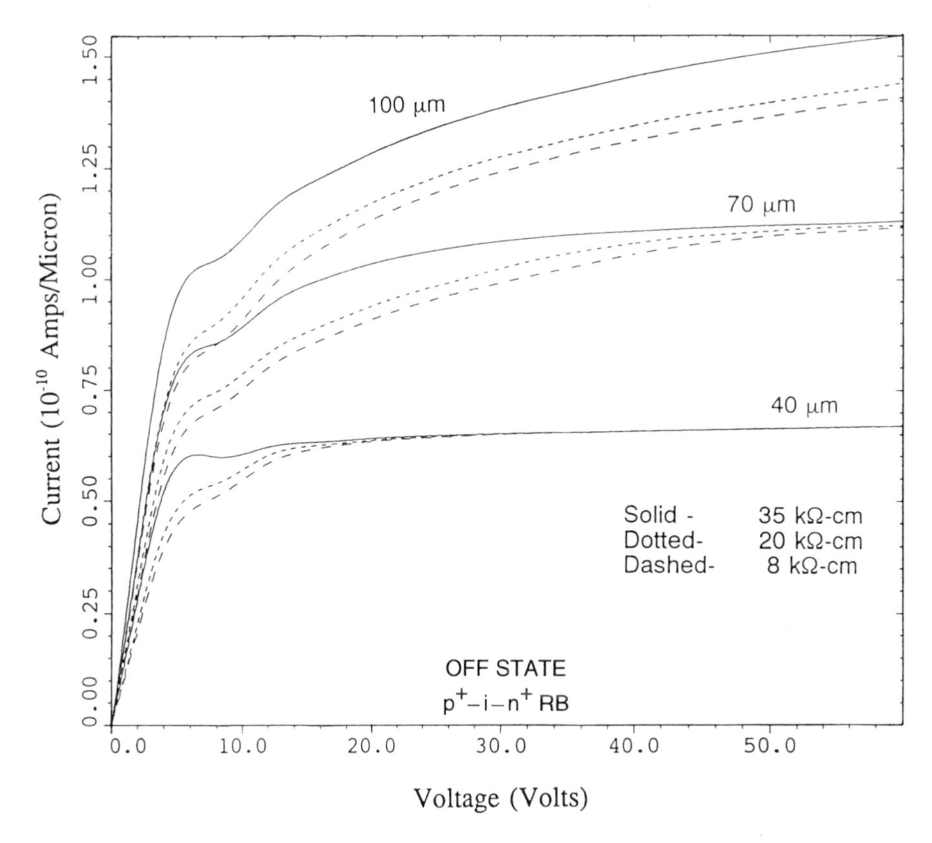

FIGURE 9.10. Current–voltage characteristics of the p^+–i–n^+ RB PCEs in the off state.

smaller if the depletion region is increased. As a result, the 40-μm reverse-biased p^+–i–n^+ PCE yields a smaller current than its 100-μm counterpart at high voltages. Note that the currents of the two PCEs are comparable at low voltages because there is no significant difference in the area of the depletion region with such bias conditions.

9.2.1.2. Light-On State In this state, the devices are illuminated with a 90-mW, 1000-nm monochrome light, and photogeneration of electron–hole pairs in the channel of the PCE is assumed uniform. Figure 9.12 shows the n^+–i–n^+ devices under illumination. The current is inversely proportional to the channel length and bulk resistivity, which are similar to the trends found in the off state, but the magnitude of the current is now much larger due to the increased electron and hole concentrations photogenerated in the channel.

The illuminated reverse-biased p^+–i–n^+ RB devices also exhibit current saturation behavior found in the off state (Fig. 9.13). However, the current has increased by as much as 340 times in the on state. The current variation due to the different bulk resistivity is very small and almost negligible in the 70- and 40-μm channel devices. This happens because the total number of free carriers in the channel is now composed predominantly

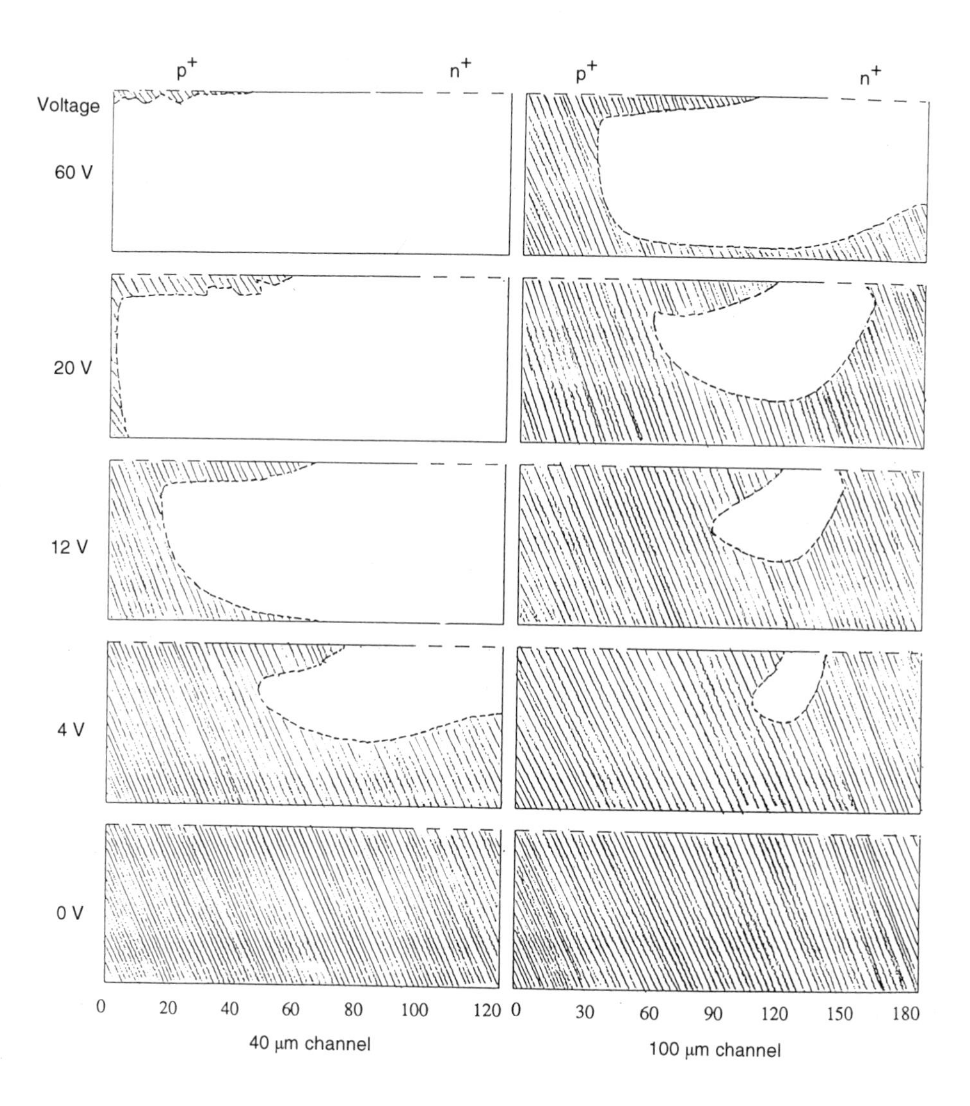

FIGURE 9.11. Areas of the depletion region (unshaded areas) simulated for the 40-μm and 100-μm p^+–i–n^+ RB PCEs at five different applied voltages.

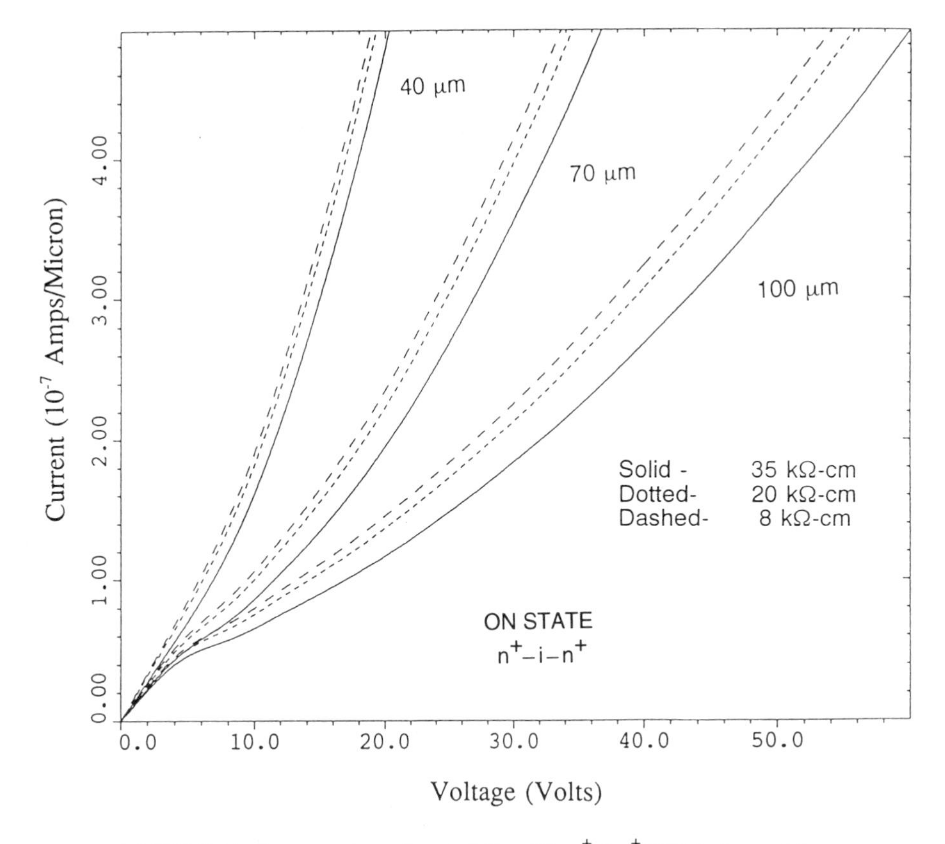

FIGURE 9.12. Current–voltage characteristics of the n^+–i–n^+ PCEs in the on state.

of those generated from illumination, and changing the bulk resistivity will not alter the total free-carrier concentration. Consequently, the bulk resistivity affects negligibly the electric field and, thus, the drift current in the channel.

The foregoing results suggest that a reverse-biased p^+–i–n^+ structure with low bulk resistivity and long channel will provide the highest ratio of light-on to light-off current and thus will be the most suitable PCE for switching in the steady state.

9.2.2. *Transient Characteristics*

When applying a constant voltage and a light pulse to the PCE, short electrical pulses can be generated (Cho *et al.*, 1994; Loubriel *et al.*, 1987). These pulses are currently used in various applications like jitter-free streak cameras and pumping sources for electrical discharge lasers (Rosen and Zutavern, 1993). For a PCE using a light source with a very fast light pulse (e.g., laser), the switching speed of the PCE is limited by the electron–hole generation time (for turn-on transient) and recombination time (for turn-off transient). Consequently, to obtain high-speed switching and, thus, fast pulses, the PCE needs to be fabricated with high-carrier-mobility such as GaAs or InP.

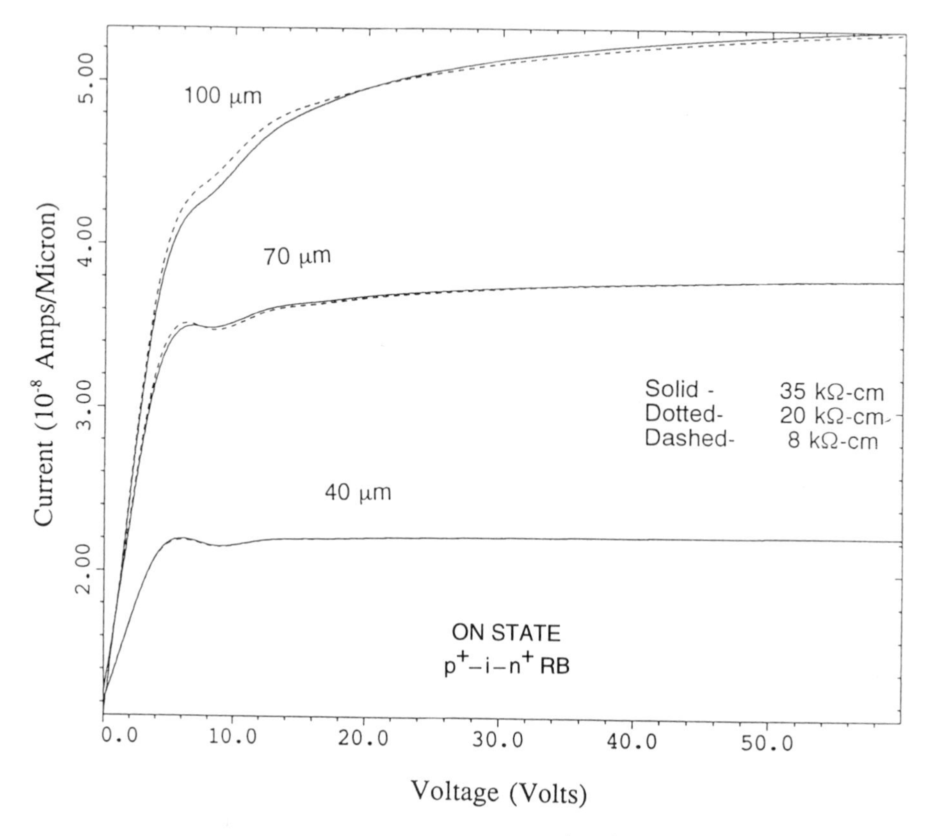

FIGURE 9.13. Current–voltage characteristics of the p^+–i–n^+ RB PCEs in the on state.

The PCE transient characteristics are presented here. In the simulation, a reverse-biased p^+–i–n^+ PCE with a channel length of 10 μm is used, and the PCE is subject to a constant voltage of 50 V and a 632-nm laser pulse. At $t = 0$, the laser is turned off, and the current versus time characteristics of the PCE are shown in Fig. 9.14a. Detailed transient response in the first 2 ps is given in Fig. 9.14b. The results indicate that, except within the initial transient, the current decays exponentially with respect to time and becomes zero at about 40 ps. The relatively long turn-off time can be attributed to the fact that the excess electrons and holes generated by the light do not recombine immediately after the light is turned off, as evidenced by the time-dependent electron and hole concentrations in the channel of the PCE in Figs. 9.15a,b, respectively. Contours of the constant total current in the PCE, from top to bottom, at 1.5, 6.1, 13.2, 23.6, 42, and 53.1 ps after the light pulse is applied are plotted in Fig. 9.16. The contours are less crowded in the channel of the PCE as time elapses, indicating that the current passing through the device is decreased. The corresponding net electron–hole recombination rates are illustrated in Fig. 9.17, which clearly shows that electron–hole recombination is less active after the light is turned off, due to the reduced electron and hole concentrations in the channel.

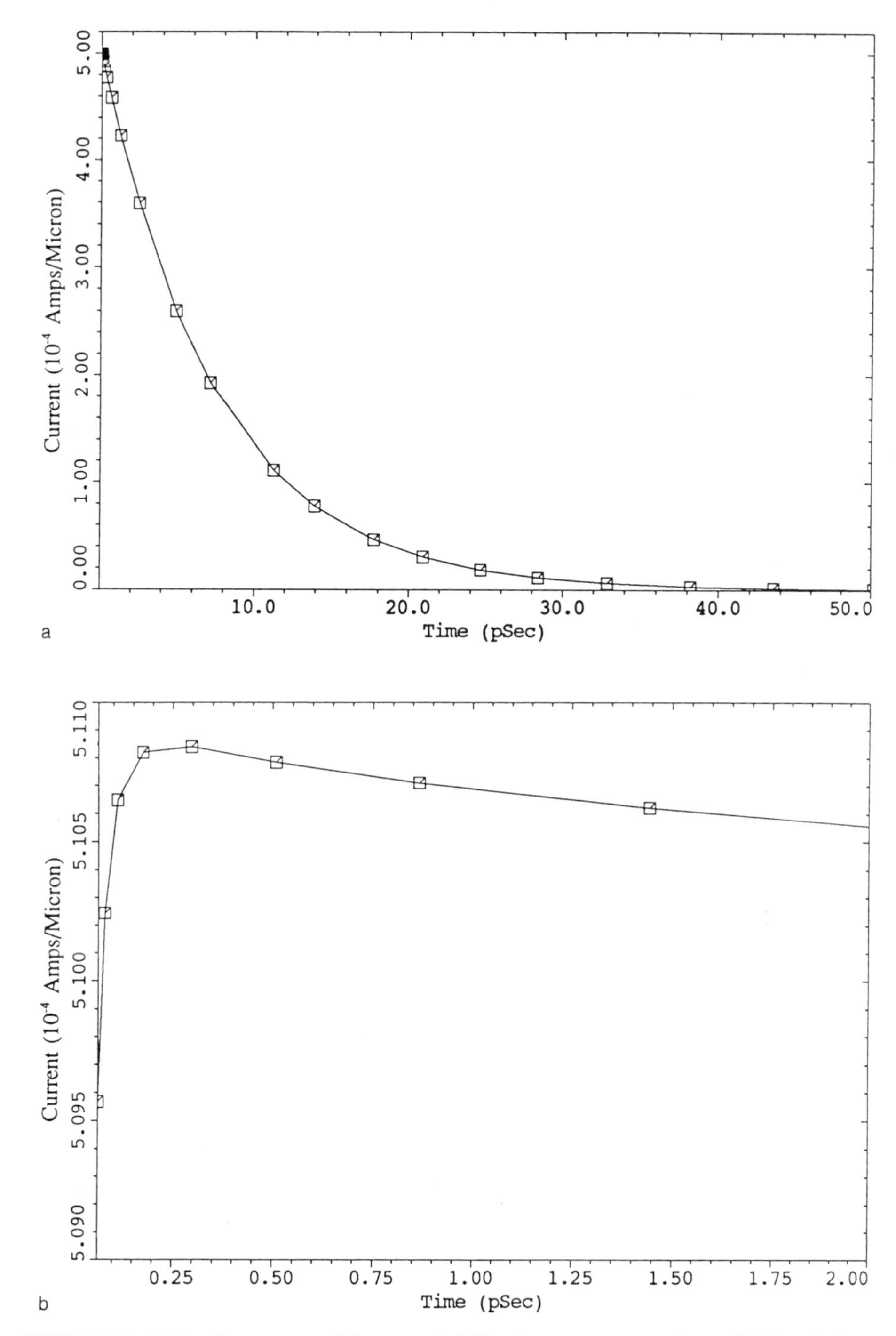

FIGURE 9.14. (a) Transient response of the current in PCE subject to a constant voltage of 50 V and a 5-ps, 632-nm laser pulse. (b) Detailed current versus time characteristics in the first two picoseconds.

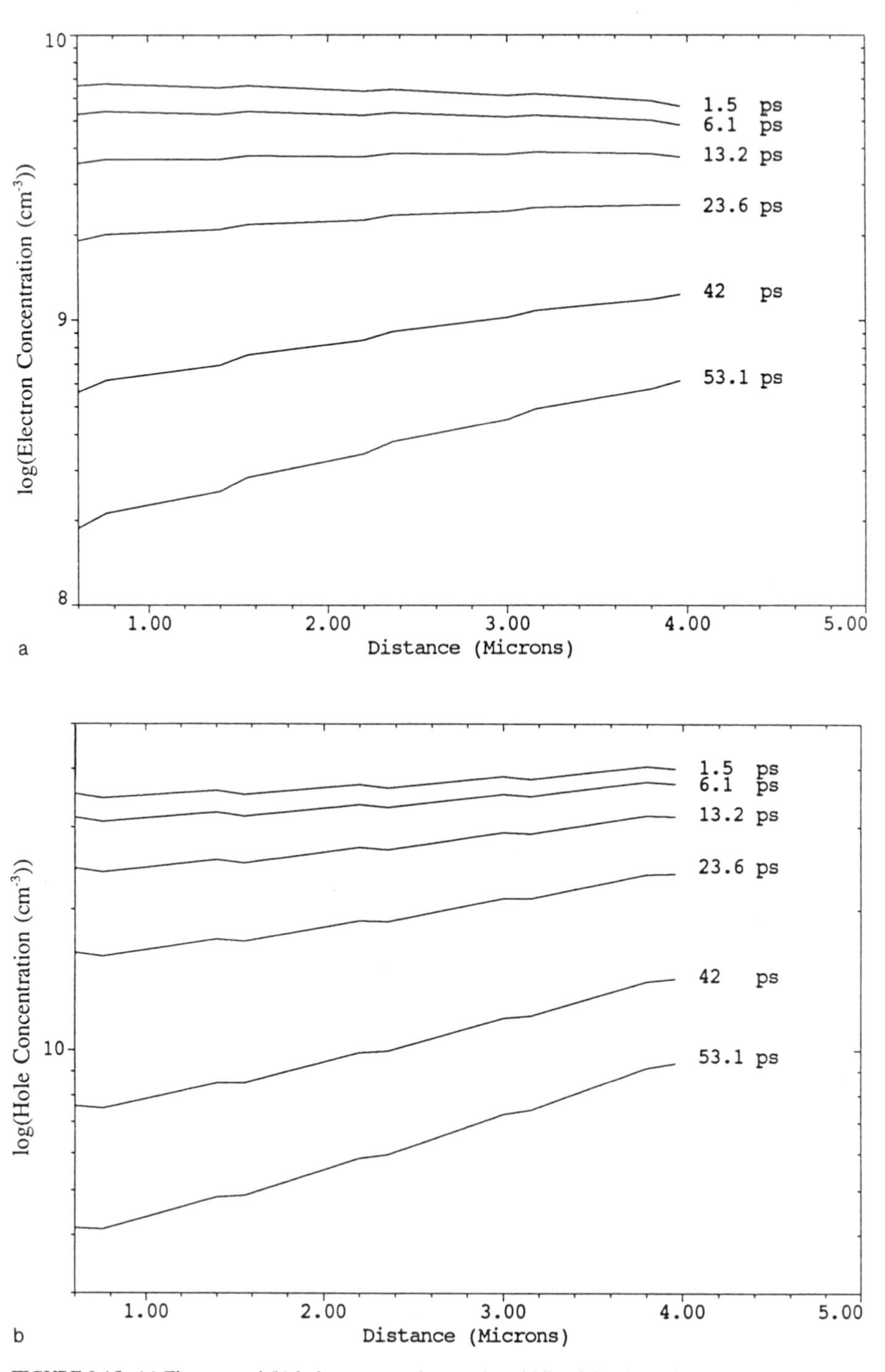

FIGURE 9.15. (a) Electron and (b) hole concentrations at the middle of the channel versus the distance (into the bulk) as a function of time.

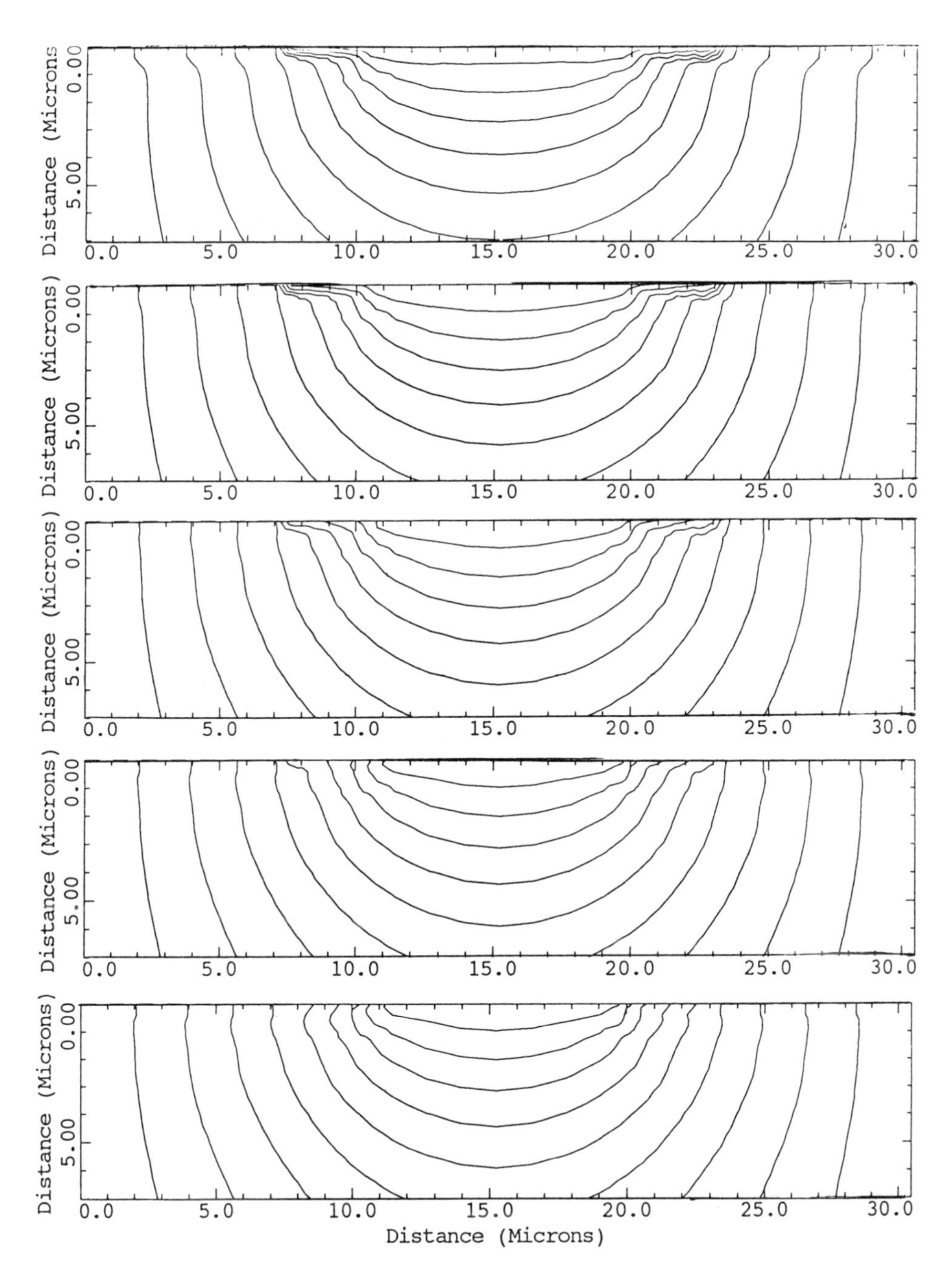

FIGURE 9.16. Contours of constant total current in the PCE at, from top to bottom, 1.5, 6.1, 13.2, 23.6, and 53.1 ps.

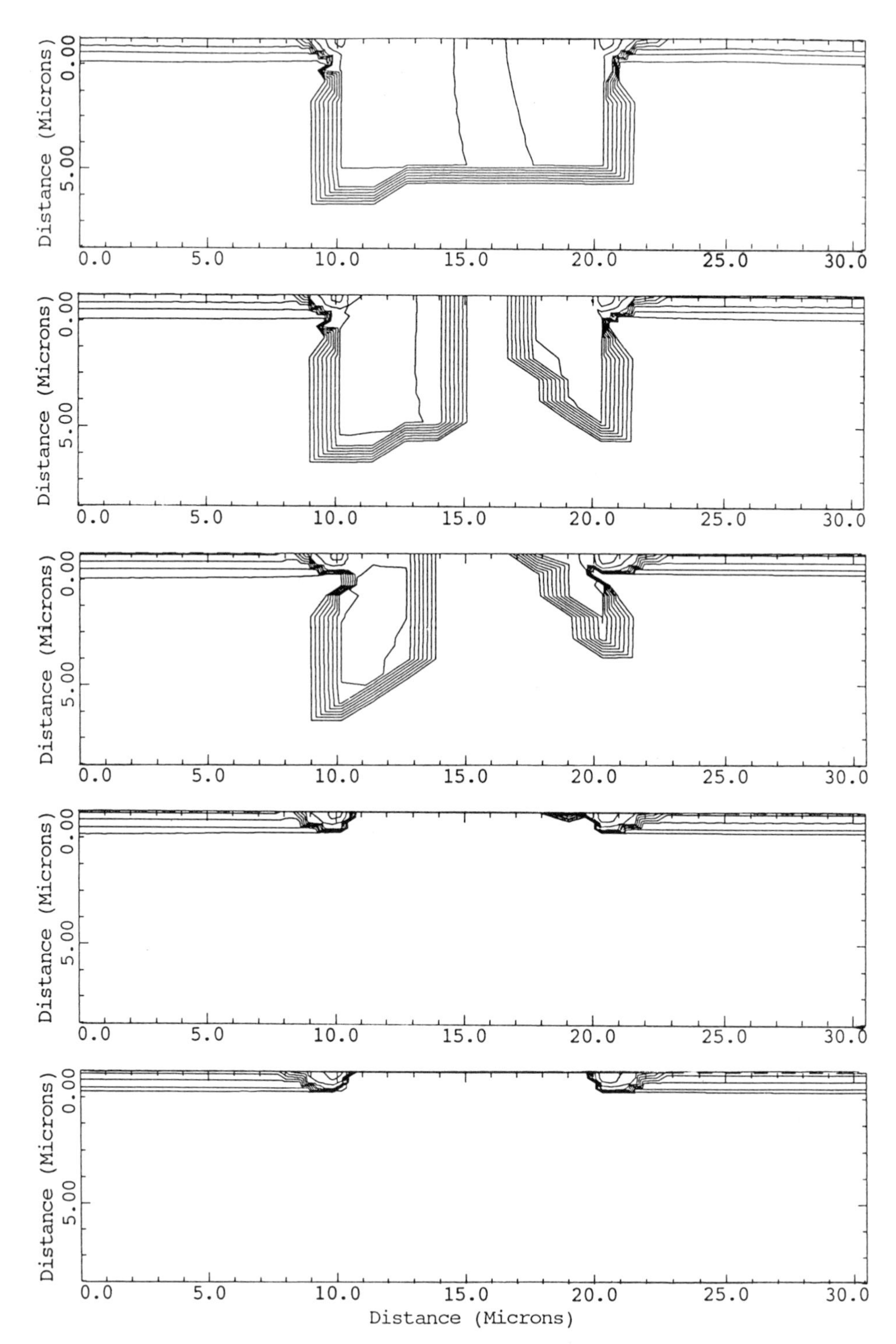

FIGURE 9.17. Contours of constant electron–hole recombination rate in the PCE at, from top to bottom, 1.5, 6.1, 13.2, 23.6, and 53.1 ps.

In general, the PCE turn-on time is approximately the free-carrier generation time. Once the excess carriers are generated, the conductivity of the channel increases drastically and the PCE is turned on. On the other hand, the turn-off time, which depends on how fast the excess free carriers are recombined, is more complicated to determine intuitively. Let us consider a reverse-biased p^+–i–n^+ PCE. Electrons and holes generated in the channel will drift toward the i–n^+ and p^+–i junctions, respectively. They can be recombined in the intrinsic region during the journey or they can reach the junction regions and recombine at the contacts, depending on the carrier recombination lifetime τ, as well as how far the carrier has to travel to reach the junction, which is a function of the channel length L. The following approach can be used to estimate the PCE turn-off time. First, divide the channel into many subregions and focus on either electrons or holes. Then calculate the electric field in the channel (V/L) and the corresponding drift velocity. Use the drift velocity to find the time t_t required for the electron (or hole) in each subregion to drift to the junction. For a particular subregion, if t_t is larger than τ, then the excess electrons in the subregion are recombined in the intrinsic region and have a lifetime equal to τ. If t_t is smaller than τ, then the excess electrons are recombined at the contact region and have a lifetime equal to t_t. The average of the lifetimes for all subregions is approximately equal to the PCE turn-off time.

9.3. VERTICAL PCE

9.3.1. DC Characteristics

A vertical n^+–i–n^+ PCE with light source and bias voltage is illustrated in Fig. 9.18. The top and bottom contact regions have a peak concentration of 10^{20} cm^{-3}, and the channel thickness between the two contacts is 500 μm (Fig. 9.19). The simulated current–voltage characteristics of the PCE for six different wavelengths are given in Fig. 9.20. At a particular voltage, the increase rate of current is decreased as the wavelength is decreased. Note that the curves consist of linear, quasi-saturation, and saturation regions. The external quantum efficiencies η in these three regions are also simulated (Fig. 9.21), and η is defined as

$$\eta = Ihcqn'\lambda'/Q_{\text{optical}} \tag{9.3.1}$$

where I is the photocurrent, h is Planck's constant, c is the velocity of light, n' is the index of refraction of air ($n' = 1$), and λ' is the wavelength. The results in Fig. 9.21 suggest that the quantum efficiency increases with voltage and decreases as the wavelength increases.

In the following analysis, a monochromatic light source with a wavelength of 950 nm and generates 10^{17} cm^{-3} electron–hole pairs in the semiconductor is used. Figure 9.22 shows the on-state I–V characteristics (up to 500 V). It is shown that the current increases rapidly at very small voltages (< 1 V) and becomes saturated at higher voltages due to the large electric field in the intrinsic bulk region. Figures 9.23a,b, 9.24a,b, and 9.25a,b plot the 3-D contours of the electric field, electron density, and hole density in the PCE, respectively, at two voltages. When the voltage is small (i.e., 1 V), the electric field peaks

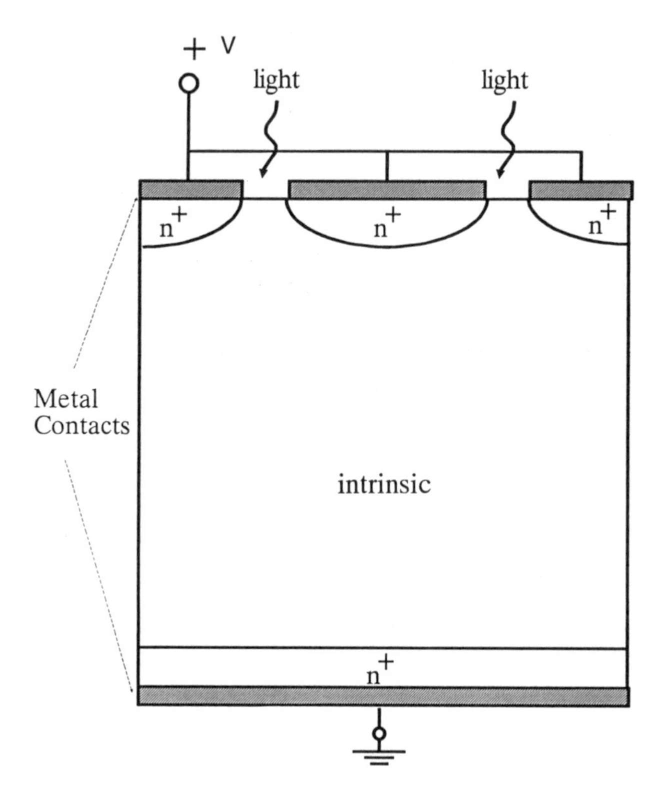

FIGURE 9.18. Simplified vertical PCE structure.

at the n^+–i junction and decreases rapidly toward the intrinsic bulk region. Since the on-state current is driven mainly by the drift tendency in the bulk region, it increases quickly with voltage because the drift velocity there increases nearly linearly with the electric field. At higher voltages, however, the high field is extended into the bulk region, which consequently causes the free-carrier drift velocity in the bulk to saturate and the on-state current to increase slowly as the voltage is increased (Fig. 9.22). Also, the effect of the electric field on the electron and hole distributions in the PCE is clearly demonstrated in Figs. 9.24 and 9.25; as the applied voltage, and thus electric field, increases, the free-carrier concentrations in the bulk near the contact regions are lowered because the carriers are being swept away from the region by the large field.

9.3.2. *Transient Characteristics*

Now the turn-off transient characteristics of the PCE are presented. In this case, the light is turned off at $t = 0$, and the device is biased with a voltage of 500 V. The transient current and voltage of the PCE versus time (i–t and v–t) curves for $0 < t < 50$ μs are plotted in Figs. 9.26a,b. The current decays to 50% of its steady-state value in about 20 μs, which coincides with the free-carrier lifetime used in simulation. In addition, it is shown that the voltage across the switch approaches to a steady-state value of about 200 V, which is considerably lower than the applied voltage of 500 V. This is due to the fact that a large voltage drop occurs in the contact regions (i.e., 10 Ω contact resistance has been used in simulation). To illustrate the physical insight during the transient, snapshots of the electric field, electron density, and hole density contours in the PCE are taken at the beginning ($t = 0$), the middle ($t = 28$ μs), and the end ($t = 50$ μs) of the transient, as shown in Figs. 9.27a–c, 9.28a,b, and 9.29a,b, respectively. It can be seen that the electric field distribution in the intrinsic region, particularly in the region close to the base contact, changes considerably for $0 < t < 28$ μs and remains nearly the same for $t > 28$ μs. Also, the electron and hole densities decrease rapidly between $t = 0$ and 28 μs due to the high electron–hole recombination rate in the device.

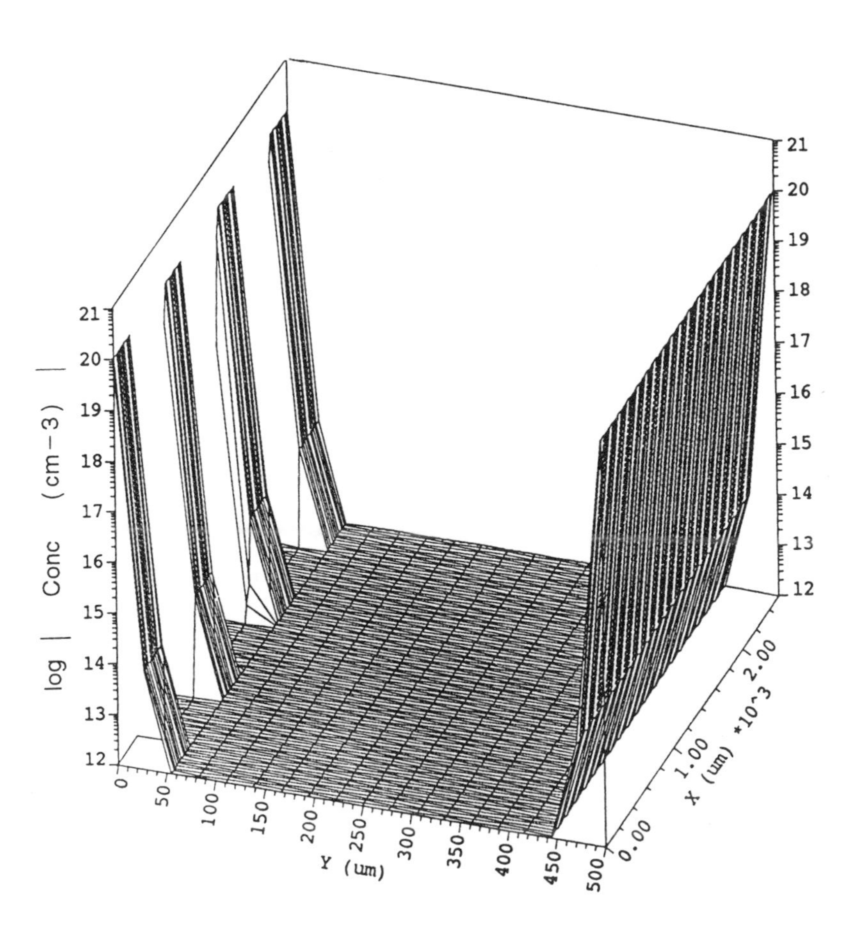

FIGURE 9.19. Impurity doping profile of the n^+–i–n^+ PCE.

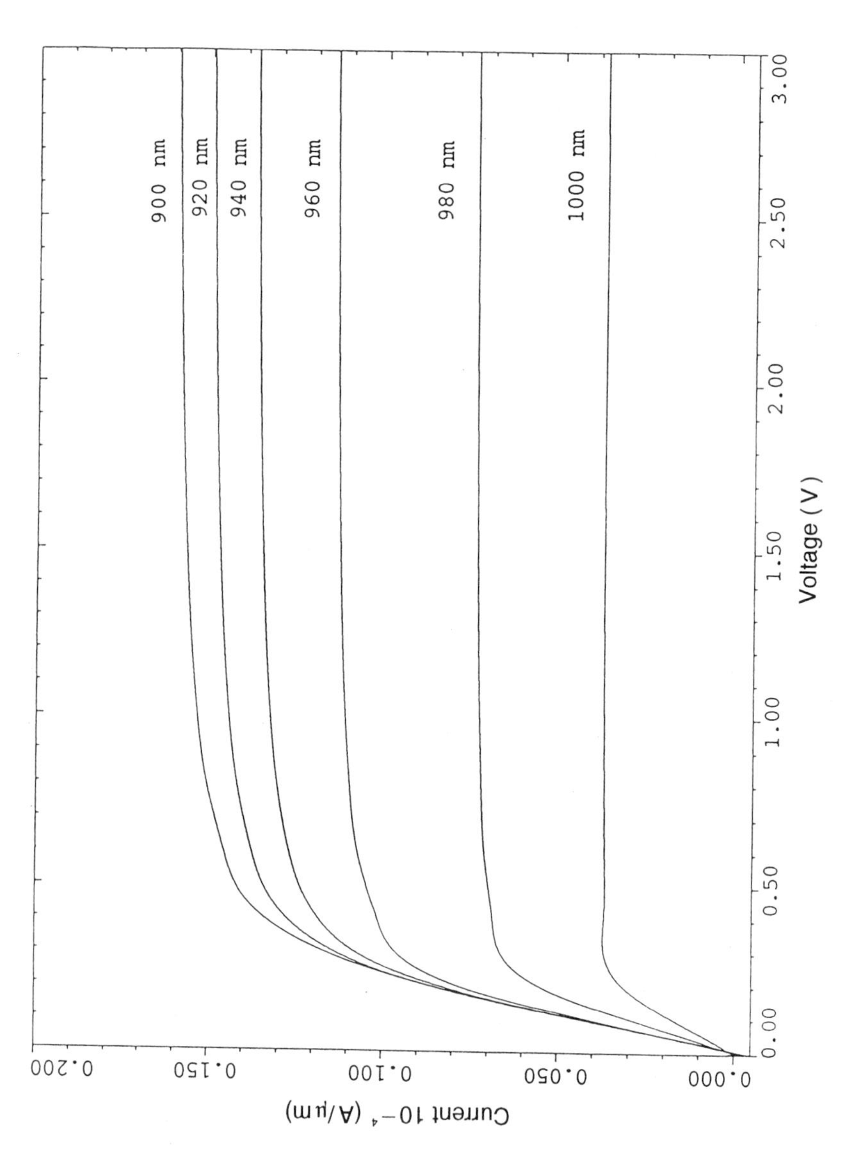

FIGURE 9.20. Current versus voltage characteristics as a function of the light wavelength.

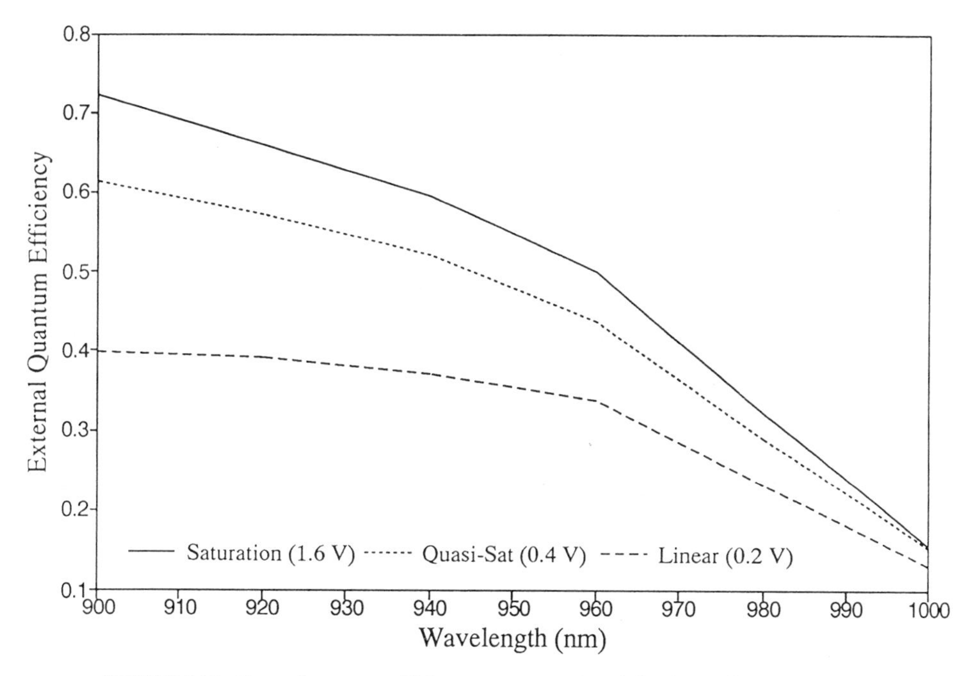

FIGURE 9.21. External quantum efficiency versus wavelength for three different voltages.

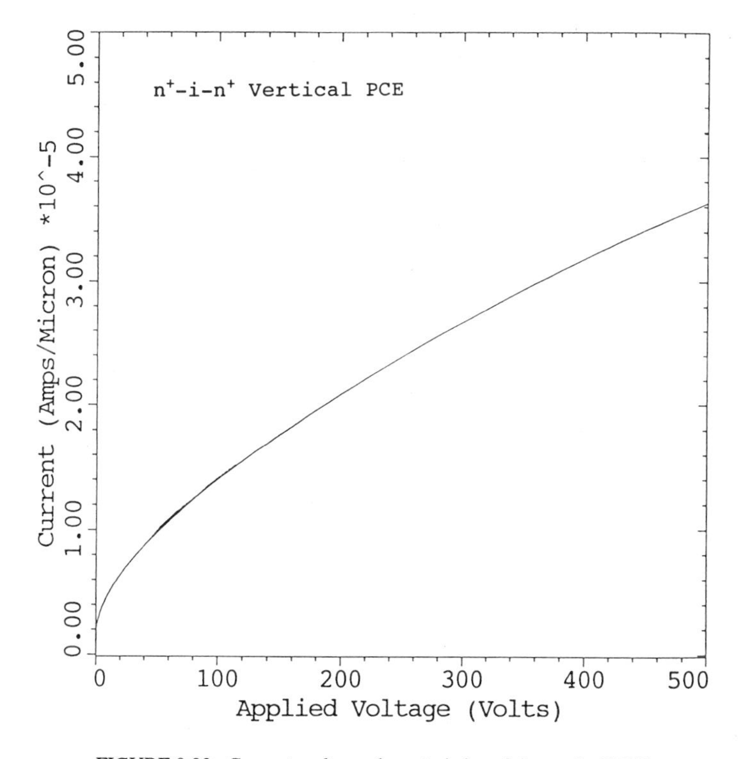

FIGURE 9.22. Current–voltage characteristics of the vertical PCE.

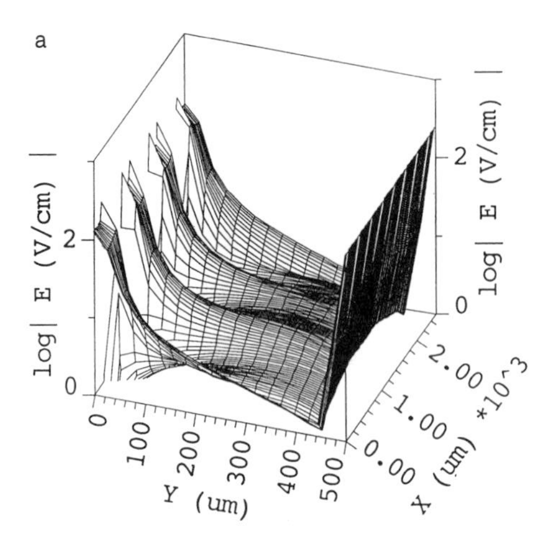

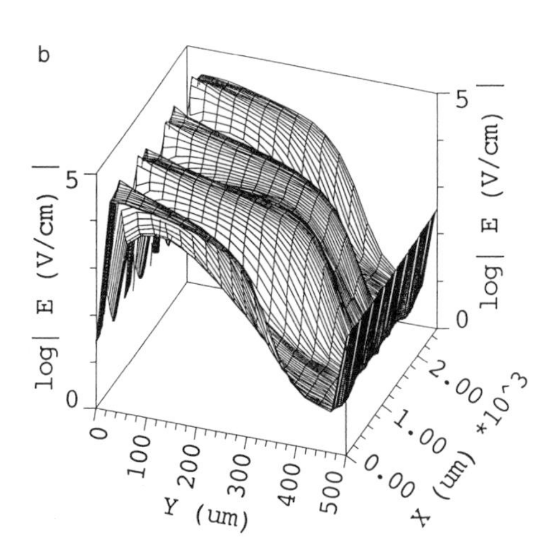

FIGURE 9.23. Three-dimensional electric field contours simulated at (a) $V = 1$ V and (b) $V = 500$ V.

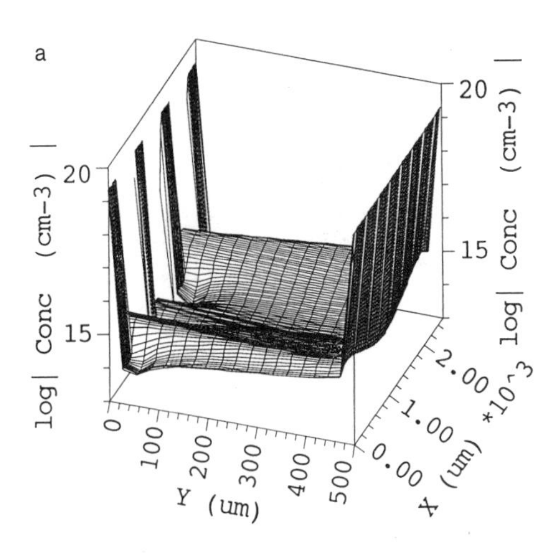

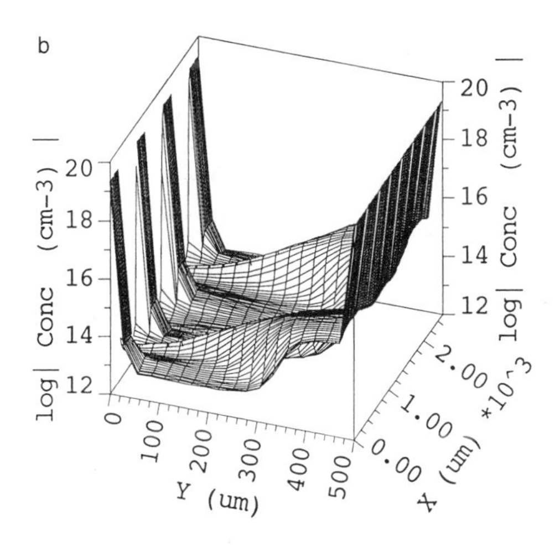

FIGURE 9.24. Three-dimensional electron density contours simulated at (a) $V = 1$ V and (b) $V = 500$ V.

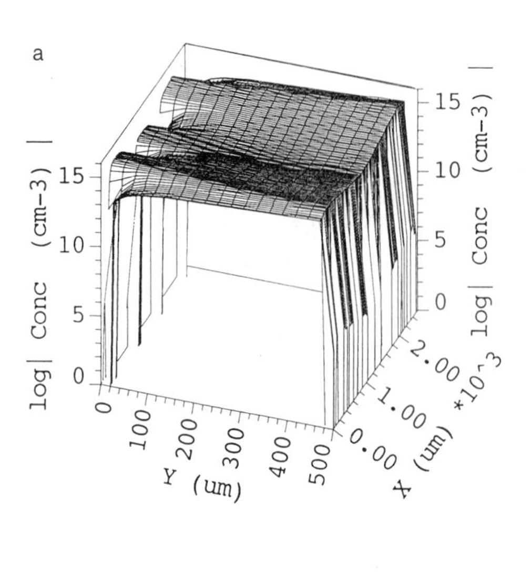

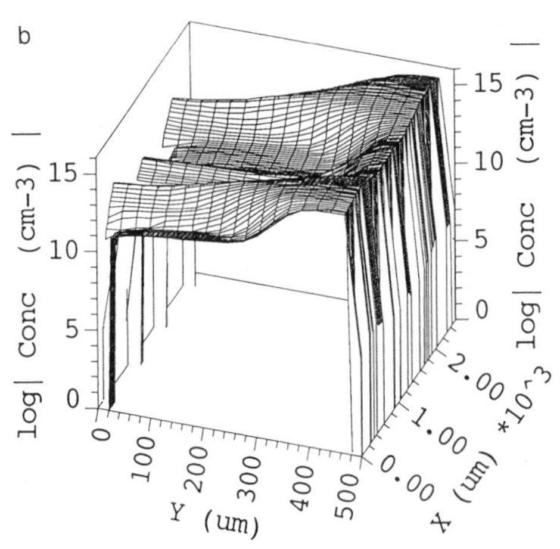

FIGURE 9.25. Three-dimensional hole density contours simulated at (a) $V = 1$ V and (b) $V = 500$ V.

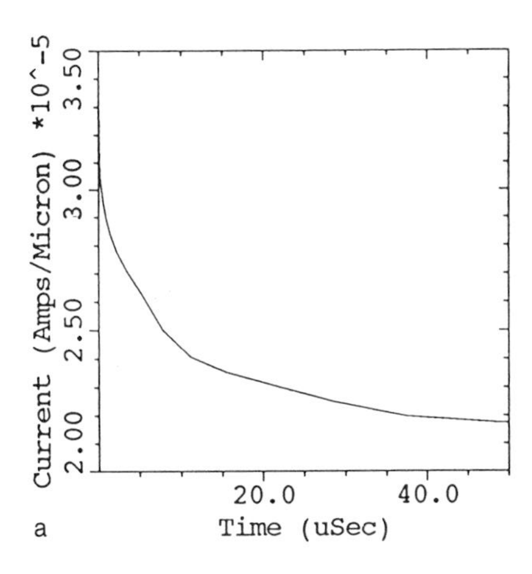

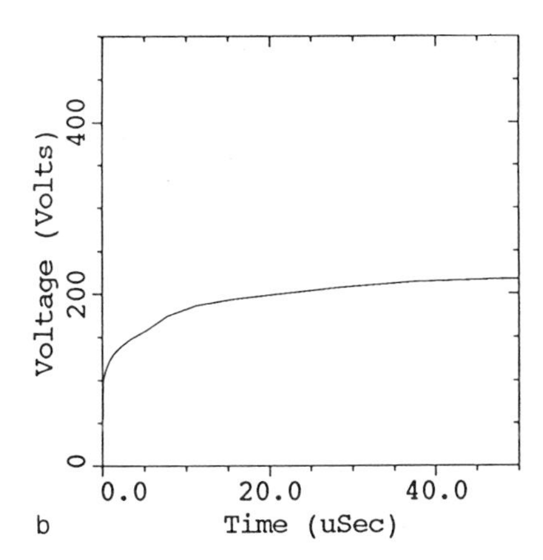

FIGURE 9.26. Turn-off transient responses of (a) current and (b) voltage of the PCE.

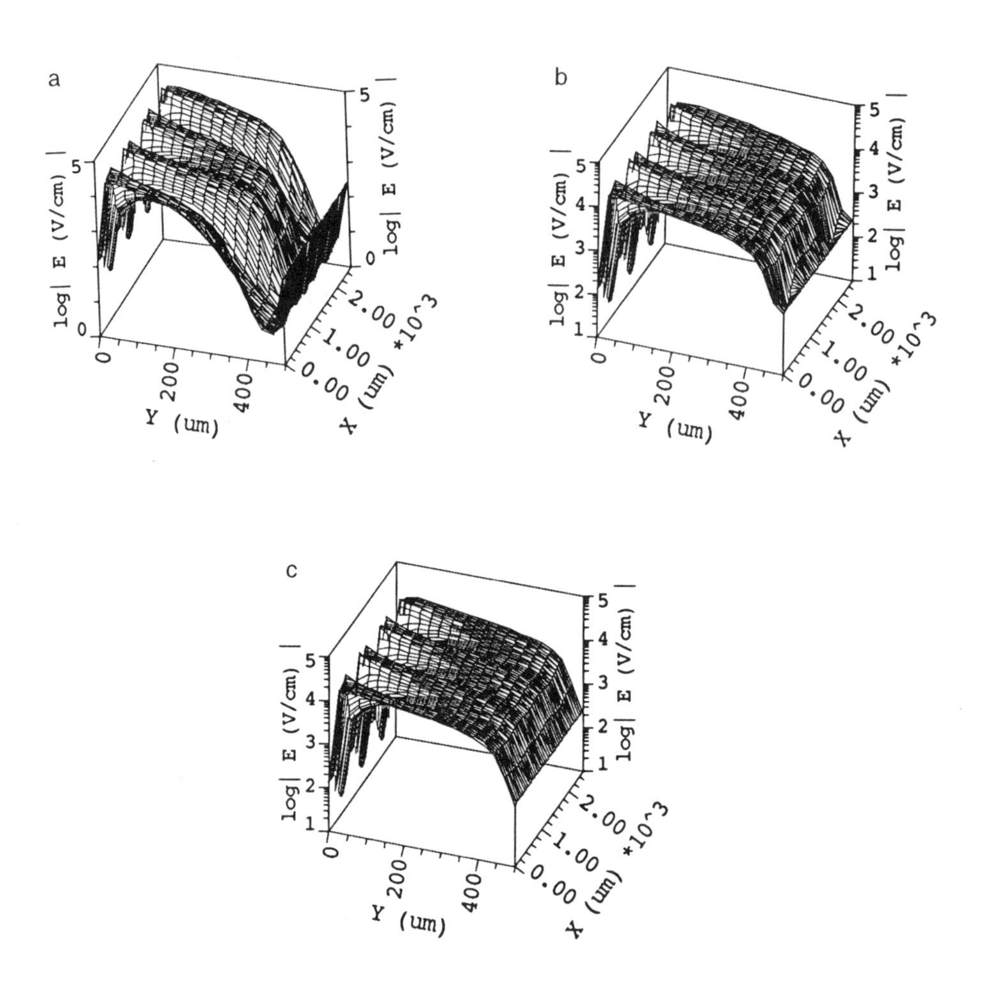

FIGURE 9.27. Three-dimensional electric field contours simulated at (a) $t = 0$, (b) $t = 28$ μs, and (c) $t = 50$ μs.

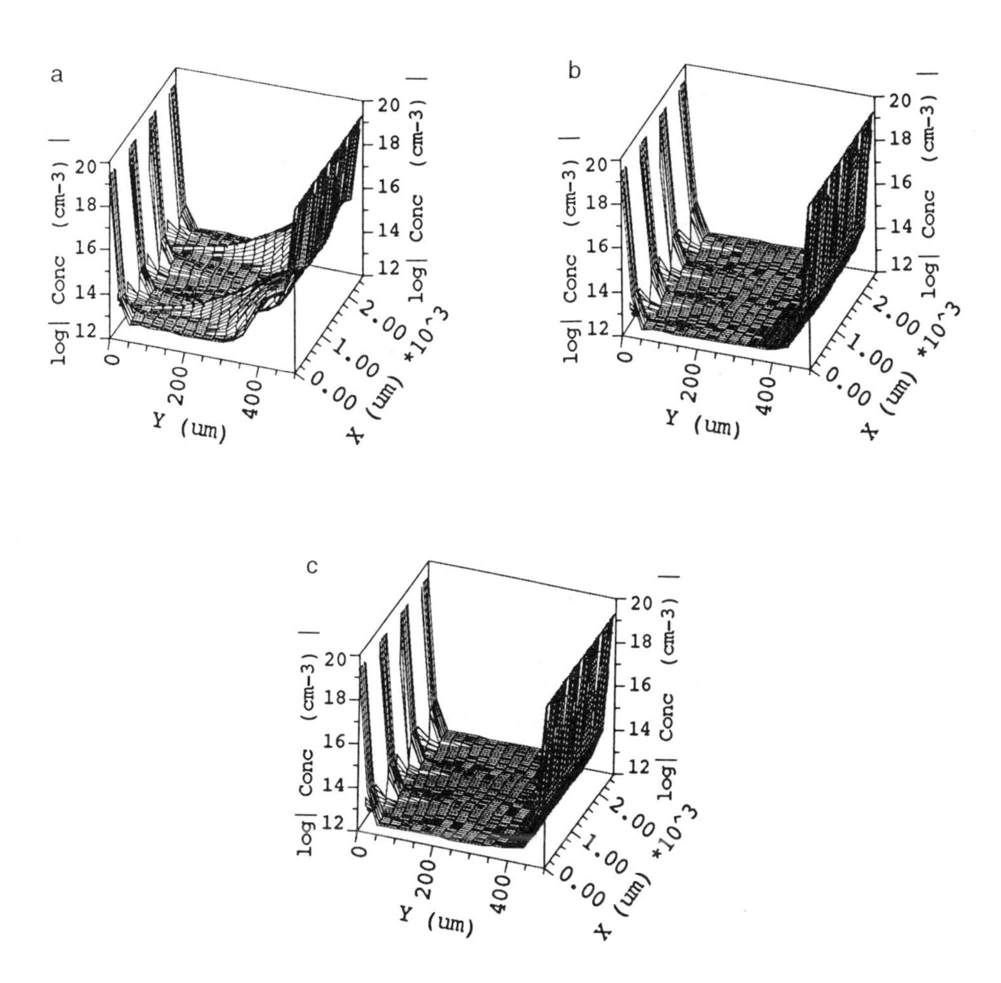

FIGURE 9.28. Three-dimensional electron density contours simulated at (a) $t = 0$, (b) $t = 28$ μs, and (c) $t = 50$ μs.

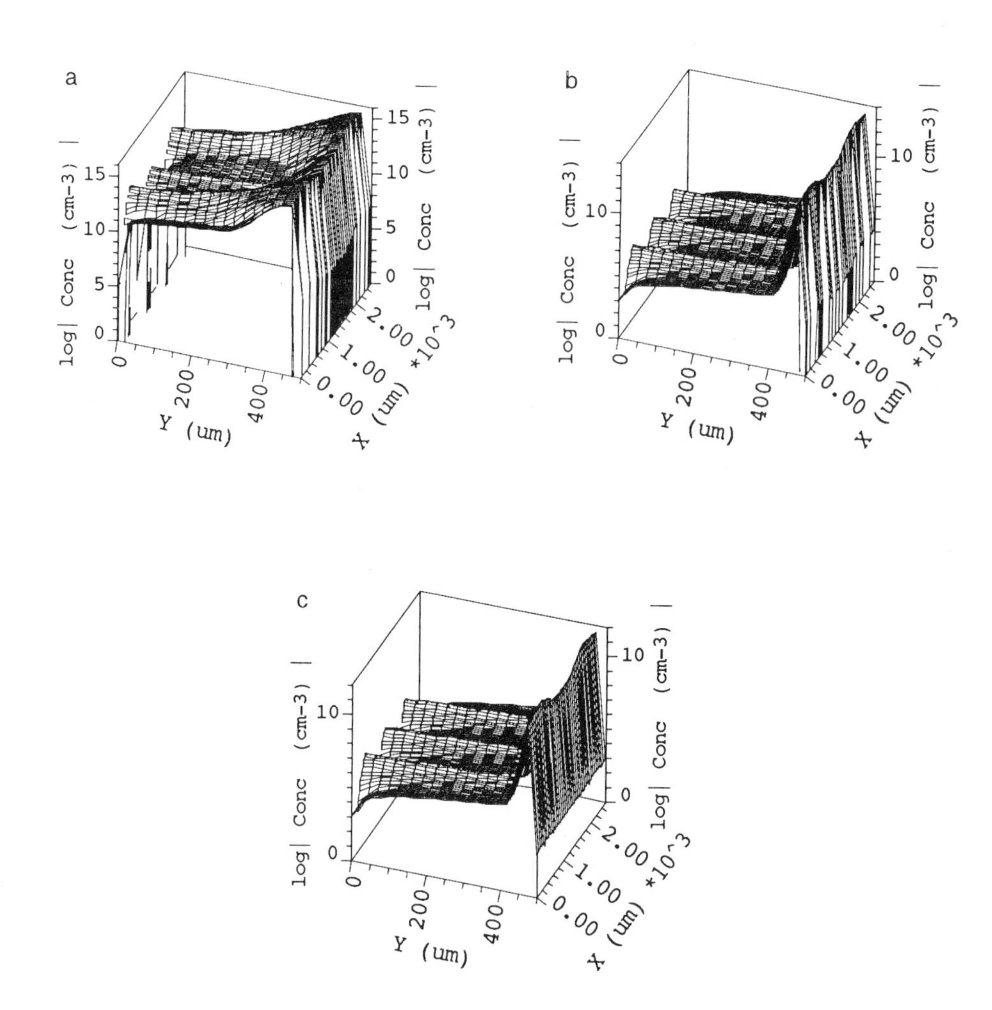

FIGURE 9.29. Three-dimensional hole density contours simulated at (a) $t = 0$, (b) $t = 28$ μs, and (c) $t = 50$ μs.

9.4. SELF-HEATING EFFECTS ON PCE PERFORMANCE

Since the PCE normally carries a very large current in the light-on condition and is subject to a large applied voltage, the heat generated in the PCE due to the lattice heating (i.e., self-heating effect) is important. A lattice temperature higher than ambient (e.g., 300 K) resulting from self-heating can degrade the PCE performance (Parthasarathy *et al.*, 1994). In this section, we present results of the vertical PCE simulated with the lattice temperature module, an option available in MEDICI that solves the coupled heat transfer equations.

Figures 9.30a–c show lattice temperature contours in the PCE at three voltages, respectively, which indicate that the lattice temperature is highly spatial dependent and considerably higher than the ambient temperature at high voltages. The current–voltage characteristics of the PCE simulated with and without the self-heating effect are compared in Fig. 9.28. Clearly, the current is reduced notably by the higher lattice temperature, and the self-heating effect is less significant at smaller voltages.

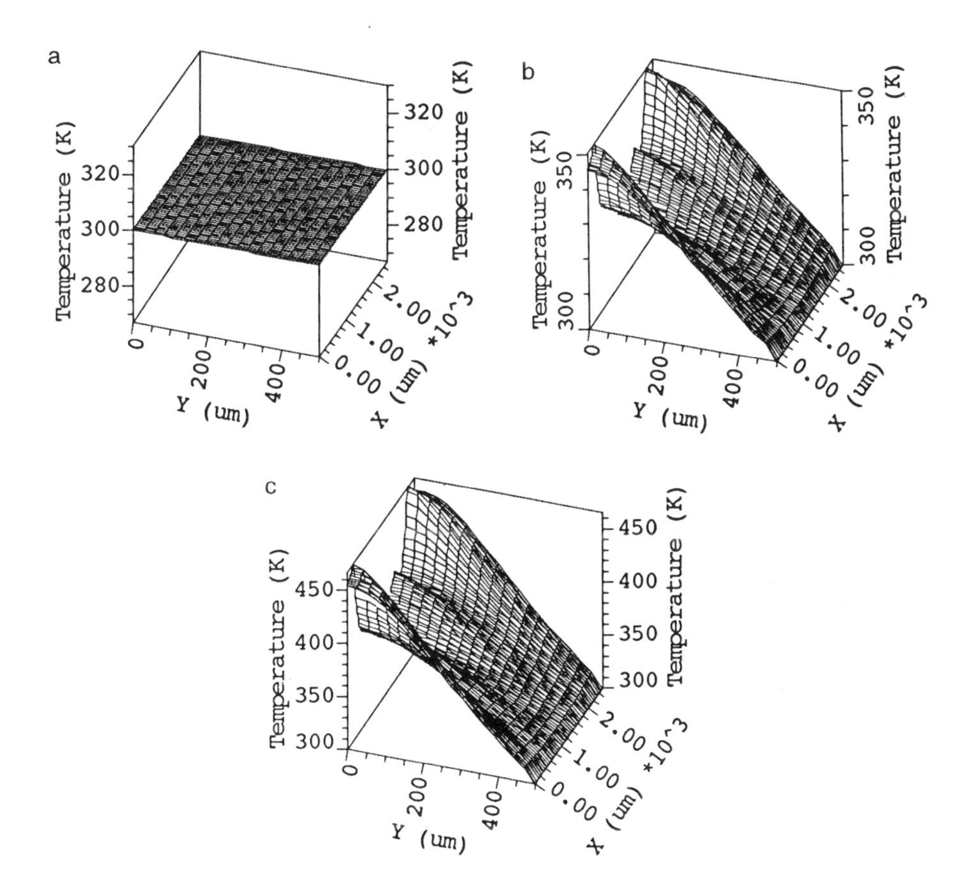

FIGURE 9.30. Three-dimensional lattice temperature contours simulated at (a) $V = 1$ V, (b) $V = 250$ V, and (c) $V = 500$ V.

Apparently, the on-state dc current flowing through the PCE would be increased, and therefore the PCE would have a large on-state conductance and be a better switching device, if the self-heating effect can be minimized. This is indeed the case for the PCE operated at cryogenic temperature (77 K), i.e., submerging the device in liquid nitrogen, where the heat generated in the PCE can be dissipated quickly outside the switch due to the very low ambient temperature. Figure 9.30 compares the current–voltage characteristics at cryogenic temperature simulated with and without the self-heating effect. The two currents are nearly the same, indicating the self-heating effect is not important. Furthermore, comparing the results in Figs. 9.31 and 9.32, it can be seen that the PCE current at 77 K is about two times higher than that at 300 K. The 3-D lattice temperature contours of the PCE at 77 K and at three different voltages are given in Figs. 9.33a–c, respectively. The electric field and total current density contours at 1 and 500 V are shown in Figs. 9.34a,b and 9.35a,b, respectively. The trends are very similar to those in the PCE operated at 300 K ambient temperature; only the magnitudes are slightly different. Comparing the electric fields of the PCE at 300 and 77 K (Figs. 9.23 and 9.33), it can be seen that the field is somewhat lower at low temperature. This contributes to the phenomenon observed in Fig. 9.32 that the current of PCE operated at 77 K saturates at a larger applied voltage than that at 300 K.

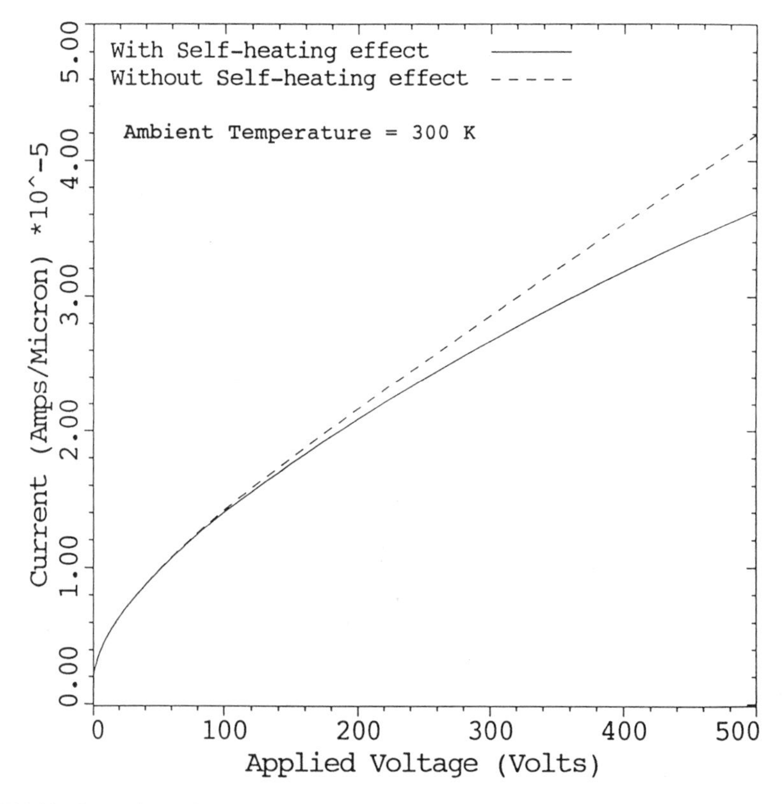

FIGURE 9.31. Comparison of the PCE I–V characteristics at room temperature simulated with and without lattice heating effect.

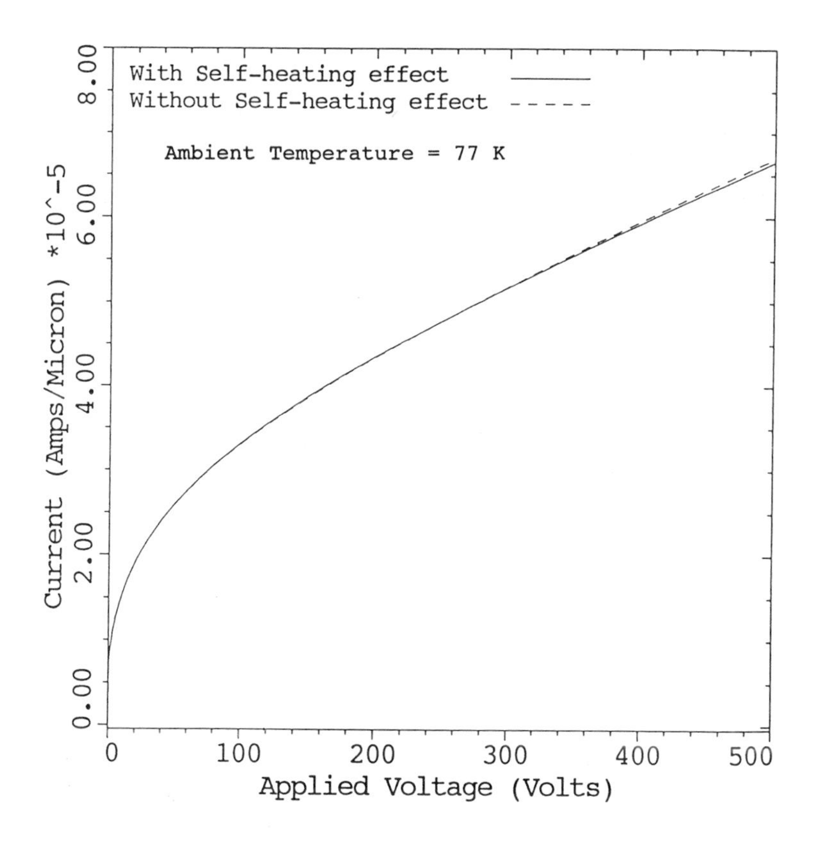

FIGURE 9.32. Comparison of the PCE I–V characteristics at cryogenic temperature simulated with and without lattice heating effect.

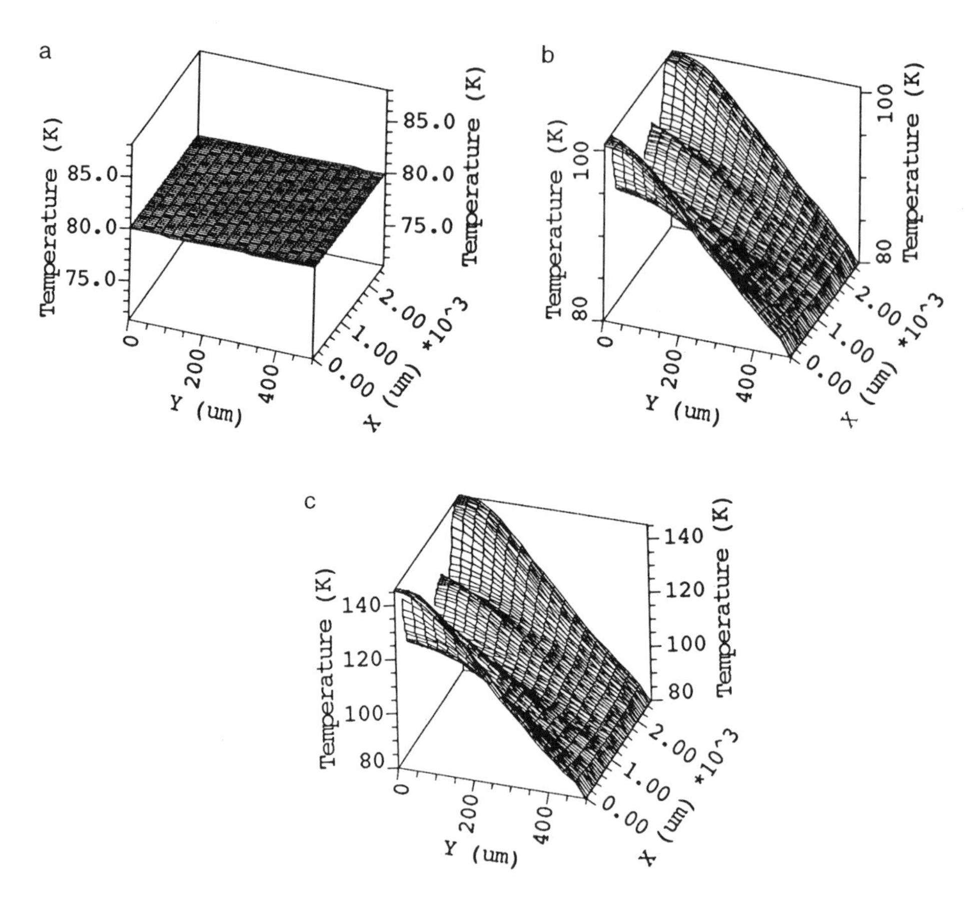

FIGURE 9.33. Three-dimensional lattice temperature contours of the PCE operated at $T = 77$ K simulated at (a) $V = 1$ V, (b) $V = 250$ V, and (c) $V = 500$ V.

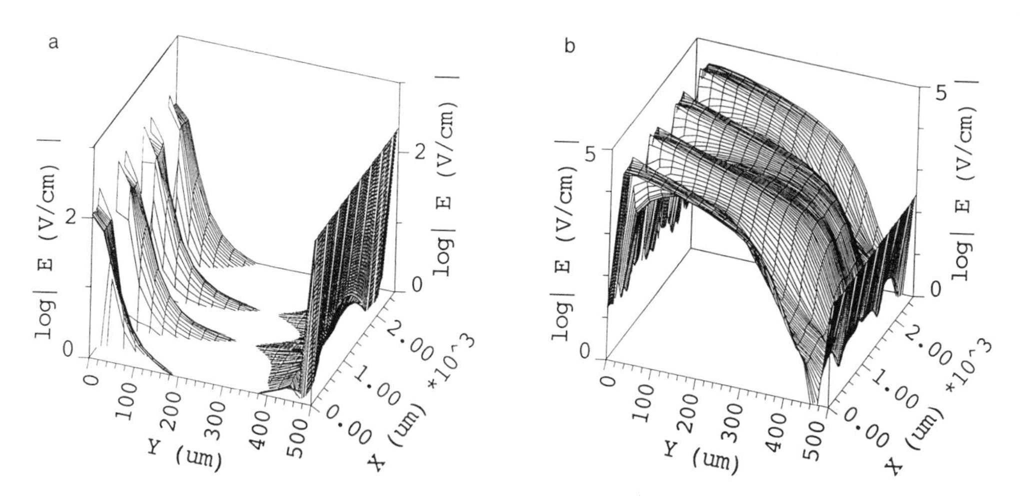

FIGURE 9.34. Three-dimensional electric field contours of the PCE operated at $T = 77$ K simulated at (a) $V = 1$ V and (b) $V = 500$ V.

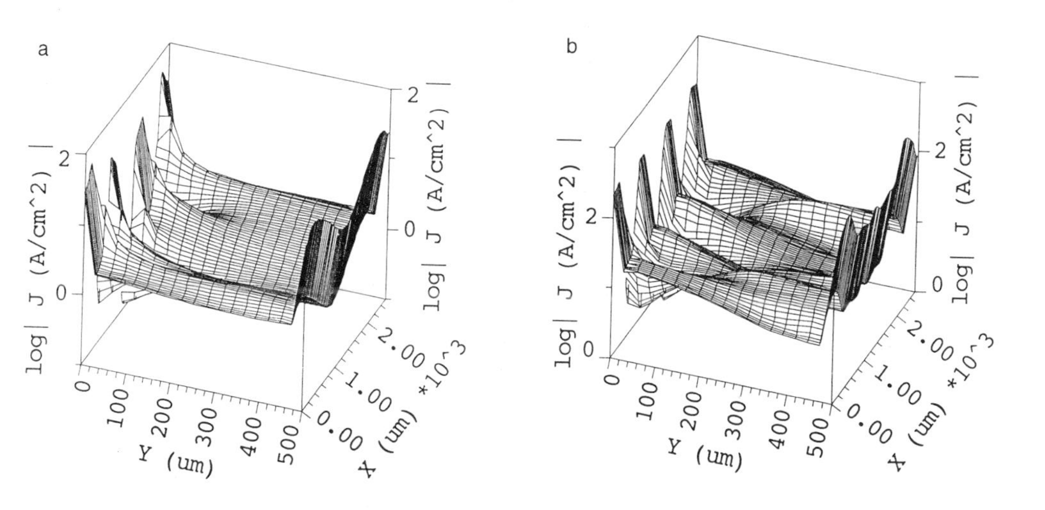

FIGURE 9.35. Three-dimensional total current density contours of the PCE operated at $T = 77$ K simulated at (a) $V = 1$ V and (b) $V = 500$ V.

EXAMPLE 9.1

```
 COMMENT  Photodiode simulation
    MESH
  X.MESH  WIDTH=2540 H1=40
  Y.MESH  DEPTH=5 H1=0.5 H2=4
  Y.MESH  Y.MAX=495 H1=25 H2=25 H3=30
  Y.MESH  Y.MAX=500 H1=3 H2=0.5
  REGION  NUM=1 SILICON
  ELECTR  NUM=1 TOP X.MAX=260
  ELECTR  NUM=2 TOP X.MIN=760 X.MAX=1020
  ELECTR  NUM=3 TOP X.MIN=1520 X.MAX=1780
  ELECTR  NUM=4 TOP X.MIN=2280
  ELECTR  NUM=5 BOTTOM
  ELECTR  NUM=9 BOTTOM THERMAL
 COMMENT  Specify profiles
 PROFILE  N-TYPE N.PEAK=1E12 UNIF OUT.FILE=mesh_dope
 PROFILE  N-TYPE N.PEAK=1E20 WIDTH=260 Y.CHAR=5 XY.RATIO=0.75
 PROFILE  N-TYPE N.PEAK=1E20 X.PEAK=760 WIDTH=260 Y.CHAR=5
       +  XY.RAT=0.75
 PROFILE  N-TYPE N.PEAK=1E20 X.PEAK=1520 WIDTH=260 Y.CHAR=5
       +  XY.RAT=0.75
 PROFILE  N-TYPE N.PEAK=1E20 X.PEAK=2280 WIDTH=260 Y.CHAR=5
       +  XY.RAT=0.75
 PROFILE  N-TYPE N.PEAK=1E20 Y.PEAK=500 Y.CHAR=5
 COMMENT  Grid refinement based on doping.
  REGRID  DOPING LOG RATIO=1 SMOOTH=1
 COMMENT  Specify contacts
 CONTACT  NUM=1 CON.RESI=0.05304
 CONTACT  NUM=2 CON.RESI=0.05304
 CONTACT  NUM=3 CON.RESI=0.05304
 CONTACT  NUM=4 CON.RESI=0.05304
 COMMENT  Specify material parameters and physical models
MATERIAL  SILICON TAUN0=2E-6 TAUP0=2E-6 bn=0 cn=0 bp=0 cp=0
MOBILITY  vsatn=3.3E7 vsatp=1.66E7 betan=0.46 betap=0.97
  MODELS  SRH BGN FLDMOB LSMMOB FERMI IMCOMPLE
 CONTACT  ALL ALUMINUM
SYMBOLIC  CARRIERS=0
 COMMENT  Specify the photogeneration
PHOTOGEN  PHOTOGEN a3=4E17 a4=-.0040 ^g.integ
       +  X.START=260 Y.START=0 X.END=0 Y.END=130 Y.MAX=500
PHOTOGEN  a3=4E17 a4=-.0040 ^clear ^g.integ
       +  X.START=1020 Y.START=0 X.END=0 Y.END=500 Y.MAX=500
PHOTOGEN  a3=4E17 a4=-.0040 ^clear ^g.integ
       +  X.START=1780 Y.START=0 X.END=760 Y.END=500 Y.MAX=500
PHOTOGEN  a3=4E17 a4=-.0040 ^clear ^g.integ
       +  X.START=2540 Y.START=0 X.END=1520 Y.END=500 Y.MAX=500
PHOTOGEN  a3=4e17 a4=-.0040 ^clear ^g.integ
       +  X.START=0 Y.START=0 X.END=1020 Y.END=500 Y.MAX=500
```

```
  PHOTOGEN  a3=4e17 a4=-.0040 ^clear ^g.integ
         +  X.START=760 Y.START=0 X.END=1780 Y.END=500 Y.MAX=500
  PHOTOGEN  a3=4e17 a4=-.0040 ^clear ^g.integ
         +  X.START=1520 Y.START=0 X.END=2540 Y.END=500 Y.MAX=500
  PHOTOGEN  a3=4e17 a4=-.0040 ^clear ^g.integ
         +  X.START=2280 Y.START=0 X.END=2540 Y.END=130 Y.MAX=500
   COMMENT  Perform steady-state solutions to find the maximum power point
    SYMBOL  NEWTON CARRIERS=2 lat.temp
    METHOD  AUTONR.PX TOLER=1E-1 CX.TOLER=1E-1 LTX.TOLER=1E+2
         +  STACK=50
       LOG  IVFILE=gen2_1_IV
     SOLVE  V5=0 V1=0 V2=0 V3=0 V4=0 ELECTRO=(1,2,3,4) VSTEP=0.5
         +  NSTEP=10 OUTF=gen2_1_000
       LOG  IVFILE=gen2_2_IV
     SOLVE  V5=0 V1=5 V2=5 V3=5 V4=5 ELECTRO=(1,2,3,4) VSTEP=1
         +  NSTEP=10 OUTF=gen2_2_000
       LOG  IVFILE=gen2_3_IV
     SOLVE  V5=0 V1=15 V2=15 V3=15 V4=15 ELECTRO=(1,2,3,4)
         +  VSTEP=5 NSTEP=7 OUTF=gen2_3_000
       LOG  IVFILE=gen2_4_IV
     SOLVE  V5=0 V1=50 V2=50 V3=50 V4=50 ELECTRO=(1,2,3,4) vstep=10
         +  NSTEP=20 OUTF=gen2_4_000
       LOG  IVFILE=gen2_5_IV
     SOLVE  V5=0 V1=250 V2=250 V3=250 V4=250 ELECTRO=(1,2,3,4)
         +  VSTEP VSTEP=10 NSTEP=25 OUTF=gen2_5_000
  PHOTOGEN  a3=0 a4=0 ^g.integ t0=0 tc=1.0
       LOG  IVFILE=trans2_IV
     SOLVE  V5=0 V1=500 V2=500 V3=500 V4=500 TSTEP=1E-12
         +  TSTOP=50E-6 OUTF=trans2_000
 CALCULATE  name=Ir A=I3 C=0.05304 PRODUCT
 CALCULATE  name=Vswitch A=V3 B=Ir DIFFEREN OUT.FILE=gen2_calc
```

REFERENCES

Auston, D. H. (1984), in *Picosecond Optoelectronic Devices*, ed. C. H. Lee (Academic Press, New York).

Cho, P. S., Y. F. Cui, J. Goldhar, and C. H. Lee (1994), *IEEE Trans. Electron. Devices* **ED-41**, 1529.

Frankel, M. Y. *et al.* (1990), *IEEE Trans. Electron. Devices* **ED-37**, 2493.

Lee, C. H. (1990), *IEEE Trans. Electron. Devices* **ED-37**, 2426.

Loubriel, G. M., M. W. O'Malley, and F. J. Zutavern (1987), in *Proc. 6th IEEE Pulsed Power Conf.*, Arlington, VA.

Parthasarathy, A., J. J. Liou, R. Petr, and A. Ortiz-Conde (1994), in *Proc. SPIE Optically Activated Switching Conf.*, Boston, MA.

Rose, A. (1963), *Concepts in Photoconductivity and Allied Problems* (Interscience, New York).

Rosen, A. and F. J. Zutavern, eds. (1993), *High Power Optically Activated Solid State Switches* (Artech House, Boston).

Shakouri, H. and J. J. Liou (1992), *IEE Proc. Part G* **139**, 343.

Thompson, S. E. and F. A. Lindholm (1990), *IEEE Electron. Devices* **ED-37**, 2542.

Triaros, C. P., D. P. Carroll, and F. A. Lindholm (1990), *IEEE Trans. Electron. Devices* **ED-37**, 2526.

Index